Advances in Intelligent Systems and Computing

Volume 763

Series editor

Janusz Kacprzyk, Polish Academy of Sciences, Warsaw, Poland
e-mail: kacprzyk@ibspan.waw.pl

The series "Advances in Intelligent Systems and Computing" contains publications on theory, applications, and design methods of Intelligent Systems and Intelligent Computing. Virtually all disciplines such as engineering, natural sciences, computer and information science, ICT, economics, business, e-commerce, environment, healthcare, life science are covered. The list of topics spans all the areas of modern intelligent systems and computing such as: computational intelligence, soft computing including neural networks, fuzzy systems, evolutionary computing and the fusion of these paradigms, social intelligence, ambient intelligence, computational neuroscience, artificial life, virtual worlds and society, cognitive science and systems, Perception and Vision, DNA and immune based systems, self-organizing and adaptive systems, e-Learning and teaching, human-centered and human-centric computing, recommender systems, intelligent control, robotics and mechatronics including human-machine teaming, knowledge-based paradigms, learning paradigms, machine ethics, intelligent data analysis, knowledge management, intelligent agents, intelligent decision making and support, intelligent network security, trust management, interactive entertainment, Web intelligence and multimedia.

The publications within "Advances in Intelligent Systems and Computing" are primarily proceedings of important conferences, symposia and congresses. They cover significant recent developments in the field, both of a foundational and applicable character. An important characteristic feature of the series is the short publication time and world-wide distribution. This permits a rapid and broad dissemination of research results.

More information about this series at http://www.springer.com/series/11156

Radek Silhavy

Editor

Software Engineering and Algorithms in Intelligent Systems

Proceedings of 7th Computer Science On-line
Conference 2018, Volume 1

 Springer

Editor
Radek Silhavy
Faculty of Applied Informatics
Tomas Bata University in Zlín
Zlín
Czech Republic

ISSN 2194-5357   ISSN 2194-5365   (electronic)
Advances in Intelligent Systems and Computing
ISBN 978-3-319-91185-4   ISBN 978-3-319-91186-1   (eBook)
https://doi.org/10.1007/978-3-319-91186-1

Library of Congress Control Number: 2018942343

This Springer imprint is published by the registered company Springer Nature Switzerland AG
The registered company address is: Gewerbestrasse 11, 6330 Cham, Switzerland

Preface

This book constitutes the refereed proceedings of the software engineering-related papers of the 7th Computer Science On-line Conference 2018 (CSOC 2018), held in April 2018.

Particular emphasis is laid on modern trends, concepts, and application of intelligent systems into a software engineering field. New algorithms, methods, and application of software engineering techniques are presented.

CSOC 2018 has received (all sections) 265 submissions, 141 of them were accepted for publication. More than 60% of accepted submissions were received from Europe, 30% from Asia, 5% from Africa, and 5% from America. Researches from 30 countries participated in CSOC 2018 conference.

CSOC 2018 conference intends to provide an international forum for the discussion of the latest high-quality research results in all areas related to computer science. The addressed topics are the theoretical aspects and applications of computer science, artificial intelligences, cybernetics, automation control theory, and software engineering.

Computer Science On-line Conference is held on-line and modern communication technology, which are broadly used improves the traditional concept of scientific conferences. It brings equal opportunity to participate to all researchers around the world.

I believe that you will find the following proceedings interesting and useful for your own research work.

March 2018

Radek Silhavy

Organization

Program Committee

Program Committee Chairs

Petr Silhavy	Tomas Bata University in Zlin, Faculty of Applied Informatics
Radek Silhavy	Tomas Bata University in Zlin, Faculty of Applied Informatics
Zdenka Prokopova	Tomas Bata University in Zlin, Faculty of Applied Informatics
Roman Senkerik	Tomas Bata University in Zlin, Faculty of Applied Informatics
Roman Prokop	Tomas Bata University in Zlin, Faculty of Applied Informatics
Viacheslav Zelentsov	Doctor of Engineering Sciences, Chief Researcher of St. Petersburg Institute for Informatics and Automation of Russian Academy of Sciences (SPIIRAS)

Program Committee Members

Boguslaw Cyganek	Department of Computer Science, University of Science and Technology, Krakow, Poland
Krzysztof Okarma	Faculty of Electrical Engineering, West Pomeranian University of Technology, Szczecin, Poland
Monika Bakosova	Institute of Information Engineering, Automation and Mathematics, Slovak University of Technology, Bratislava, Slovak Republic

Pavel Vaclavek	Faculty of Electrical Engineering and Communication, Brno University of Technology, Brno, Czech Republic
Miroslaw Ochodek	Faculty of Computing, Poznan University of Technology, Poznan, Poland
Olga Brovkina	Global Change Research Centre Academy of Science of the Czech Republic, Brno, Czech Republic and Mendel University of Brno, Czech Republic
Elarbi Badidi	College of Information Technology, United Arab Emirates University, Al Ain, United Arab Emirates
Luis Alberto Morales Rosales	Head of the Master Program in Computer Science, Superior Technological Institute of Misantla, Mexico
Mariana Lobato Baes	Superior Technological of Libres, Mexico
Abdessattar Chaâri	Laboratory of Sciences and Techniques of Automatic control and Computer engineering, University of Sfax, Tunisian Republic
Gopal Sakarkar	Shri. Ramdeobaba College of Engineering and Management, Republic of India
V. V. Krishna Maddinala	GD Rungta College of Engineering & Technology, Republic of India
Anand N. Khobragade	Maharashtra Remote Sensing Applications Centre, Republic of India
Abdallah Handoura	Computer and Communication Laboratory, Telecom Bretagne, France

Technical Program Committee Members

Ivo Bukovsky	Roman Senkerik
Maciej Majewski	Petr Silhavy
Miroslaw Ochodek	Radek Silhavy
Bronislav Chramcov	Jiri Vojtesek
Eric Afful Dazie	Eva Volna
Michal Bliznak	Janez Brest
Donald Davendra	Ales Zamuda
Radim Farana	Roman Prokop
Martin Kotyrba	Boguslaw Cyganek
Erik Kral	Krzysztof Okarma
David Malanik	Monika Bakosova
Michal Pluhacek	Pavel Vaclavek
Zdenka Prokopova	Olga Brovkina
Martin Sysel	Elarbi Badidi

Organizing Committee Chair

Radek Silhavy   Tomas Bata University in Zlin, Faculty of Applied
                              Informatics

Conference Organizer (Production)

OpenPublish.eu s.r.o.
Web: http://www.openpublish.eu
Email: csoc@openpublish.eu

Conference Website, Call for Papers

http://www.openpublish.eu

Contents

The Digitization of Society – Case of Specific Chosen State Alliance of Four Central European States

Michal Beno[1(✉)] and Darina Saxunova[2]

[1] VSM/CITY University of Seattle, Panonska cesta 17, 85104 Bratislava, Slovakia
michal.beno@vsm-student.sk
[2] FM CU, Odbojarov 10, P. O. Box 95, 820 05 Bratislava, Slovakia
darina.saxunova@fm.uniba.sk

Abstract. The purpose of this research is to study the process of the Visegrad Group citizens being gradually incorporated in the digital environment. The present digital world needs digital sovereign residents who are competent i.e. capable of dealing with Information Age systems; electronic devices, digital computers and modern scientific technologies and are open i.e. who are open and proactively ready to benefit from advantages and challenge possible risks of intended usage. Further requirements on the direction through the entire digital world are digital access and the spreader usage of distinctive products and services. An online survey was conducted examining this development, and respondents were questioned on certain issues and aspects of the digital world in the Visegrad Group; including countries as the Czech Republic, Hungary, Poland and Slovakia

Keywords: Internet of things · Digitalization · ICT · Electronic devices
Services

1 Introduction

The digitization is believed to be the next step, more significant, greater acclaim than the Internet, crucial for positive change in society. Technological progress, innovation and engagement, greater consumer power and increased competition are some of the prerequisites and enablers to render for further development. Digital technology can offer a myriad of tools to maximize human skills.

The term "digitization" designates the analogue information conversion into digital information. The term "digitalization" refers to the adoption or increase in the use of digital computer technology by all voluntary interrelations within society. As Communications scholar Feldman [1] argues, unlike analogue data with "continuously varying values, digital information is based on the two distinct states. In the digital world, things are there or not there, on or off. There are no in-betweens." Pepperell [2] states that "digital information is discrete and clean, whilst analogue information is continuous and noisy".

© Springer International Publishing AG, part of Springer Nature 2019
R. Silhavy (Ed.): CSOC 2018, AISC 763, pp. 1–11, 2019.
https://doi.org/10.1007/978-3-319-91186-1_1

Life in every country of the world has changed dramatically due to what is referred to as "the digital world" – the extensive communication among human beings through creating digital artefacts that can be shared mutually among them via an Internet connection, and the ability of searching and finding information. The Internet and especially social technologies are used for various purposes by different groups of people.

A study in Australia shows that parents, students and other stakeholders in education expect students to participate in the digital world for multiple reasons, for instance, ensuring life-long learning and being digitally prepared for future technologies [3].

The digital economy is growing at seven times the rate of the rest of the economy, but Europe is lagging behind many other regions when it comes to fast, reliable and connected digital networks [4]. The Internet usage represents fundamental statistics monitored worldwide, the results in 2016 were as follows: 11.2% in the European Union compared to 88.8% of the rest of the world. The highest user rate was evidenced in Germany (17.8%), UK (14.7%), France (13.8%). The Visegrad Group countries (further abbreviated V4 Group) together reported 11.5% user rate [5]. The European Union digital economy grows by 12% each year, with 7 million jobs in the ICT sector, estimating half of the productivity growth from investment in ICT. It is common knowledge that there are more mobile phone subscriptions than people in Europe [4].

This scientific paper examines the accession and entrenching of the V4 Group citizens in the digital environment. The present digital world needs digital sovereign residents who are competent i.e. capable of dealing with Information Age systems; electronic devices, digital computers and modern scientific technologies, concurrently they are open i.e. who are open and proactively ready to benefit from advantages and challenge possible risks of intended usage. The digital access and spreader utilization of different products and services are further significant requirements on the way through the digital world.

Some years ago, those who had their own email account were labelled "digital pioneers". In 2015, there were over 863 105 652 websites on the World Wide Web [6] which was invented in March 1989 by Berners-Lee [7]. In August 1998, Google was officially registered [8] and, after 14 years, there was an annual even growth of daily searches. In 2012, Google Zeitgeist reported 1.2 trillion searches worldwide [9].

The current digitalization is not characterised only by changes in communication behaviour and media usage, in the meantime, digital innovations have penetrated almost all areas of our lives, sometimes causing disruptive changes and reformation. It affected the re-establishment of our purchasing behaviour, daily work routine and economic sectors, e.g. Web 2.0 applications, car and accommodation sharing, flight ticket reservations, etc. This utilization requires more implicit knowledge and competences than merely surfing the Internet.

An online survey was aimed to overlook these developments, to examine and ascertain the opinion of a group of respondents being questioned about these issues and aspects of the digital world in the group of the four countries, called the Visegrad Group, which includes the Czech Republic, Hungary, Poland and Slovakia.

2   Methodology

A two-step research project was conducted to focus on the main topic of this paper. Firstly, the results of the collected survey data were examined. The final step is aimed at investigating the issues and advances and identifying the benefits and advantages of a digital society.

Electronic surveys have distinctive technological, demographic and response characteristics that have an impact on many aspects. A survey design, subject privacy and confidentiality, subject, distribution methods and response rates are critical methodical components that must be addressed in order to conduct online research.

Surveys are generally perceived not to be the perfect vehicles for data collection, because surveys require subjects to recall past behavior and, as suggested by others, survey questions lead to biased subjective judgements and answers [10].

However, research costs (observations, focus groups, interviews, telephone surveys) and the scope of the research may make it impractical or financially impossible to use more than one data collection approach. Electronic surveys provide the ability to conduct large-scale data collection by other than organizations at the centers of power in society [11].

Using new technology offers an inexpensive possibility of conducting a survey online, instead of using other options. Email/web surveys are widely considered to be cheaper than postal surveys, responses are faster, more complete, are less likely to be ignored and more accurate [12, 13].

With the advent of WWW and electronic mail, the Internet has opened up new prospects in surveying. Rather than being mailed a paper survey, a respondent can now be given a hyperlink to a website containing the survey. Alternatively, in an email survey, a questionnaire can be sent to a respondent via email, possibly as an attachment [14].

This section continues describing the scope of the study. The research only covers the domain of the V4 Group, in addition having guided the authors' choice. Respondents were chosen from different age groups, as we realized that a certain level of comprehension of the digital world was crucial to answer the questionnaire, so only an educated audience was selected for the research. The questionnaire of the survey was prepared in the V4 Group's national languages to avoid misunderstanding an English text. A general viewpoint of the digital world was solicited from V4 Group citizens. There were 11 questions based on the 4 determinants of competence, openness, access and the variety of accepted practices in the digital world.

We have created our own Index called **COVA** which includes these determinants: **Competence,** in this area, the V4 Group population is questioned on their knowledge of digital topics. **Openness** as the next determinant regarding citizens' digital attitudes, openness to digital topics and innovations, fears, opportunities and benefits is studied. **Variety of uses** i.e. the regular use of online/computer applications is monitored. **Access**, this determinant considers users' access to the digital infrastructure, i.e. Internet usage, usage of devices and broadband function.

These areas have a different influence on the digital self-determination of an individual. Owing to this reason, we divided the questionnaire into 4 parts of varying importance, which are included in the calculation of the total index in %.

In this case, we calculated the COVA as follows:

$$
\begin{aligned}
\text{COVA} = (& \frac{Competence\ (\mathbf{40\%}) * Competence\ of\ V4\ Group\ country\ in\ \%}{\mathbf{100}} + \\
& \frac{Openness\ (\mathbf{30\%}) * Openness\ of\ V4\ Group\ country\ in\ \%}{\mathbf{100}} + \\
& \frac{Variety\ of\ uses\ (\mathbf{10\%}) * Variety\ of\ uses\ of\ V4\ Group\ country\ in\ \%}{\mathbf{100}} + \\
& \frac{Access\ (\mathbf{20\%}) * Access\ of\ V4\ Group\ country\ in\ \%}{\mathbf{100}})
\end{aligned}
\tag{1}
$$

The survey questionnaire contained several types of questions for respondents to answer consisting of open questions, questions allowing respondents to note their answers on a different scale, with ranges varying from "disagree" to "agree". A few of the questions required respondents to check appropriate replies.

Data were collected from February 2, 2017 to March 10, 2017 through online sampling on the web of survio.com. Survio is an online survey service using the free and easy tools for any type of online survey.

The findings from an online survey examine the viability of advanced operations in the digital world. The survey was an appropriate research strategy, because the purpose of the study was to describe the incidence of the phenomenon under investigation. The analysis of the study assisted researchers to refine the scale.

The questionnaire was mailed to people from the 4 different countries, using a modern online survey system, a comprehensive survey developed by the Survio company, which provides an effective solution for data collection, analysis and sharing results. 1 017 out of 1 501 questionnaires were completed. The response rate in Slovakia and the Czech Republic is 53.00%, in Poland 83.60% and in Hungary 100%. The drop-out rate is only 2.90%. Most drop-outs occurred already at the beginning of the survey, probably due to disinterest. The average processing time is between 2–5 min (55.97%). This is within the calculated time requirement of 8 min.

The survey was coded, so that it was possible to determine the type of a respondent, including gender, age, education and occupation. We used the codes as control variables in the analyses. The sample consists of 49.56% of the male and 50.44% female respondents; the majority (55.75%) are university graduates and 30.09% are high school (secondary education) graduates (GCE level). Most of the examinees are represented by full-time employees (56.54%); almost 10% are self-employed persons and businesspersons.

3   Results

To accompany society towards the digital age means to enable every human being to move freely within the digitalized world. An important aspect of the digitization of the population is Internet usage. The total number of users worldwide on December 31, 2016 was 3 696 238 430 [15].

In our opinion, a detailed picture of the digitization status of the population can be given, by a much more complex and broader view of the subject. It is crucial to pay attention not only to the pure utilization rate, but also to the dimensions that deal with usage behavior, knowledge of digital media, and attitude.

Digital society needs "digital citizens" self-confident, who can cope well with modern products, devices and technologies (competence) and be proactive with the advantages and possible risks of utilization (openness). Further requirements are digital access and a wide range of different usages of products and services. There is almost no profession for which a computer and digital technology are not needed. Inhabitants who are unable to access this spectrum; become increasingly limited in social and economic progression. The ratio of off-liners become increasingly smaller in numbers. Thus, the objective is to enable every individual to be able to move independently and freely in a digitalized world, regardless of gender, education and age.

Firstly, we calculated our 2017 COVA, which is a composite index of the digitalization rate of countries, based on specific information which we summarized as access, competence, openness, and the variety of usages in the digital world. The results of our estimation demonstrate that citizens of V4 group countries with access to this digital environment are at a pre-medium level, as presented in Table 1.

Table 1. COVA in % (N = 1017).

Country	Competence in %	Openness in %	Variety of uses in %	Access in %	COVA
Czech Republic	64.95	60.21	39.34	44.95	56.37
Hungary	69.97	55.60	34.33	49.44	57.99
Poland	66.13	68.54	40.04	48.47	60.78
Slovakia	67.25	63.72	35.68	45.82	58.75

These indexes highlight that a large proportion of Czech, Slovak, Hungarian and Polish inhabitants participate in the digitalized world. But the COVA value of each country still shows a development potential. Further analysis revealed that competence and openness shows above average results. This means that the infrastructure and equipment are provided in many places to enable residents to participate in the digital world. The necessary ability and skills have increased in these countries and most respondents are aware of the capabilities required in the virtual world.

However, the digital world affords just as many difficulties as it does have solutions to the questions of the variety of uses and access. Three major factors in the use of digital technologies continue to differentiate the citizens of V4 Group and to serve as barriers to those two indexes: access to technology, individual competence in the variety of uses, and security. The access index remains under the total index value, with growth potential. It is fundamental to note that the majority of the V4 Group population uses the Internet frequently.

The most remarkable result is the "variety of uses" sector which is below the other pillars. It appears that ordering online services, posting articles, utilization of smart home services and personal data have little importance for most citizens. It can be reasonably assumed that there are issues linked to the deficiencies regarding Internet security and the absence of building trust in newer technologies. We believe that citizens are not correctly informed about the benefits of these digital facilities and proving the fact that the small share of the capacity of digital facilities is utilized.

Competence

Digital competence could benefit citizens, communities, the economy and society in general in various areas. Van Deursen [16] defined five areas in which citizens can gain personal benefit from Internet usage: social, economic, political, health and cultural. All these areas contribute to specific and societal-level benefits. This categorization has also been used in this paper.

Society is above average in the awareness of basic digital terms e.g. homepage (90.44%), App (86.71%), Cloud (61.13%), website (96.86%) and social networks (96.49%). Obvious arguments are that English terms are technically less known than expressions in national languages, showing that a small proportion of respondents are familiar with words e.g. Industry 4.0 (13.16%), Smart metering (16.15%), e-health (42.79%) and e-government (51.16%).

We identify the following applications for smart homes, e.g. energy saving, service for monitoring and improving energy efficiency of the end-user; home automation, remote control devices for security monitoring.

Further evidence depicts that a very high percentage (99.76%) of the V4 Group population possesses basic knowledge of computers (including installation of equipment 87.77%) and electronic communications, have Internet skills (99.76%), knowledge of the use of social media (90.87%) and the operation of specialized software for Internet calling (94.43%) and Internet banking (93.21%). Creating web applications (21.66%) is the only fundamental component which lags presently behind the remaining computer knowledge, because this requires specialized skills.

Openness

It is well-known fact that the Internet is one of the greatest inventions of mankind's and it is understood to be a worldwide network of computerized devices and servers. The vast opportunities and potential of digital media are explained in this section. Many advantages of Internet utilization are confirmed by 99.19% of respondents and a first search for required information on the Internet was conducted by 93.49% of the interviewees. Due to this fact, the Internet should be used in a way that creates harmony both at work and home.

As mentioned in the Introduction, skills are vital for the further development of the population in order to benefit from the knowledge and experience of others, from participation, sharing ideas, solutions and gathering much diverse information. It is believed that schools have commenced to take a different approach by introducing digital media into the educational system itself, 89.63% of respondents agree that digital media should be part of the school syllabus. 78.83% positively regard the work-life balance provided by Internet usage.

Almost 49% of respondents prefer usage of the Internet instead of contacts via telephone or personal encounter/meeting. The analysis identifies the significant result that only 37% of interviewees would prefer to be offline more, and 35.44% are reluctant to use digital devices like the computer, tablet or smartphone, for fear of reaching their determined restricted limit. The most remarkable result emerging from the data is that only 14.02% of respondents prefer avoiding the Internet as much as possible e.g. limiting computer time of usage, calling instead of sending instant messages or texts, using an alarm clock or timer, etc.

Variety of Uses

The way people interact, sharing information, and even do business is changing, to keep up with the continuous evolution of this global network. There are several actual and potential barriers to the sustained use of ICT by the population.

Respondents were asked about their regular usage of online/computer applications. It is obvious that Internet users mostly search through search engines for information on the net (99%), use social networks (81%) e.g. Facebook, Xing, Google+, WhatsApp and others to contact one another, buying/purchasing online (77%) and do Internet banking (76%). Ordering online services is used by only 53% of respondents, with 45% using forums, blogs and searching for help. 18% utilize modern technologies like smart home services. Most of population is not interested in posting messages, articles or saving personal data for that matter.

The service trade has been a rapidly growing phenomenon worldwide, apart from during the years of the financial crisis and slightly worsening in recent years. It is a remarkably fast developing sector, attracting the attention of researchers, who strive to reveal and remove shortcomings leading to service quality improvement and enhancement of investor returns. Our results offer invaluable evidence for the future development of online services in these four countries. As anticipated, our survey proves that courier service (75%) is the top sector among the rest. Remarkably, cab reservation and car rental are among the less frequently utilized categories (total 29%). We found a high ratio of holiday and private accommodation bookings (total 87%). Only 10% of respondents refuse to use any of the listed services online.

The impact of IT on the consumer sector is difficult to measure. Our argument is the consumers' receipt of a considerable fraction of the benefits from Internet usage. How can one improve the efficiency? One option could be a detailed time study of how people spend their free time.

Internet usage is covered by several different time studies. In 2000, a study by the Stanford Institute for the Quantitative Study of Society (SIQSS) [17] found that increased Internet usage correlated with isolating behaviors such as spending less time with friends and family. The more time people spend on the Internet, the less time they spend on interacting with people in the "real world". Dryburgh [18] finds that Canadians who use the Internet spend less time watching TV, reading, sleeping and doing leisure activities at home. A Swiss study found that the Internet did not cause people to spend less time on social activities but did cause less TV watching [19]. Anderson and Tracey [20] analyze a UK time study, finding that people who gain access to the Internet spend significantly less time on hobbies, games and musical instruments.

As far as we are aware, this is the only section to which countries should pay attention for future development, especially of modern technologies like Smart metering, e-government and e-health. All nations have the potential for growth. The population in general must be more inclusive in the digitalization of each country.

Access

We believe the use of the Internet has changed our way of thinking, as well as having made a unique contribution by providing our society with immediate and convenient access to an extraordinary range of ideas and information. Access to information and ideas has always been important for both personal development and the progress of a community and nation.

The proliferation of devices has changed the way people interact with the world around them. In 2010, experts estimated that 17.6 million tablets were sold — a number that was expected to increase more than three-fold in 2011. Market projections predicted that there could be more than 300 million tablets sold worldwide in 2015, with more than 80 million tablet users in the U.S. alone [21, 22].

Business or home users can choose from many types of portable devices, with modern products constantly arriving on the market. This paper focuses on four portable devices and one desktop PC. Smart media devices are those that can transfer data by means of a wired non-cellular or cellular wireless connection: mobile phone, smartphone, tablet and laptop.

Most residents are familiar with the use of modems and telephone lines to access the Internet, but there are many other options available, e.g. Wi-Fi, Mobile Internet, Cable TV, as illustrated in Table 2.

Table 2.   Internet connections in % (N = 1017).

Internet connections	%
Dial-up modem	9.51
ISDN	8.48
ADSL	28.82
WiFi	86.63
Cable TV	25.86
Optical cable	27.14
Mobile Internet	71.04

The analysis identifies discrepancies among various types of Internet connections, from traditional dial-up access (9.51%) to the many different broadband options, especially Wi-Fi (86.63%) and Mobile Internet (71.04%).

The findings in this paper are based on the real behavior of a sample of users. It reveals which devices the users use. We live in a mobile world. 75% of all users utilize the smartphone and 82% laptop, rather than the desktop PC (64%), as shown in Fig. 1. Our study provides further evidence for a potential mobile environment, digitalization and further utilization. Mobile phones present only 36%, which means that users prefer the Internet via smartphone or tablet (52%).

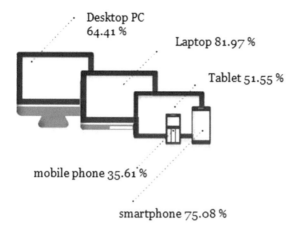

Fig. 1. Devices (N = 1017), author's own elaboration.

4   Conclusion

Digital Information Technology has become increasingly intensive and rapidly entrenched in our society. Digitization has major consequences for the way people live, work and study, for promoting health and combating diseases, and for dealing with freedom and security.

The Visegrad Group have a strong potential to lead in the digital area, but only if the stakeholders manage to coordinate efficiently grasping this once-in-a-lifetime opportunity. This was one of the recommendations resulting from the Think Visegrad Conference that took place on November 30, 2016 in Brussels, to debate the Visegrad Group Contribution to the EU Digital Single Market [23].

Although access to ICT cannot be considered a basic human right, it has earned its place in the discussions of governments, major corporations and civil society groups as being a crucial component to the successful development of any society. ICT is one of the most innovative and growing sectors in the global economy. According to the OECD data of 2011, 4.9% of people worked in this sector, 3.3% in Slovakia and 3.2% in the Czech Republic [24].

The different results of the Boston Consulting Group e-intensity Index suggest that V4 Group have different expertise, and each of them has its own areas of success. In 2015, the Czech Republic had the highest Index, followed by Poland, then Slovakia and Hungary. This means that, for example, e-commerce has not become as prevalent and popular as in other countries e.g. UK, South Korea, Denmark, China [25].

A person who can read clearly has the advantage. The Internet is a powerful communication tool and information resource which can provide many benefits to the population, from searching, using Skype, making reservations, to online banking. Initiatives promoting the benefits of the Internet and programs usage assist in teaching the population. Citizens are trained skills needed to navigate through many goods and services offers online, are important in encouraging people of all ages to access and use it.

However, it is also important to recognize that many people, particularly in the older age groups, will never participate totally in the digital world. The majority in this group commented that they prefer being offline more frequently, visit friends or call them, rather than using modern Internet tools. They avoid using this medium as much as possible.

The possibilities of modern technology may be benefitted by people when they are knowledgeable about the manners of digital media usage in an increasingly digitalized world. The citizens' participation in the area of digital healthcare for seniors, economic opportunities in business and dating platforms enables to bring this process in social and economic life. The research results in our conclusion that those who can recognize the opportunities of digitalization and the appropriate competences, can only gain advantage in the digital transformation.

Our experiment has reviewed the necessary requirements for a digital world – competence, openness, variety of uses and access. Taking them into consideration, a conceptual Index is projected for developing a digital society. The COVA Index proposes level aspects. The objective is to highlight all different areas that should be considered when comparing and developing a digital world. Many respondents share the opinion of growing openness and skills ability in the three elements –technical, economic and social. Acquired skills and openness deliver numerous, substantial benefits to individuals, enterprises, governments and societies. The evidence implying from this paper encourages the idea that the variety of digital world usage must be observed carefully. Modern technology, such as Smart metering, e-government and e-health, is not shaped by individuals.

We have demonstrated that V4 Group citizens access the digital world through modern technologies. The results show that there are numerous usages regarding digital instruments, mostly portable devices connected via Wi-Fi. The research results lead to highlighting the potential growth of myriad applications, especially online reservation services and Smart metering applications.

References

1. Feldman, T.: An Introduction to Digital Media. Routledge, New York (1997). ISBN 978-0415154239
2. Pepperell, R.: The Posthuman Condition: Consciousness Beyond the Brain. Intellect Books, Portland (2003)
3. Howell, J.: Teaching with ICT: Digital Pedagogies for Collaboration and Creativity. Oxford University Press, South Melbourne (2014)
4. Europäische Kommission: Digital Agenda for Europe: rebooting Europe's economy, Manuscript updated in November 2014. Ed. The European Union explained. Publ. Off. of the Europ. Union, Luxembourg (2014)
5. Internet World Stats. http://www.internetworldstats.com/stats9.htm. Accessed 15 Mar 2017
6. Internet Live Stats. http://www.internetlivestats.com/total-number-of-websites/. Accessed 23 Mar 2017
7. Berners-Lee, T.: Information management: a proposal. http://faculty.georgetown.edu/irvinem/theory/Berners-Lee-HTTP-proposal.pdf. Accessed 14 Mar 2017

8. History.com. http://www.history.com/this-day-in-history/google-is-incorporated. Accessed 23 Mar 2017

9. Google Zeitgeist: ein Jahr Suche. https://www.google.com/zeitgeist/2012/#the-world. Accessed 14 Mar 2017

10. Schwarz, N.: Self-reports: how the questions shape the answers. Am. Psychol. **54**(2), 93–105 (1999)

11. Couper, M.P.: Web-based surveys: a review of issues and approaches. Publ. Opin. Q. **64**, 464–494 (2000)

12. McCoy, S., Marks, Jr., P.V.: Using electronic surveys to collect data: experiences from the field. In: Proceedings of Americas Conference on Information Systems 2001, Boston, USA (2001)

13. Tse, A.C.B.: Comparing the response rates response speed and response quality of two methods of sending questionnaires: email vs mail. J. Market Res. Soc. **40**(4), 353–362 (1998)

14. Fricker, Jr., R.D., Schonlau, M.: Advantages and disadvantages of internet research surveys: evidence from literature. Field methods **14**(4), 347–367 (2002). Sage publications Quarterly 64

15. Internet World Stats, World Internet users and 2017 Population Stats. http://www.internetworldstats.com/stats.htm. Accessed 04 Apr 2017

16. van Deursen, A.J.A.M.: Internet skills. vital assets in an information society. university of twente. http://doc.utwente.nl/75133/. Accessed 15 Mar 2017

17. SIQSS, Stanford SIQSS Study. http://cs.stanford.edu/people/eroberts/cs201/projects/personal-lives/stanford.html. Accessed 07 Apr 2017

18. Dryburgh, H.: Changing our ways: why and how Canadians use the internet. statistics Canada report no. 56F0006XI. http://citeseer.ist.psu.edu/viewdoc/download;jsessionid=2533748515B47DFC7AD79DF86E868C4F?doi=10.1.1.556.4068&rep=rep1&type=pdf. Accessed 07 Apr 2017

19. Franzen, A.: Social capital and the internet: evidence from swiss panel data. Kykios **56**, 341–360 (2003)

20. Anderson, B., Tracey, K.: Digital living: the impact (or otherwise) of the internet on everyday life. Am. Behav. Sci. **45**(3), 456–475 (2001)

21. CNN: Tablets sales may hit $75 billion by 2015. http://money.cnn.com/2011/04/19/technology/tablet_forecasts/index.htm. Accessed 04 Jul 2017

22. Gartner: Gartner Says Apple Will Have a Free Run in Tablet Market Holiday Season as Competitors Continue to Lag. http://www.gartner.com/newsroom/id/1800514. Accessed 04 Jul 2017

23. ThinkVisegradfund, Internet users (per 100 people). https://think.visegradfund.org/wp-content/uploads/R_30_11_2016_Contribution_Single_Market.pdf. Accessed 07 Jul 2017

24. OECD, ICT employment (indicator). https://doi.org/10.1787/0938c4a0-en. Accessed 07 Apr 2017

25. BCG.perspectives, The 2015 BCG e-Intensity Index. https://www.bcgperspectives.com/content/interactive/telecommunications_media_entertainment_bcg_e_intensity_index/. Accessed 07 Apr 2017

An Efficient Security System in Wireless Local Area Network (WLAN) Against Network Intrusion

P. H. Latha[1(✉)] and R. Vasantha[2]

[1] Atria Institute of Technology, Bengaluru, KR, India
phdlatha2017@gmail.com
[2] Sambharm Institute of Technology, Bengaluru, KR, India

Abstract. The computer network faces any kind of unauthorized activities i.e. Network Intrusion (NI). The detection of these NI needs a better understanding of how the attacks work. The NI detection is necessary to protect the system information in current activities of the cyber attacks. This paper is intended to improve the security aspect in the Wireless Local Area Network (WLAN) by implementing a machine learning approach i.e. Support Vector Machines (SVMs). In this, the computer lab generated data are used for experimentation. The SVM detects the NI by recognizing the patterns of attack. The simulation outcome of the proposed security framework recognizes the NI and bells the alarm. The analysis of this security system is performed by considering the efficiency of detection and false alarm rate that offers significant coverage and effective detection.

Keywords: Coverage · Efficiency · Network Intrusion (NI) · Security
Wireless Local Area Network (WLAN) · Support Vector Machines (SVMs)

1 Introduction

The systems for detection of Network Intrusion (NI) are fundamental tools used by the system administrators to distinguish different security violation inside system administrators. A system of NI detection monitors and examines the system traffic from the network devices of an organization and gives the cautions if an intrusion is observed. In view of the strategies for interruption detection, NI detection systems are classified as: (a) Signature based NI detection (SNID) system and (b) Anomaly detection based NI detection (ANID) system. The system SNID like Snort [1], the attack signatures are pre-introduced in the NI detection system. A pattern matching can be done for traffic against these introduced signatures to recognize the NI. Conversely, an ANID system categorizes the network traffic as an intrusion when it finds a deviation from the typical traffic design [2]. The system of SNID is significant in the identification of known attacks and shows higher accuracy in detection with less false detection rates. Also, the performance of SNID suffers during the identification of latest or unknown attacks due to attack signature constraint which can be introduced in an NI detection system [3]. The ANID

© Springer International Publishing AG, part of Springer Nature 2019
R. Silhavy (Ed.): CSOC 2018, AISC 763, pp. 12–19, 2019.
https://doi.org/10.1007/978-3-319-91186-1_2

system is appropriate for the identification of unfamiliar or latest attacks. But, ANID system delivers high false-positive rates; its hypothetical potential in the distinguishing proof of novel attacks has caused its acceptance among research area [1, 4].

The NI faces two major challenges which take place while developing significant intrusion detection. The selection of proper features from the network dataset to detect the anomaly is difficult [5]. The selection of one kind of attack may not perform well for other kind of attacks because of regular changes in the attack scenarios. The next challenge is development of NI detection system under non-presence of labeled traffic data set in real time applications. Their needs high efforts to generate the labeled data set by using the real time traffic traces. Also, to maintain the network confidentiality of internal organizational structure and privacy preservation of different users are reluctant towards reporting all the intrusion which may occurs in their networks [6].

The bases to develop ADNIDS are various machine learning methodologies like Self-Organized Maps (SOM) Artificial Neural Network (ANN), Naïve-Bayesian (NB), Support Vector Machines (SVM), Random Forest (RF) and Artificial Neural Network (ANN). The process of distinguishing the usual traffic from the anomalous traffic is done by utilizing NIDs as they are treated as classifiers [7, 8]. A role to choose the particular feature is performed by NIDSs to extract respective feature set to improvise the results from the dataset relevant to traffic. As the redundant noise and features are removed, feature selection is employed in discarding the wrong training [9, 10]. The application of speech, audio and image processing can be successfully expanded using deep-learning techniques. These schemes goal to learn an efficient feature framework acquired from a massive amount of unlabeled information and parallel execute these trained features on a restricted amount of labeled information in the enhanced version of classification. Non-identical distributions are responsible to maintain unlabeled and labeled information and hence they might persist some relevance among one another [11, 12]. The current paper is organized as different sections: Sect. 2 idealizes the background of the network intrusion. The Sect. 3, discusses a some of the closely related work. Then, Sect. 4 introduces proposed system (to detect the network intrusion) using SVM. Section 5 gives the analysis of the obtained results and finally conclusion for this paper along with future work direction in Sect. 6.

2 Background

As the reliance on the Internet, extranet network and Intranet is expanding day by day, the problem of computer intrusion by unauthorized access is an increasing difficulty. An intrusion is an attempted or an unauthorized user access to exhibit activity that happens to be not approved. A software application or device named Intrusion Detection System (IDS) monitors (is shown in Fig. 1) the system or network connectivity for the detection of malicious events or violations in the policies. Any activity which is detected is reported typically to the central resource of administration via a system of security management. A tool that is defined either in hardware or software form used to monitor the internal events or external cyber events is called as the Intrusion Detection System (IDS) [1, 3, 4, 6, 8]. The main of aim of IDS lies within detecting attacks, network evidence collection, dealing with awareness of situation and connection policy enforce-ment. Basically, it consists of four main components namely, the sensor, detector, data base or knowledge base and the response component. The sensor gathers data from the system monitoring it. The detector or the Analyze engine analyses the gathered infor-mation to identify the chance of any intrusions. The component for response handles the attacks response actions. IDSs are sub-classified as Host Based Intrusion Detection System (HIDS) and Network Based Intrusion Detection System (NIDS). The classifi-cation is based on the monitoring the system technique. The basic approaches that are involved in the detection of intrusion are misuse detection and anomaly detection [9–12].

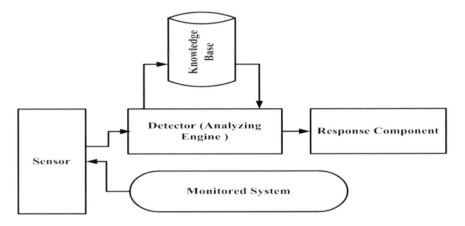

Fig. 1. Structure of intrusion detection system

3   Related Works

This section foregrounds the current status of work being carried out in developing approaches to overcome the issues of attacks, faults discovered and the vulnerabilities which makes the research area of Wireless Local Area Network (WLAN) even more challenging. Wireless Intrusion Detection System (WIDS) was designed to implement the protection parameter of a multi-servicing wireless network. To provide protection at multi-levels on the network, which could efficiently update against new kinds of attacks the WIDS along a new algorithm for searching executing at the platform of Ubicom was initiated. Various kinds of attacks could be detected effectively for a WLAN persisting standard of IEEE 802.1b [13]. An additional technique of data-mining based on the method Extra Trees that featured the Aegean WiFi Intrusion Detection (AWID) for the data set publically was studied. This involved WIDS and was found that the algorithms that performed exceptionally were Bagging, Random Forests and Extra Trees. The precision rate was improved to 96.31% with a time enhancement of 390 s in case of a voting classifier [14].

The conceptualization of null data frames that are accommodated in the WLAN of standard IEEE 802.11 can also be manipulated to identify the vulnerabilities bought in the network. Frames in a WLAN are forged with an extreme ease and hence this results in low overhead. Incorporating the execution of null data frames eliminates the chance of attacks, malicious activities or intrusion mechanisms over the network of WLAN [15]. A new type of attack introduced was Stealth Man-in-Middle (SMITM) that is derived from poisoning the Address Resolution Protocol (ARP). Here, directly the victim is affected by forging its frame. WIDS simulated over the network simulator of NS-3 is used to detect these attacks and rectifies it. When in the static condition, the scheme has been observed to cover the single sensor use in the complete attack period [16]. The approach of Anomaly Detection (AD) implied in medical field of WSNs is proposed. In order to perceive regarding the abnormal variations and to deal with malicious or unreliable instrument injections, the scheme of AD was introduced. Also, to minimize the triggering of false alarms induced by abnormal measurements. One-step forecasting is attained by implying the Exponentially Weighted Moving Average (EWMA). An outcome of less than 4% was obtained with a good detection rate for false alarm value [17]. The approach named Position-Aware, Secure and Efficient mesh Routing (PASER) prevented more number of attacks when compared to those with the traditional routing protocols. The wireless network can be easily sabotaged or the intruder may make changes in the payload information or even acquire data from Unmanned Aerial Vehicles (UAVs). For the real-time scenarios of UAVs, PASER obtains same performance result as that of protocol Analyzing security of Authenticated Routing Protocol (ARAN) [18]. The drawback of constrained resources in WSNs makes it highly susceptible for types of intrusions like Denial of Service (DoS) attack that produces a result of massive compromised nodes. A system of self-recovery and detection of intruder which is agent based was proposed, which monitors the unusual event processing and intrusion activity in the localized set of nodes. The resultant indicated that the nodes that compromised can self-recover efficiently and also the total network power consumption is also reduced [19]. A communication model for establishing the control and frame attachment in the process of transmitting the data is proposed. The perception of *h*Jam that includes the utilization of OFDM derived from WLANs is highlighted. The technique explores the physical features of the OFDM modulation, which requires single data packet and a group pf messages to be forwarded together. This technique improvises the throughput by a factor of 200% in comparison with existent family protocols of 802.11 [20].

4   Proposed System

The purpose of improving the intrusion detection system (IDS) of a large-scale Wireless Local Area Network (WLAN). It also offers a robust security model which incorporates machine learning (ML) based classifier namely support vector machine (SVM) to recognize different attack patterns by means of prediction results. The study also further conducted an experimental simulation which validates the proposed system in terms of different performance metrics. It also has exhibited that our proposed network anomaly detection model recognizes suspicious activities very efficiently and also generates a control signal to the notify the controller about it. The prime aim of the proposed system is to present an analytical modeling with the extensive system modeling and its description includes association of different operational entities which take active participation in the intrusion detection scenario. The proposed system basically operates on different user sequence entity. It also incorporated a sequential pattern profiling scenario by tracking down the daily activities and workflows generated by individual users in a sequence γ_{Seq}. The following Fig. 2, exhibits the tentative architecture of the proposed SVM based intrusion detection model.

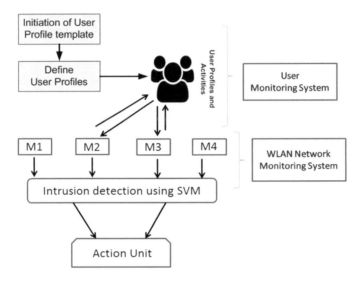

Fig. 2. Design of the proposed system

However, the prime reason behind incorporating SVM classifier is, the currently deployed SVM classifiers are claimed to be well-capable of handling multiclass classification problems. The SVM classifier basically transforms any input stream of data into a feature space f which usually has higher dimension. SVM basically uses the geometrical characteristics and patterns of training data rather not considering the dimension of input space every time. In the process of normal user profiling scenario the study considered a set of users' $u \in S_{\alpha}$, here S_{α} denotes the system components which include a sequence of users. The system performs a set of transaction denoted by T_i which are basically being tracked down by the proposed system's user monitoring component. The set of transactions in this system can be represented with $S_{\alpha} \in T_i \in S_{log}$. Where S_{log} denotes system log which contains a set of user profiling and system oriented information. The occurrence of first user profile can establish a set of transaction, therefore an effective a dynamic user profiling is needed. Therefore, the study introduced a continuous monitoring system. The user monitors M1, M2, Mn assist the system to keep track of false positive events in each profile, in this way it can provide efficient intrusion detection. The monitors later used to update each profiles. The system user sequence can be defined with a set of items such as {set1, set2, .., setn}. Therefore the system also intended to view cumulative activities of users' altogether in a sequence which assists the system to detect the suspicious event with their corresponding set of items and list of different activities. It also monitors the user's attempt towards increasing period of transaction time.

5   Results Discussion

This section of the study presents the experimental outcomes obtained after simulating the prototyped IDS under different test case scenarios. The study considered performing the experiment considering a WLAN architecture operated on a real time environment.

There are 41 various features extracted in the presence of normal TCP/IP connection. The SVM classifier has been trained with different classes of data such as normal and attacked data attributes. The prime objective of the classifier is to distinguish the normal classes and attack classes by objectifying different set of patterns. Data points were randomly generated with the normal data usage and attack patterns. The study also defines a future space where the training set examples will be classified.

The study also performed a test case procedure which helps verifying the proposed system. In the initial phase the system takes input datasets and converts into different frequency pf attributes. The study also performed a comparative analysis to ensure the superiority of the proposed system in terms of accuracy and CPU runtime. The experimental outcomes for the proposed and existing IDS are summarized into the following Fig. 3.

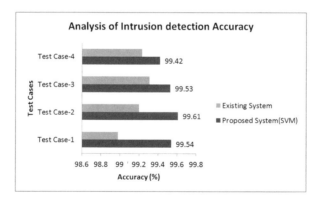

Fig. 3. Comparative analysis with respect to Accuracy

The training data are separated into attack and normal data attributes. Further the SVM was fed with the training datasets. During the training procedure SVM predictive models are added into operations. The study considers a test dataset which consist approximately 7000–8000 randomly generated patterns.

The above stated Fig. 3 clearly depicts the fact that the proposed SVM based method provides robustness by achieving higher accuracy in network anomaly detection as compared to the proposed system. The study performed another mode of comparative analysis which includes comparison of these two systems with respect to CPU execution time. The summarized outcomes are compared with the following Fig. 4.

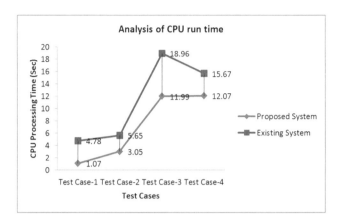

Fig. 4. Comparative analysis with respect to processing time

    The above figure clearly shows that the proposed system excels better performance with negligible computational complexity thus exhibit effectiveness as compared to the existing intrusion detection algorithms for WLAN.

6   Conclusion

The study proposed a SVM classifier based security model with an aid of efficient computational aspect to recognize the malicious events taking place in distributed WLAN. The study conceptualized and analytical methodology where a monitoring system for both user and network has been introduced. The study also exhibited the matter of fact that how SVM assists the system to distinguish between different set of attack or normal workflow patterns. The proposed system can be used as benchmarking one to improve the security of WLAN by using different machine learning approaches. Also, this study can be implemented to find the intrusion detection and response time. The experimental analysis further conveyed the effectiveness of the proposed system in terms of accuracy and computation time.

References

1. Patel, A., Qassim, Q., Wills, C.: A survey of intrusion detection and prevention systems. Inf. Manag. Comput. Secur. J. **18**(4), 277–290 (2010)
2. Kim, J., Bentley, P.J., Aickelin, U., Greensmith, J., Tedesco, G., Twycross, J.: Immune system approaches to intrusion detection – a review. Nat. Comput. **6**(4), 413–466 (2007)
3. Loo, C.E., et al.: Intrusion detection for routing attacks in sensor networks. Int. J. Distrib. Sens. Netw. **2**(4), 313–332 (2006)
4. Wu, H., Freedman, J., Ivory, C.J.: Network intrusion detection and analysis system and method. US Patent 7,493,659, 17 Feb 2009
5. Guerrero-Zapata, M., et al.: The future of security in wireless multimedia sensor networks. Telecommun. Syst. **45**(1), 77–91 (2010)

6. Uppuluri, P., Sekar, R.: Experiences with specification-based intrusion detection. In: Lee, W., Wespi, L.M.A. (eds.) Recent Advances in Intrusion Detection, Lecture Notes in Computer Science, vol. 2212, pp. 172–189 (2001)

7. Sekar, R., Gupta, A., Frullo, J., Shanbhag, T., Tiwari, A., Yang, H., Zhou, S.: Specification-based anomaly detection: a new approach for detecting network intrusions. In: Proceedings of the 9th Conference on Computer and Communications Security, CCS 2002, Washington, DC, USA, pp. 265–274 (2002)

8. Mitchell, R., Chen, I.R.: Behavior rule based intrusion detection for supporting secure medical cyber physical systems. In: International Conference on Computer Communication Networks, Munich, Germany (2012)

9. Berthier, R., Sanders, W.: Specification-based intrusion detection for advanced metering infrastructures. In: Proceedings of the 17th Pacific Rim International Symposium on Dependable Computing, Pasadena, CA, USA, pp. 184–193 (2011)

10. Cárdenas, A.A., Amin, S., Lin, Z.-S., Huang, Y.-L., Huang, C.-Y., Sastry, S.: Attacks against process control systems: risk assessment, detection, and response. In: The 6th Symposium on Information, Computer and Communications Security, Hong Kong, China, pp. 355–366 (2011)

11. Chen, Y., Luo, B.: S2a: secure smart household appliances. In: The Second Conference on Data and Application Security and Privacy, San Antonio, TX, USA, pp. 217–228 (2012)

12. Jokar, P., Nicanfar, H., Leung, V.: Specification-based intrusion detection for home area networks in smart grids. In: International Conference on Smart Grid Communications, Brussels, Belgium, pp. 208–213 (2011)

13. Ali, Q.I.: Design and implementation of an embedded intrusion detection system for wireless applications. IET Inf. Secur. **6**(3), 171–182 (2012)

14. Alotaibi, B.: A majority voting technique for wireless intrusion detection systems (2016)

15. Gu, W., et al.: Null data frame: a double-edged sword in IEEE 802.11 WLANs. IEEE Trans. Parallel Distrib. Syst. **21**(7), 897–910 (2010)

16. Kumar, V., et al.: Detection of stealth man-in-the-middle attack in wireless LAN. In: Proceedings of the 2nd IEEE International Conference on Parallel Distributed and Grid Computing (PDGC). IEEE (2012)

17. Salem, O., Yaning, L., Ahmed, M.: Anomaly detection in medical WSNs using enclosing ellipse and chi-square distance. In: 2014 IEEE International Conference on Communications (ICC). IEEE (2014)

18. Sbeiti, M., et al.: Paser: secure and efficient routing approach for airborne mesh networks. IEEE Trans. Wirel. Commun. **15**(3), 1950–1964 (2016)

19. Sun, T., Liu, X.: Agent-based intrusion detection and self-recovery system for wireless sensor networks. In: Proceedings of the 5th IEEE International Conference on Broadband Network & Multimedia Technology (IC-BNMT). IEEE (2013)

20. Wu, K., et al.: hJam: attachment transmission in WLANs. IEEE Trans. Mob. Comput. **12**(12), 2334–2345 (2013)

FMHT: A Novel Framework for Mitigating Hidden Terminal Issue in Wireless Mesh Network

T. H. Naveen[1(✉)] and G. Vasanth[2]

[1] Visvesvaraya Technological University, Belagavi, Karnataka, India
thnaveen.rsrch@gmail.com
[2] Department of Computer Science and Engineering, Government Engineering College,
Krishnarajapet, Mandya District, Karnataka, India

Abstract. Various researchers have theoretically discussed the Hidden terminal problem in Wireless Mesh Network (WMN), but almost no standard implementation or modeling is being explored in recent years when WMN is used in large-scale communication in the presence of interference and collision. Therefore, this paper presents a framework called as Framework for Mitigating Hidden Terminal FMHT that contributes to the identification of hidden terminal followed by compensating the communication loss by introducing a novel algorithm for enhancing channel capacity. The sole target was to improve the Quality-of-Service in WMN over different conditions of dynamic traffic. The proposed system also assists in formulating an effective decision for traffic flow admission where the simulation outcome shows better improvement in data delivery performance in contrast to existing techniques.

Keywords: Wireless mesh network · Hidden terminal issue
Routing performance · Throughput · Collision · Dynamic traffic

1 Introduction

A Wireless Mesh Network (WMN) uses mesh topology for establishing communication among the nodes and is characterized by the client node, access point, and gateway [1]. Although, nodes could go the mobile majority of the research-based approaches doesn't consider mobility as a dominant characteristic of a node in WMN. The probable reason behind this is the inclusion of mobility concept incorporates more work towards routing update rather than performing actual communication. The conventional WMN is characterized by limited scalability with the presence of intermittent routing in multi-hop problems. Such problems often reduce its eligibility towards its usage in large-scale implementation [2]. Another significant problem is a hidden terminal which is particularly less emphasized in WMN. Irrespective of various literature [3–6], there is no significant discussion of any solution in regards to WMN. The adverse effect of a hidden node in WMN could be high latency network with increasing probability of interference and collision. The theoretical solution to hidden node problem is by maximizing the

© Springer International Publishing AG, part of Springer Nature 2019
R. Silhavy (Ed.): CSOC 2018, AISC 763, pp. 20–29, 2019.
https://doi.org/10.1007/978-3-319-91186-1_3

transmit power, eliminating obstacles, software for improving protocol, usage of a unidirectional antenna, adoption of antenna diversity. Some of the frequently evolved solutions to address this problem are channel assignment [7] and scheduling [8, 9]. At present, there is less number of effective solution to address the issues of identifying the hidden nodes as well as to improve the communication. Therefore, the principal contribution of the proposed system is to introduce simplified modeling of a framework meant for identification of hidden node followed by improvement of the quality of service regarding channel capacity. Section 2 briefs about existing research methodology while extracted problems are briefed in Sect. 3. Adopted methodology as a solution is outlined in Sect. 4 followed by an illustration of algorithm implementation in Sect. 5. Obtained results are discussed in Sect. 6 and conclusion of the proposed manuscript is provided in Sect. 7.

2 Related Work

This section outlines the existing methodologies utilized for improving the communication system in WMN particularly associated with the hidden terminal issue. Discussion about existing methodologies can be seen in our prior study [10]. Problems associated with selection of best route are witnessed in the work of Boushaba et al. [11]. Aoudia et al. [12] have introduced an opportunistic approach for developing MAC protocol for energy-efficient routing in WMN. Avallone et al. [13] have presented a heuristic method to address routing issue and problem of channel assignment in WMN. Igarashi et al. [14] have presented a transmission scheme for mitigating delay problems. Study towards fault-tolerant routing is presented by Peng et al. [15] using the software-defined network. Roh and Lee [16] have used Benders decomposition for solving the rate control problem along with channel assignment issue in a mesh network. Zhou et al. [17] have used cross-layer approach for enhancing the network performance concerning scheduling in a mesh network. Shao et al. [18] have introduced an attribute for enhancing load balancing for enhancing the coding performance in multipath communication. Usage of the bio-inspired algorithm was seen in work of Nugroho et al. [19] for improving communication performance in a mesh network. Luo et al. [20] have presented a rate control mechanism using fairness modeling approach for addressing energy issue in a mesh network. Adoption of network coding-based approach was seen in work of Gu et al. [21] for efficient route discovery. Darehshoorzadeh et al. [22] have used random modeling concept for assessing the performance of opportunistic routing in WMN. Sheshadri and Koutsonikolas [23] have used experimental approach for improving the MAC/PHY layers for better routing in WMN. Alrayes et al. [24] have used dynamic interface allocation approach for multi-radio communication in WMN. Chen et al. [25] have used adaptive network coding-based approach for addressing collision problems in traffic flow. Kim et al. [26] have used distributed tree-based approach for enhancing the performance of on-demand routing in WMN. Most recently, study towards allocation of a channel for mitigating collision problem in WMN is presented by Yoshihiro and Noi [27] using a unique channel assignment approach. All the schemes mentioned above mainly target to improve communication performance of WMN.

3   Problem Description

The prominent problems identified are as following (i) Existing research approaches mainly focuses on channel allocation techniques or network coding techniques in order to improve the communication without any consideration of dynamic environment, (ii) few studies have addressed hidden terminal problems in WMN, (iii) there is a lack of benchmarked studies towards addressing communication issues, (iv) consideration of channel capacity and its possible impact in throughput enhancement as a solution towards hidden node issue is less found in existing studies, (v). Applicability of existing studies towards real-time traffic has been less investigated in the existing system. Therefore, the problem statement is *"Designing modeling towards addressing the detection and mitigation of hidden terminal issue in WMN."* The next section outlines the methodology implemented to address this problem.

4   Proposed Methodology

The prime motive of the proposed study is to introduce a framework that can effectively identify and mitigate the problem of the hidden terminal in WMN. The schematic diagram of proposed system FMHT is showcased in Fig. 1 as follows.

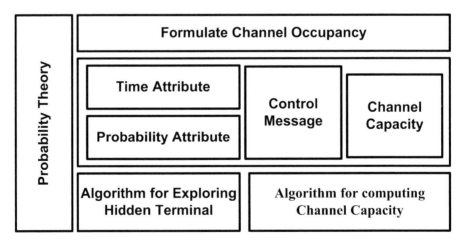

Fig. 1. Proposed architecture of FMHT

FMHT uses probability theory to perform modeling where it contributes by introducing channel occupancy with time and probability attribute as an essential role. A different form of the control message (e.g., route request and response) is used with a unique usage of channel capacity considering problems associated with multi-hop networks. The design of algorithm also emphasizes on the obtaining better probability of successful transmission over dynamic traffic scenario. The proposed FMHT potentially assists in leveraging the admission control of traffic flow as a solution to overcome the hidden node problem. The next section illustrates algorithm implementation.

5   Algorithm Implementation

The algorithm implemented for the proposed FMHT system not only targets hidden terminal but also aimed at estimating channel capacity that finally leads to the formation of admission control strategy to address the hidden terminal problem in the wireless mesh network. This section briefs about the prominent algorithms and their contribution.

5.1   Algorithm for Exploring Hidden Terminal

This algorithm takes the input of dimension of simulation area A and number of nodes N which computes and finally determine hidden terminal in the wireless mesh network. The core steps of the algorithm are as follows:

> **Algorithm for Exploring Hidden Terminal**
> **Input**: N (Number of nodes), A (Simulation Area), R (Sensing Range)
> **Output**: H_{pool} (hidden terminal)
> **Start**
> 1. init N, A
> 2. deploy rand(x, y)
> 3. **For**i=1: N
> 4. h_1=explore($\omega(i)$<R)
> 5. $h_1(h_1$==$i)$=[]
> 6.   **For**j=1:size(h_1)
> 7. nh_1→explore($\omega(h_1(j)$<R)
> 8. c_{node}→intersect(h_1, nh_1)
> 9.   **For**k=1:size(c_{node})
> 10. $nh_1(nh_1$==$c_{node}(k))$=[]
> 11.   **End**
> 12.   **For**k=1:size($sens_1$)
> 13.   **For**nn=1:size(h_1t)
> 14.   If $(h_1t(nn), sens_1)$ && $(h_1t(nn), h_1)$
> 15. H_{pool}→uniq[$h_1t(nn)$]
> 16.**End**
> 17. **End**
> 18. **End**
>
> End

All the nodes are deployed randomly within A (Line-2). The algorithm constructs a matrix ω to reposit all the distance (Line-4). The first phase of the algorithm performs neighbor discovery where a structure H_{list} is constructed to store a list of hops. For all the hops (Line-3), the algorithm extracts one hop h_1 neighbor by comparing it with the communication radius R (Line-4 & Line-5). The next step is to explore the nodes within two hops with the aid of its adjacent nodes (Line-6). It initially obtains the index of the nodes with neighbors single hop (Line-7) followed by computing the common node between h_1 and nh_1 (Line-8). To ensure better computational performance, the common nodes are further investigated to find the nodes that reside within single hop (Line-9).

Upon finding such nodes, they are removed from single hop (Line-10), and the residual nodes present in double hops are then added, i.e., h_2 and nh_1. The matrix $sens_1$ stores all the nodes within a sensing range (Line-12) and initiates a loop with it to obtain the single hop node, i.e., h_{1t} followed by an internal loop for all the nodes within sensing range (Line-12). This mechanism also obtains one-hop node h_1t and checks if h_1t is a member of $sens_1$ as well as single hop h_1 (Line-14). A matrix H_{pool} is constructed using h_1t that ultimately offers the list of the hidden terminal.

5.2 Algorithm for Computing Channel Capacity

This algorithm computes channel capacity that is required to compensate the data trans-mission process that may be possibly hampered due to the hidden terminal problem. Hence, it takes in the input of hidden nodes explored from the prior algorithm and performs processing. The core steps involved in this algorithm is as follows:

> **Algorithm for computing Channel Capacity**
> **Input**: Channel Occupancy Rate (C_o), node identity (n_{id})
> **Output**: Cc (Channel Capacity)
> 1. init S_{node}, $H_{node}(=H_{pool})$
> 2. $n \rightarrow S_{node}(n_{id})$ & $n_1 \rightarrow H_{node}(n_{id})$
> 3. **For** $P_t = p:a:q$
> 4.   **For** $i=1: T_{REQ}$
> 5.   **If** $i==1$
> 6. $op \rightarrow (1-P_t)^{n+n1-1}$
> 7.   **Else**
> 8. $op \rightarrow (1-P_t)^{n1}$
> 9.   **End**
> 10. $M_1=[op]$
> 11. **For** $i=0:(r-1)$
> 12. $M=M+(1-M_1)^w$
> 13. **End**
> 14. Compute $T_s=M+q_1$, $T_c=q_2+q_3$, & $p_{spacket} \rightarrow M_0 * P_{SDATA}$
> 15. Compute P_i, P_s, P_c
> 16. $C_o=[C_i, C_b, C_s]$ & $s \rightarrow R_s * q_4$
> 17. $C_c \rightarrow \text{argmax}(y_{vec})-y_{vec}(\min(x_{vec}-R_{b1}))$;
> **End**

This algorithm introduces a unique variable called as channel occupancy rate Co which represents its state of engagement with the current network regarding probability, i.e. [0, 1] range. The algorithm also takes the input of hidden nodes H_{node} (Line-1) and other nodes within sensing range S_{node} (Line-1). The complete analysis is carried out considering mean probability of data transmission P_t with p and q corresponds to the minimum and maximum range of transmission with a difference of a (Line-3). For the entire request packet T_{REQ}, the probability operator op is further computed (Line-6 and Line-8) concerning n and n1. The product of probability operator op is used to compute the successful probability of transmission request M1 (Line-10). The algorithm also computes the probability of collision of transmission of packet **M3** which is computed as $(1-M0)$, where $M0 \rightarrow M0 + (1-M0)M1$. The algorithm further computes time required to perform request and response for transmitting data packet M (Line-12), where w represents a weight factor depending on network parameters. Finally, the duration required for data transmission with success T_s and collision T_c is computed (Line-14). The variable q_1 and q_2 are the dependable attributes of T_s and T_c respectively using time and probability based parameters. Finally, the probability of idle state P_i, successful state P_s, and collision state P_c is computed (Line-15) using the probability of transmission and number of Packets. A variable s is computed by multiplying channel utilization R_s with another network coefficient q_4 that is computed by dividing time required for processing the data with T_s, i.e., the time required to successfully forwarding the data (Line-16). Finally, channel capacity is computed using the maximum value of y_{vec} obtained from s matrix subtracted from the utilized value of channel x_{vec}.

The next part of the implementation mainly performs admission control for enhancing the data delivery services in WMN. For this purpose, the proposed system applies on-demand routing algorithm to explore the optimal path leading to its sink node concerning cost and route. The algorithm computes the route as well as a buffer if the flow start point is found to be less than some iteration. The throughput is calculated as twice the bandwidth subtracted by initialized throughput. This empirically initialized throughput can be represented as tp $\rightarrow$ constant*rand. The throughput values more than initialized bandwidth is only considered for future computing. The algorithm also checks for interference and any form of rejected traffic flows and eliminates any such possibilities in routing in WMN. One of the interesting contributions of the proposed system is also its support to mobility owing to the highly flexible resource allocation scheme for better admission control. This, in turn, enhances the communication process to a greater extent.

6   Results Discussion

The implementation of the proposed system is carried out considering on-demand routing technique on MAC protocol of IEEE 802.11 standard to assess the performance. The scripting of proposed FMHT is carried out using MATLAB considering bandwidth of 150 kbps. The size of the MAC header and Physical header is considered as 224 bits and 192 bits respective during the analysis. Figure 2 highlights the trend of some nodes being identified as the nodes within the sensing range and hidden nodes. Found that node 15 of both the sensing node and hidden nodes meeting node ID 17. The outcome shows a smoother trend of identification which is also in good agreement with mobility management.

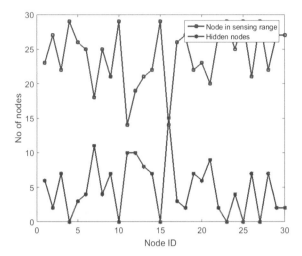

Fig. 2. Exploring hidden nodes in FMHT

The Fig. 3 shows the performance of throughput on added four different flows after a specific period. However, the analysis found a 4th flow to offer less quality of service that adversely affects the throughput. This phenomenon leads to the generation of retransmission effect thereby minimizing the channel capacity, and therefore the algorithm does not accept the 4th flow to balance the throughput performance. Hence, the proposed algorithm offers efficient conservation of the channel capacity for better use in the different form of traffic demands in WMN. Another significant observation is the algorithm can maintain the better stability of throughput over a longer period and thereby it offers guaranteed quality of service in WMN under dynamic traffic.

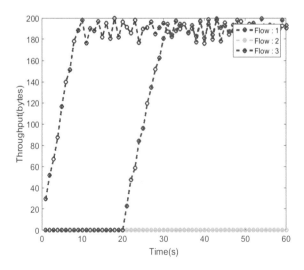

Fig. 3. Analysis of throughput

The proposed scheme is also theoretically compared with some relevant existing schemes with similar mission of performing routing in the presence of hidden terminal in WMN (Table 1).

Table 1. Theoretical comparison of existing techniques

Approaches	Average throughput trend	Supportability of mobility	Identification of hidden terminal
Yoshihiro [28]	Increase and then decrease	No	No
Hammash [29]	Increasing	No	No
Matoba [30]	decreases	No	No
Proposed	Stiff increase and then uniform stabilize	Yes	Yes

7   Conclusion

Ensuring an effective Quality of Service in WMN is quite a challenging task in real-time application owing to the higher degree of dynamicity associated with it, the absence of any form of centralized authority, vagueness in mobility concept implementation, the restricted energy of nodes, and multi-hop communication. We strongly feel that WMN will have to deal with various problems to obtained better Quality of Service performance, e.g., a diversified requirement of quality of Service, communication of unreliable medium, unpredictable delay, etc. Therefore, the solution presented by proposed system invokes a balance between two problems, i.e., identification of hidden terminal and enhancing the quality of service concerning channel capacity. The study outcome proves that proposed system offers a simplified technique to address the problem in comparison to the existing system.

References

1. Ma, L., Khreishah, A., Zhang, Y., Yan, M.: Wireless Algorithms, Systems, and Applications. Springer, Boston (2017)
2. Sinha, K., Ghosh, S.C., Sinha, B.P.: Wireless Networks and Mobile Computing. CRC Press, Boca Raton (2016)
3. Boroumand, L., Khokhar, R.H., Bakhtiar, L.A., Pourvahab, M.: A review of techniques to resolve the hidden node problem in wireless networks. Smart Comput. Rev. **2**(2) (2012)
4. Al-Saadi, A., Setchi, R., Hicks, Y., Allen, S.M.: Routing protocol for heterogeneous wireless mesh networks. IEEE Trans. Veh. Technol. **65**(12), 9773–9786 (2016)
5. Karia, D.C., Jadiya, A., Kapuskar, R.: Review of routing metrics for wireless mesh networks. In: 2013 International Conference on Machine Intelligence and Research Advancement, Katra, pp. 47–52 (2013)
6. Ahmeda, S.S., Esseid, E.A.: Review of routing metrics and protocols for wireless mesh network. In: 2010 Second Pacific-Asia Conference on Circuits, Communications and System, Beijing, pp. 27–30 (2010)

7. Lee, C., Shin, D., Choi, S.: Weighted conflict-aware channel assignment in 802.11-based mesh networks. In: 2016 18th International Conference on Advanced Communication Technology (ICACT), Pyeongchang, p. 1 (2016)
8. Takeda, T., Yoshihiro, T.: A queue-length based distributed scheduling for CSMA-driven wireless mesh networks. In: 2016 International Conference on Computing, Networking and Communications (ICNC), Kauai, HI, pp. 1–6 (2016)
9. Yoshihiro, T., Nishimae, T.: Practical fast scheduling and routing over slotted CSMA for wireless mesh networks. In: 2016 IEEE/ACM 24th International Symposium on Quality of Service (IWQoS), Beijing, pp. 1–10 (2016)
10. Naveen, T.H., Vasanth, G.: Qualitative study of existing research techniques on wireless mesh network. Int. J. Adv. Comput. Sci. Appl. **8**(3), 49–57 (2017)
11. Boushaba, M., Hafid, A., Gendreau, M.: Source-based routing in wireless mesh networks. IEEE Syst. J. **10**(1), 262–270 (2016)
12. Aoudia, F.A., Gautier, M., Berder, O.: OPWUM: opportunistic MAC protocol leveraging wake-up receivers in WSNs. J. Sens. **2016**, 1–9 (2016)
13. Avallone, S., Banchs, A.: A channel assignment and routing algorithm for energy harvesting multiradio wireless mesh networks. IEEE J. Sel. Areas Commun. **34**(5), 1463–1476 (2016)
14. Igarashi, Y., Matsuura, Y., Koizumi, M., Wakamiya, N.: Priority-based dynamic multichannel transmission scheme for industrial wireless networks. Wirel. Commun. Mob. Comput. **2017**, 1–14 (2017)
15. Peng, Y., Gong, X., Guo, L., Kong, D.: A survivability routing mechanism in SDN enabled wireless mesh networks: design and evaluation. China Commun. **13**(7), 32–38 (2016)
16. Roh, H.T., Lee, J.W.: Channel assignment, link scheduling, routing, and rate control for multi-channel wireless mesh networks with directional antennas. J. Commun. Netw. **18**(6), 884–891 (2016)
17. Zhou, A., Liu, M., Li, Z., Dutkiewicz, E.: Joint traffic splitting, rate control, routing, and scheduling algorithm for maximizing network utility in wireless mesh networks. IEEE Trans. Veh. Technol. **65**(4), 2688–2702 (2016)
18. Shao, X., Wang, R., Huang, H., Sun, L.: Load balanced coding-aware multipath routing for wireless mesh networks. Chin. J. Electron. **24**(1), 8–12 (2015)
19. Nugroho, D.A., Prasetiadi, A., Kim, D.S.: Male-silkmoth-inspired routing algorithm for large-scale wireless mesh networks. J. Commun. Netw. **17**(4), 384–393 (2015)
20. Luo, C., Guo, S., Guo, S., Yang, L.T., Min, G., Xie, X.: Green communication in energy renewable wireless mesh networks: routing, rate control, and power allocation. IEEE Trans. Parallel Distrib. Syst. **25**(12), 3211–3220 (2014)
21. Gu, Y., Han, H., Li, X., Guo, J.: Network coding-aware routing protocol in wireless mesh networks. Tsinghua Sci. Technol. **20**(1), 40–49 (2015)
22. Darehshoorzadeh, A., De Grande, R.E., Boukerche, A.: Toward a comprehensive model for performance analysis of opportunistic routing in wireless mesh networks. IEEE Trans. Veh. Technol. **65**(7), 5424–5438 (2016)
23. Sheshadri, R.K., Koutsonikolas, D.: An experimental study of routing metrics in 802.11n wireless mesh networks. IEEE Trans. Mob. Comput. **13**(12), 2719–2733 (2014)
24. Alrayes, M.M., Biswash, S.K., Tyagi, N., Tripathi, R.: An enhancement of AODV with multi-radio in hybrid wireless mesh network. ISRN Electron. **2013**, 1–13 (2013)
25. Chen, J., He, K., Du, R., Zheng, M., Xiang, Y., Yuan, Q.: Dominating set and network coding-based routing in wireless mesh networks. IEEE Trans. Parallel Distrib. Syst. **26**(2), 423–433 (2015)

26. Kim, S.H., Chong, P.K., Kim, D.: A location-free semi-directional-flooding technique for on-demand routing in low-rate wireless mesh networks. IEEE Trans. Parallel Distrib. Syst. **25**(12), 3066–3075 (2014)
27. Yoshihiro, T., Noi, T.: Collision-free channel assignment is possible in IEEE802.11-based wireless mesh networks. In: 2017 IEEE Wireless Communications and Networking Conference (WCNC), San Francisco, CA, pp. 1–6 (2017)
28. Yoshihiro, T., Nishimae, T.: Practical fast scheduling and routing over slotted CSMA for wireless mesh networks. In: 2016 IEEE/ACM 24th International Symposium on Quality of Service (IWQoS), Beijing, pp. 1–10 (2016)
29. Hammash, D., Kim, M., Lee, B.: HIAM: hidden node and interference aware routing metric for multi-channel multi-radio mesh networks. In: ACM Conference (2013)
30. Matoba, A., Hanada, M., Kanemitsu, H., Kim, M.W.: Asymmetric RTS/CTS for exposed node reduction in IEEE802.11 ad hoc networks. J. Comput. Sci. Eng. **8**(2), 107–118 (2014)

Recursive Algorithm for Exhaustive Search of Possible Multiversion Software Realizations with the Choice of the Optimal Versions Set

Roman Yu Tsarev, Denis V. Gruzenkin[✉], and Galina V. Grishina

Siberian Federal University, Krasnoyarsk, Russia
tsarev.sfu@mail.ru, gruzenkin.denis@good-look.su,
ggv-09@inbox.ru

Abstract. N-version software is used all over the world as one of the approaches that can provide with the high level of reliability and software fault tolerance. The application of redundant module versions of software allows to obtain a correct result even if there is an error in the separate module versions. However, the program redundancy that can increase software reliability needs extra resources. It results in an optimization problem. There is a necessity for a certain variant of multiversion software realization i.e. such a modules versions set is required that demands less resources and guarantees high level of reliability simultaneously. The exhaustive search of all possible multiversion software realizations is carried out by the recursive algorithm proposed in the article.

Keywords: Multiversion software · N-version software · Reliability
Optimization · Exhaustive search · Recursion · Recursive algorithm

1 Introduction

The problem of software reliability has been in existence as long as software exists. There is a demand for the software that guarantees a high level of reliability. It forces software designers to resort to such methods and tools that allow creating error and fault tolerance software. Since the beginning of 1960s when the software industry started to develop, a vast amount of approaches and methods of assessment and increasing of software reliability have been suggested [1].

There are some of the approaches that can be distinguished. They are based on time, information and program redundancy. The introduction of redundancy enables both to increase reliability and to guarantee fault tolerance of software. The program redundancy is implemented by two main approaches. They are N-version programming [2, 3] and recovery blocks [4, 5].

N-version programming has successfully proved itself particularly in such spheres as fault-tolerant control software for communications satellite system [6], railway interlocking systems [7], producing an architectural framework to automate and enhance

© Springer International Publishing AG, part of Springer Nature 2019
R. Silhavy (Ed.): CSOC 2018, AISC 763, pp. 30–36, 2019.
https://doi.org/10.1007/978-3-319-91186-1_4

application security [8], developing a N-version programming-based protection scheme for microgrids [9], web services systems [10].

The idea of N-version programming can be understood in the following way. The developing software has to solve a certain problem. The solution of this problem is the achievement of a certain goal. The problem is divided into some subtasks and the goal is achieved by finding solutions to them. On the conceptual level every subtask is solved by an appropriate module. The module is realized by means of several versions (multi-versions) in order to ensure high reliability and fault tolerance of software. So, the modules and the software as a whole are becoming multiversion.

The introduction of the program redundancy in the form of the redundant modules versions ensures high reliability. It happens due to the fact that if one (or several) of the modules versions returns an incorrect result, the other versions return correct results nevertheless. When all results are obtained, it is necessary to analyze them and choose the one that is correct. It will be sent to all versions of the next module as input data. The analysis is carried out by a decision-making unit. The process of decision-making is usually realized by a voting algorithm [11, 12]. During the implementation of N-version software, the voting algorithm is implemented after every modules versions implementation. This algorithm defines the correct result of the operation of the whole module (i.e. all its versions).

The main problem of the application of any type of redundancy is the requirement for extra resources. The problem is connected with optimizing that could correspond to a higher level of reliability and at the same time to less amount of resources [13]. While developing N-version software the problem is defined as the choice of a modules versions set that could achieve the goals. The formal description of the problem is presented below.

2   The Generation Model of Optimal Versions Set of N-Version Software

The conventional signs that are used are as follows

n – the number of subtasks that are required to be solved to achieve the goal;
i – a subtask number, $i = 1, 2, \ldots, n$;
m_i – the number of multiversions that are available for the solution of i-subtask;
j – an available multiversion number for the solution of i-subtask, $j = 1, 2, \ldots, m_i$;
R_{ij} – the reliability (the possibility of no-failure operation) of j-multiversion, solving i-subtask;
N_i – the set of all multiversions subsets with the power range from 1 to m_i;
N_i^* – the multiversions subset, $N_i^* \in N_i$;
$|N_i^*|$ – power N_i^*;
RN_i^* – the N_i^* reliability during i-subtask solution. It is equal to the possibility that no less than $|N_i^*|/2$ multiversions from a large number of multiversions N_i return a similar result;
v_i – the voting algorithm that analyzes the results of i-subtask solution, $i = 1, 2, \ldots, n$;
R_{vi} – the voting algorithm reliability v_i, $i = 1, 2, \ldots, n$;

C_{ij} – the cost of j-multiversion that solves i-subtask;
C_{vi} – the cost of the voting algorithm development v_i, $i = 1, 2, \ldots, n$;
C_i – the total cost of multiversions selected for the i-subtask solution;
x_{ij} – Boolean variable is equal to1, if j-multiversion is selected for i-subtask solution, and it is equal to 0 in an opposite case.

The selection problem of the optimal versions set of N-version software can be presented by dual-purpose nonlinear task of integer programming with Boolean variables:

$$\max f_1(x) = \prod_{i=1}^{n} R_i R_{vi} \tag{1}$$

$$\min f_2(x) = \sum_{i=1}^{n} \sum_{j=1}^{m_i} C_{ij} x_{ij} + \sum_{i=1}^{n} C_{vi} \tag{2}$$

on conditions that:

$$\sum_{j=1}^{m_i} x_{ij} \geq 1, i = \overline{1, n},$$

$$R_i = \sum_{N_i^* \in N_i} \left[\prod_{j \in N_i^*} x_{ij} \prod_{k \in N_i - N_i^*} (1 - x_{ik}) \right] RN_i^*,$$

$$RN_i^* = \sum_{j=|N_i^*|/2}^{|N_i^*|} \left[\sum_{M \in N_i^* ||M|=j} \left\{ \prod_{k \in M} R_{ik} \prod_{l \in N_i^* - M} (1 - R_{il}) \right\} \right],$$

The objective function (1) maximizes the software reliability while the objective function (2) minimizes the software cost. As a rule, these two purposes conflict with each other.

The software cost includes the cost of every multiversion or decision-making unit realizing the voting algorithm only once. So, if one and the same voting algorithm is applied after several subtasks solutions, its cost is included only once.

The realization variants of N-version software can be a solution to this problem, i.e. a certain modules versions set (or a version set) of N-version software. It is possible to choose the variant after the exhaustive search of all possible variants of N-version software. In order to solve the task the algorithm is proposed.

3   The Exhaustive Search Algorithm of All Possible Realizations of N-Version Software

If the i-subtask is solved by the i-module realized in the form of some multiversions then the number of i-module multiversions is equal to m_i. The i-module multiversions set is presented by the series $x_{i1}, x_{i2}, \ldots, x_{imi}$, where x_{ij} is equal to 1, if j-multiversion is used in the i-module of N-version software and it is equal to 0 in the opposite case (see Fig. 1).

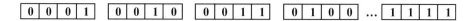

Fig. 1. Possible module version sets realized in four multiversions

There is an algorithm which is used to generate module version sets (Listing 1). The input data for this algorithm is the natural number N that varies in the range from 1 to 2mi − 1. Every N number in this range corresponds to one of the i-module version sets. The number of i-module version sets that is equal to 2mi − 1 is exhaustive. The i-module version sets is stored in a binary form in a one-dimensional array Bin, whose size is equal to mi.

```
assign for record the rightmost cell of the array Bin
execute in a cycle:
  record in the array Bin record cell the remainder of N
on division by 2;
  assign for record the cell that precedes the current
record cell;
  divide N by 2;
until the integer part from division is equal to 0;
record zeros in the remaining cells of the array Bin.
```

Listing 1. A binary array generation algorithm corresponding to the module version set

An exhaustive search algorithm of different realizations of N-version software has been developed. The algorithm is presented as a recursive function in pseudolanguage (Listing 2). On the basis of multiversions multitude the algorithm allows to consider all possible variants of the modules version sets of N-version software.

```
recursive_function (module number i, binary array Alt)
  for natural number N from 1 to 2mi - 1 execute in a cy-
cle:
    generate a binary array Bin for the current value N;
    copy the array Bin values in the i-line of the array
    Alt;
    if i-module is not the last then
      call for the recursive function (with the number of
      the next module (i+1) and the binary array Alt);
    carry out a required action on the current variant;
  the end of the cycle;
the end of the recursive function.
```

Listing 2. The algorithm of the recursive function performing an exhaustive search of all possible version set variants of N-version software.

The first function call is accompanied by sending the module number i = 1 as the first argument of the function. The current variant of the modules version set of N-version

software is stored in a two-dimensional binary array Alt with the size n by max mi, i =
1, 2, ..., n.

The cycle is executed in the recursive function and the last line of the cycle implies
any required actions that can be carried out on the obtained variant of version set of N-
version software. There are some examples of such actions. They are the calculation of
the reliability (1) or the cost (2) of the current variant of N-version software generation,
the record of both the obtained variant version set and the values of characteristics
corresponding to the variant into separate arrays for further application without the
recurrence of the exhaustive search.

Figure 2 shows the example of the recursive function execution during the
exhaust search of different realizations of N-version software consisting of six
modules that are realized by the following number of multiversions: $m_1 = 5$, $m_2 = 4$,
$m_3 = 4$, $m_4 = 3$, $m_5 = 4$, $m_6 = 5$.

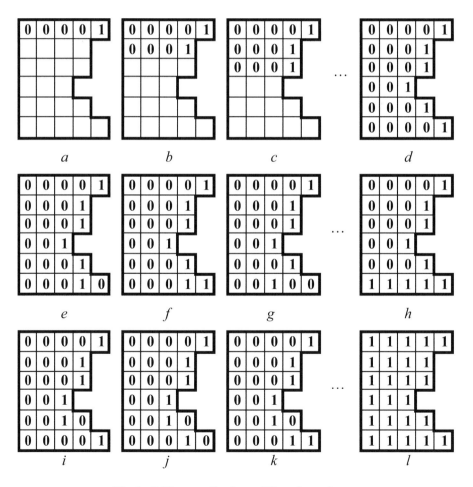

Fig. 2. Different realizations of N-version software

Every picture in Fig. 2 corresponds to one of the version sets of N-version software. The lines mean modules, the number of cells in a line mean the maximum possible number of the current module, the values in cells mean the values of the variable x_{ij} that reflects whether the current module version is applied ($x_{ij} = 1$) or not ($x_{ij} = 0$).

Figure 2a–d show the generation of the first variant of realization for every module from one multiversion according to the algorithm in Listing 1. A new recursive function copy is called for every module. Figure 2d shows the first variant of the version set of N-version software.

After that the sixth module version set is searched in the last recursive function copy (Fig. 2e–h).

Then there is a return to the previous recursive function copy and another version is selected for the fifth module. The fifth line in Fig. 2i corresponds to this case. And again a new recursive function copy is called and the sixth module multiversions are searched (Fig. 2i–k).

The last possible variant of the version set of N-version software is in Fig. 2l. All available multiversions are selected.

So, the proposed recursive algorithm allows to make a complete exhaustive search of all possible realizations of N-version software.

4   Conclusion

N-version software has a high level of fault tolerance and reliability due to the realization of program redundancy principle. In practice, reliable software modules are realized as a series of functionally equivalent versions. A software designer can include one or another module version into N-version software. The selection of versions is caused by the necessity to ensure a high level of reliability and to reduce the applied resources. The proposed recursive algorithm enables to make the exhaustive search of all possible realizations of N-version software that allows a decision maker to select an optimal variant.

References

1. Sommerville, I.: Software Engineering, 9th edn. Addison-Wesley, Wokingham/Reading (2010)
2. Avizienis, A., Chen, L.: On the implementation of N-version programming for software fault-tolerance during program execution. In: Proceedings of the IEEE Computer Society International Conference on Computers, Software and Applications, COMPSAC 1977, pp. 149–155 (1977)
3. Gruzenkin, D.V., Chernigovskiy, A.S., Tsarev, R.Y.: N-version software module requirements to grant the software execution fault-tolerance. Adv. Intell. Syst. Comput. **661**, 293–303 (2018)
4. Randell, B., Jie, X.: The evolution of the recovery block concept. In: Lyu, M.R. (ed.) Software Fault Tolerance, pp. 1–21. Wiley (1995)

5. Kaur, R., Arora, S., Jha, P.C., Madan, S.: Fuzzy multi-criteria approach for component selection of fault tolerant software system under Consensus Recovery Block Scheme. Procedia Comput. Sci. **45**(C), 842–851 (2015)
6. Kulyagin, V.A., Tsarev, R.Y., Prokopenko, A.V., Nikiforov, A.Y., Kovalev, I.V.: N-version design of fault-tolerant control software for communications satellite system. In: 2015 International Siberian Conference on Control and Communications, SIBCON 2015 - Proceedings, Article no 7147116 (2015)
7. Eriş, O., Yildirim, U., Durmuş, M.S., Söylemez, M.T., Kurtulan, S.: N-version programming for railway interlocking systems: Synchronization and voting strategy. In: IFAC Proceedings Volumes (IFAC-PapersOnline), pp. 177–180 (2012)
8. Malaika, M., Nair, S., Coyle, F.: N-version architectural framework for application security automation (NVASA). CrossTalk **27**(5), 30–34 (2014)
9. Hussain, A., Aslam, M., Arif, S.M.: N-version programming-based protection scheme for microgrids: a multi-agent system based approach. Sustain. Energy Grids Netw. **6**, 35–45 (2016)
10. Peng, K.-L., Huang, C.-Y., Wang, P.-H., Hsu, C.-J.: Enhanced N-version programming and recovery block techniques for web service systems. In: International Workshop on Innovative Software Development Methodologies and Practices, InnoSWDev 2014 - Proceedings, pp. 11–20 (2014)
11. Durmuş, M.S., Eriş, O., Yildirim, U., Söylemez, M.T.: A new voting strategy in Diverse programming for railway interlocking systems. In: Proceedings 2011 International Conference on Transportation, Mechanical, and Electrical Engineering, TMEE 2011, Article no 6199304, pp. 723–726 (2011)
12. Rezaee, M., Sedaghat, Y., Farmad, M.K.: A confidence-based software voter for safety-critical systems. In: Proceedings - 2014 World Ubiquitous Science Congress: 2014 IEEE 12th International Conference on Dependable, Autonomic and Secure Computing, DASC 2014, Article no 6945688, pp. 196–201 (2014)
13. Gruzenkin, D.V., Grishina, G.V., Durmuş, M.S., Üstoğlu, I., Tsarev, R.Y.: Compensation model of multi-attribute decision making and its application to N-version software choice. Adv. Intell. Syst. Comput. **575**, 148–157 (2017)

Noise Detection and Elimination
by Using Analytical Approach

N. Rekha[1](✉) and Fathima Jabeen[2]

[1] Department of Electronics and Communication Engineering, K.S. Institute of Technology,
Bangalore, India
rekhaphd2014@gmail.com
[2] Islamiah Institute of Technology, Bangalore, India

Abstract. The existence of noise periodically decreases the total contains required to transmit over the GSM channel. Also, due to increased challenges in the upcoming communication system for noise cancellations in the voice as well as in data transmitted. The existing filters were not able to perform the effective noise cancellation, thus this paper introduces analytical approach to detect both the GSM transient signals of superior or inferior form and then to cancel its noise level. The proposed approach considers the probability theory to perform the modeling of the system and allocate the power of the transitive device along with noise level. The result analysis of the proposed system gives that it offers the detection of the different noise forms and also can significantly determines the both the superior as well as inferior signal quality. These outcomes suggest that the design of accurate filter can be efficient for noise cancellation in GSM signal.

Keywords: Analytical approach · Power allocation · Noise cancellation
Noise level · GSM signal

1 Introduction

The noise is surfacing of unwanted signal which leads to original channel degradation [1]. Conventionally, any electrical or electronic circuit's experiences noises like short noise, flicker noise, avalanche noise etc. [2]. Also, some other kinds of noise are exist which mainly originated from the certain variation in the original quantity and these noises only eliminated through bandwidth enhancement and signal averaging [3]. The GSM channel faces some critical challenges in which the user's experiences different forms of noises and interferences over the microphone signal [4]. This situation takes place only when the user device will be in transmission mode along with mobility stage and this phenomenon is common in the GSM based communication system [5]. Also, some noise segments are exist which is generated through crystal and are particularly periodic in nature and will generate the potential noise level. Due to this the GSM channel faces critical issue with the network. Currently the services over 3G or 4G network, the users prefer multimedia data with higher quality than the voice data. But, the presence of noise is leading to quality degradation and also, the existing techniques

© Springer International Publishing AG, part of Springer Nature 2019
R. Silhavy (Ed.): CSOC 2018, AISC 763, pp. 37–45, 2019.
https://doi.org/10.1007/978-3-319-91186-1_5

can filter only noise in case the information of the particular noise is provided. Thus, this paper a simple analytical approach is presented which analyzes the impact of power allocation in noise minimization by using probability theory. The complete paper is planned with different sections like: Sect. 2 review of existing techniques, Sect. 3 problem statement, Sect. 4 System design & Sect. 5 implementation, Sect. 6 results analysis and end up with Sect. 7 conclusion.

2   Review of Existing Techniques

This section elaborates the existing techniques evolved in the noise level control, elimination during data transmission through wireless communication. This paper is the extension of stochastic approach presented in Rekha et al. [6]. The work of Mostafa [7] addressed the impulsive noises of conventional OFDM signals. The work of Zhang et al. [8], expressed the selection method for noise minimization with low transmission cost. In Antonanzas et al. [9] introduced the controller design by using least mean square (LMS) mechanism.

The work of Ferras et al. [10], have described the noise resistant system over a medium. Author Liu et al. [11] explained the optimized scheme to enhance the scheduling scheme for noisy wireless network. Prominent works of Ruder et al. [12] have discussed the interference cancellation scheme for GSM related problems. Demissie [13] has explained the cluster noise cancellation mechanisms for GSM signals by using LMS mechanism. In Goosens et al. [14], a unique filter is designed to perform the denoising of various forms of noises.

Jiang et al. [15] focused on the flickering noise and introduced noise minimization technique which offers enhanced linearity. The neural network based noise minimization concept is introduced in Salmasi et al. [16]. From the above techniques it is found that most of them are directly or indirectly worked for noise elimination. Few are considered the GSM based technique for noise elimination and these techniques have their own significances as well as limitations.

3   Problem Statement

The following issues were found in the analysis of current existing techniques:

- Most of the techniques were evolved with detection of noises generated by the static sources and are highly incapable of detecting the fast moving source generated noises.
- The recently available researches were considered the multimedia of acoustic noises but not the image, text based data.
- Also very rare researches have been witnessed to overcome the issues of the GSM based noise level before subjecting to any other noise cancellation techniques.
- The allocated power impact over the GSM device for noise generation is studied in existing systems.

Thus it is quite difficult task to detect the undefined types of noises generated by the impact of *power which is allocated in GSM signals and then cancel the noise effectively.*

4   System Design

The proposed system is the extended work presented in [] which is also meant for harmonics elimination by using stochastic approach. In this, probability theory is been adopted for the analytical approach to detect the noise on the basis of power allocation performed. Figure 1, illustrates the schematic diagram of system design.

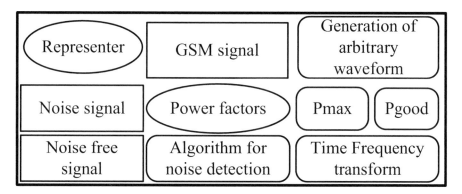

Fig. 1.   Schema of proposed methodology

The proposed system considers an analytical approach a GSM signal generated from the fast moving vehicle and is incorporated with undefined noises along with significant interference level. Therefore, the system introduced a mechanism where such undefined noise is represented from the GSM signal as well as from arbitrary waveform generator to perform modeling on undefined form of noises in GSM signal. Then a power factors is designed where power of good signals as well as maximum power required is computed. An algorithm for detection of noise is developed which again segregates good from bad signal in GSM communication system. The proposed system contributes with a simple design with faster response time to perform detection of good signals for the vehicles moving at faster velocity. The algorithm implementation is explained as below.

5   Algorithm Implementation

The algorithm aims to perform a precise detection of all form of radio-frequency noises present in the acoustic GSM signal. The basic idea behind the proposed study is to precisely represent a space where the detection of the GSM signals is required to be carried out. Such exploration of the space is always indexed with an aid of certain attributes from the surveilled GSM signal with the help of a new variable called as *represent or*. Although, the detection of various GSM signals can be done through the approaches of discriminative, stochastic, neuronal type and in this work a probabilistic theory to do the same. The *represent or* is the amount which can be estimated through the GSM signals aggregated from the transmitter and is the collection of all the attributes impacting communication quality. Are present R(t) can be gathered by power of constructive signals (P_{good}) computing level divided with radio-frequency (P_{noise}) interference obtained by GSM signal as follows:

But, the prime intension of the research is to classify the different types of different noisy GSM signals subjected to the extent of allocated power over a specific GSM channel. This is resolved through harnessing the characteristics of the broadband meant for this form of transient GSM signal. Thus, a fixed band of frequency is considered for GSM which varies from minimum of 921 MHz frequency and maximum of 925 MHz frequency. Thus, there is no much difference among the P_{good} and P_{noise}. Hence, it is feasible for approximating the noises generating from radio-frequency interference by considering the highest feasible extent of power (P_{noise}) keeping the similar range of frequency band (P_{max}).

The initiation of the proposed system begins with the computation of the time-frequency transform followed by utilization of coefficients to estimate the *represent or*. The represent or can be computed by assessing the patterns of time and frequency pattern and then followed by P_{good} and P_{max}. This computation is performed at a regular time interval (t). It is considered that, if transient factor are exist during t time then all the GSM signals are subjected to mixing so that it is challenging to distinguish the signal on the their power attributes basis. Thus, the system considers ($t - 1$) time allocated for P_{good} so that $P_{good}(t)$ is equivalent to $P_{good}(t - 1)$. The P_{max} computation is performed by considering the highest value on the frequency band equivalent to GSM signal standard form. In same manner, the proportion between the extents of power provides an empirical value of *represent or R* at t time followed by repetition of same method over all time duration correlating to time and frequency form of data. From conventional GSM signal, the symbol rate is minimum of 270 Kbps while the time duration and frequency pattern is about 1 ms. The algorithm is introduced by associating the descriptor with the error rate value which is need to operate at 3.7 µs in the scale of transmission bit for noisy GSM signal. This work has aggregated the estimation of P_{good}/P_{noise} on multiple windows that successfully overlap with each other. The steps involved in the algorithm are described below:

Algorithm to generate transient signal
Input: transmitter (Tr), Additive white Gaussian (ϕ_g), signal (s_g)
Output: transient signal (λs_g)
Start
1. Get $Tr \rightarrow \phi_g(s_g)$
2. **If** e<thr
3.   flag q++
4. **Else**
5.   flag q—
6. $\lambda s_g(t) \rightarrow \tau.(e^{-1/x} - e^{-1/y \cdot t}).\psi(t).\sin(2\pi pt)$

End

Once they represent or pattern is extracted then numerous second-order attributes will be extracted. The preliminary parameter represents a minimum that depicts various transients' occurrences in the GSM signal and the second attribute depicts corrupted level of signal to noise ratio (SNR) during the transient stage. This is observed that these occurrences are associated with the transient exhibit a deeper impact on transmitting GSM signal. The minima are computed by the cut-off value in which the minimum indicates all the numerical value less than the cut-off value. This cut-off value (*thr*) is needed as the cut-off value is less than it could result in missing of certain minima whereas if the numerical value of cut-off is greater than the results in incorporating falsified detection. Therefore, there is always a generation of errors within the summation of all the minima values. The proposed system considers assessing the GSM signal that is characterized by 2.4 million points along with the frequency of sampling corresponding to 2.4 GHz. This is an evident depiction of window of 1 ms time duration. The proposed algorithm initially assesses the GSM signal evolution over a increased duration of time considering increasing number of points. The algorithm than focuses on capturing the amount of transient signals that is characterized by the shortest duration owing to the implication of frequency of finer sampling origin. The study considers shortest frequency to be 20 ns. The numerical value is computed through multiple operations. The first step is to establish the communication system among the mobile devices (i.e. transmitter) so that the GSM signal level is configured properly. The algorithm also uses additive Gaussian noise to incorporate transient noise in radio-frequencies of GSM signal followed by estimating all the error rates value and detection of class in order to index the extracted signal. The assessment of the noise detection is carried out for >1 ms. Then a conditional statement is introduced for both the category of noises to perform better indexing of the measurement signals of the radio-frequencies associated with the transient noises that are found to have a significant impact on the GSM signal quality. The first conditional statement suggests that if the error rate is found less than the cut-off percentage than it is considered to be better signal quality of GSM signal whereas if the error rate is found more than the cut-off percentage than it is considered to be inferior quality of GSM signal. The proposed study considers the error rate as the rate of error that has been evolved up prior to any other form of correction of error. After the minima attribute as well as summation of the second minima attribute is extracted from single assessment, all these values are finally evaluated to check its preciseness. These results in storage of such numerical values temporarily that will further be acted as orientation based. The GSM signal segregation is carried out in two ways where the first classifier recons to superior signal quality while other for inferior signal quality. The proposed system empirically expresses a recursive transient signal as follows,

$$\lambda(t) \rightarrow \tau.(e^{-1/x} - e^{-1/y^t}).\psi(t). \sin(2\pi\rho t) \tag{1}$$

In the above expression, where λ represents amplitude, x represents time duration, y represents rinse time, ψ represents unit step function, and ρ represents frequency. The validation of the proposed algorithm is performed by using considering the monitored GSM signals which are considered by a hypothetical form of transient signals. The system considers two different form of transient GSM signal which are captured through the fast moving vehicle. The algorithm considers similar variation of time interval of orientation base GSM signal and similar computation is also applicable to the SNR fluctuation. The algorithm implementation scenario is as shown below (Fig. 2):

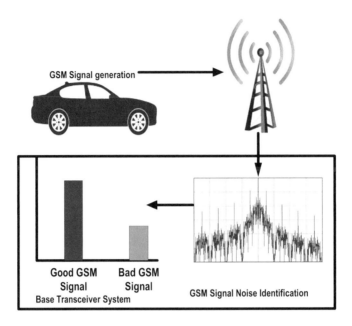

Fig. 2. Implementation of algorithm with signal generation (GSM), base transceiver system and noise identification

6   Result Analysis

The proposed system is implemented using Matlab the test values are done once with the preliminary round of monitoring, the orientation base of the GSM signal is found to consist of maximum good signal and less bad signals. The numerical scenarios assist us to encapsulate the complete GSM signal inflicted by noise. Also, it assists the defining of spectrum among two forms of signals with a special significance to the spectrum points which are close to the corresponding GSM signals peak values with different form of random noises mixed with additive white Gaussian noise.

Figure 3 illustrates the proposed system signal analysis. The algorithm does the computation to obtain the obtaining the minima quantity and summation of it to be computed simultaneously by cut-off concept as explained above. This is observed that the quantity of the true minimal can over-estimated in case the cut-off value is configured to higher value. Also, in case the cut-off value is configured to very low than it could result in non-detection of true minima. Basically, a correct version of minima could result in minimal factor of the transient signals in GSM. Also, a scatter diagram is used to analyze the consequences. For effectiveness, the threshold is given with percentage of the mean value of the representation considering 1 ms duration of window.

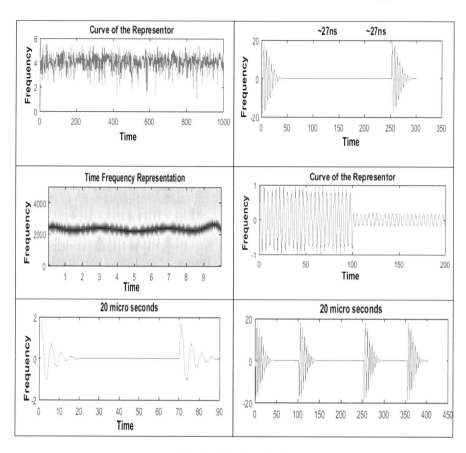

Fig. 3. Signal analysis

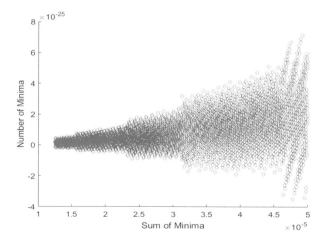

Fig. 4. Scatter plot for noise detection

Figure 4 illustrated the outlines the scatter plot of the noise detection and is observed that the dense area of scatter indicates the greater extent of noise while the scare area indicates the lower extent of transient noise. The complete algorithm is processed in 0.1662 s and can offer cost effective solution to detect the noise detection in GSM signal.

7   Conclusion

The paper mainly presents an analytical approach to detect the undefined transient noises generated by GSM signals radio frequencies. This proposed tool of noise cancellation can be deployed over any moving (terrestrial area) vehicle to detect the transmission quality of GSM signal. The proposed analytical approach considers relation among the allocated power and the level of undefined (noise) level. With performance analysis of the proposed system found the high level ability in detection and classifying the different kinds of noises generated from GSM signals. In future, the proposed system can be considered for further improvement in noise detection with different algorithms.

References

1. Munegowda, B.: Denoising Audio Signal from Various Realistic Noise using Wavelet Transform. GRIN Verlag (2016)
2. Silver, N.: The Signal and the Noise: The Art and Science of Prediction. Penguin, UK (2013)
3. Li, R.C.: RF Circuit Design. Wiley, Hoboken (2012)
4. Wilamowski, B.M., Irwin, J.D.: Fundamentals of Industrial Electronics. CRC Press, Boca Raton (2016)
5. Middlestead, R.W.: Digital Communications with Emphasis on Data Modems: Theory, Analysis, Design, Simulation, Testing, and Applications. Wiley, Hoboken (2017)
6. Rekha, N., Jabeen, F.: Study on approaches of noise cancellation in GSM communication channel. Commun. Appl. Electron. 3(5), 5–11 (2015)
7. Mostafa, M.: Stability proof of iterative interference cancellation for OFDM signals with blanking nonlinearity in impulsive noise channels. IEEE Sig. Process. Lett. 24(2), 201–205 (2017)
8. Zhang, J., Chepuri, S.P., Hendriks, R.C., Heusdens, R.: Microphone subset selection for MVDR beamformer based noise reduction. IEEE/ACM Trans. Acoust. Speech Lang. Process. 26, 550–563 (2017)
9. Antoñanzas, C., Ferrer, M., Diego, M., Gonzalez, A.: Blockwise frequency domain active noise controller over distributed networks. Appl. Sci. J. 6, 124 (2016)
10. Ferràs, M., Madikeri, S., Motlicek, P., Dey, S., Bourlard, H.: A large-scale open-source acoustic simulator for speaker recognition. IEEE Sig. Process. Lett. 23(4), 527–531 (2016)
11. Liu, J.S., Lin, C.H.R., Tsai, J.: Delay and energy tradeoff in energy harvesting multi-hop wireless networks with inter-session network coding and successive interference cancellation. IEEE Access 5, 544–564 (2017)
12. Ruder, M.A., Lehmann, A.M., Schober, R., Gerstacker, W.H.: Single antenna interference cancellation for GSM/VAMOS/EDGE using L_p-norm detection and decoding. IEEE Trans. Wirel. Commun. 14(5), 2413–2425 (2015)
13. Demissie, B.: Clutter cancellation in passive radar using GSM broadcast channels. IET Radar Sonar Navig. 8, 787–796 (2014)

14. Goossens, B., Luong, H., Pižurica, A., Philips, W.: An improved non-local denoising algorithm. Research Gate (2008)
15. Jiang, P.C., Yan, T.T., Jinand, J., Zhou, J.J.: A low flicker noise and high IIP2 downconversion mixer for zero-IF GSM receiver. IEEE (2014)
16. Salmasi, M., M-Nasab, H., Pourghassem, H.: Comparison of feed-forward and recurrent neural networks in active cancellation of sound noise. In: International Conference on Multimedia and Signal Processing, Guilin, China, pp. 25–29 (2011)

A Novel Experimental Prototype for Assessing IoT Performance on Real-Time Analytics

B. C. Manujakshi[1(✉)] and K. B. Ramesh[2]

[1] Department of Computer Science and Engineering, Acharya Institute of Technology,
Bengaluru, India
manujakshibc2014@gmail.com
[2] Department of Electronics and Instrumentation Engineering,
RV College of Engineering, Bengaluru, India

Abstract. Internet-of-Things (IoT) is one of the stepping stone to future ubiquitous computing with the aid of cloud environment. We reviewed the existing literature to find that there are more theoretical-based study and less standard and established modeling approach to claim the efficiency of the IoT application. Therefore, we present simple and novel prototyping of our experimental framework that not only offers real-time analysis of heterogeneous and dynamic sensory data captured from different IoT nodes but also offer a very user-friendly experience to carry out any form of an analytical operation on the top of it. The study outcome shows good streaming of real-time data of different physical attributes with better capability to read and analyze the real-time information. The prototype will offer simpler experience to handle IoT-based data and open avenues of various researches on IoT.

Keywords: Internet-of-Things · Ubiquitous computing · Sensor nodes
Sensor network · Data aggregation · Prototyping

1 Introduction

With the advancement of the mobile network and cloud computing, Internet-of-Things (IoT) is gaining fair attention for constructing smart city [1]. The formation of the IoT environment consists of the integration of various sensor nodes with different sensing capabilities [2]. However, it is interesting to know that it is a combination of various forms of heterogeneous wireless technologies where mainly wireless sensor network and cloud computing plays the major role [3, 4]. Owing to various forms of impediments in a wireless network, it is still uncertain to say that data captured from the sensor nodes are reliable [5]. Such reliability issue degrades the performance of various applications in IoT [6, 7]. At present, there is various research work being carried out towards IoT where still the research is just a beginning of identification of various problems [8, 9]. The major challenge in IoT-based research work is mainly the large-scale data that incorporates the higher amount of redundancies and thereby degrading data quality that significantly affects the IoT performance. The amount of complexity associated with the large-scale data generated from the IoT is completely based on the potential resource as

© Springer International Publishing AG, part of Springer Nature 2019
R. Silhavy (Ed.): CSOC 2018, AISC 763, pp. 46–55, 2019.
https://doi.org/10.1007/978-3-319-91186-1_6

well as computational cost needed to process the data ignoring the data size. Another bigger problem is the process of data transmission in IoT that also results in the narrowed scope of data structuring that slow down the process of data analysis for complex data. Therefore, this paper presents a novel and simple experimental prototype designed to incorporate the reliability of the data generated by the IoT devices using customized computational IoT platform and Raspberry Pi. Section 2 discusses the existing research work followed by problem identification from the existing research contribution in IoT in Sect. 3. Section 4 discusses proposed methodology followed by an elaborated discussion of algorithm implementation in Sect. 5. Analysis of accomplished result is discussed under Sect. 6 followed by conclusion in Sect. 7.

2   Related Work

This section briefs up the research contribution as a continuation of our prior work [10]. Most recently, the predictive approach has been primarily being emphasized for analyzing sophisticated data in IoT with the aid of machine learning as seen in the work of Akbar et al. [11]. Study towards analyzing real-time sensory data has been emphasized by Cao et al. [12] with higher supportability of context awareness and optimization. Adoption of the system on a chip has been presented by Conti et al. [13]. The work carried out by Mariani et al. [14] has discussed the relationship between IoT and big data analytics. Discussion on a different form of applicable intelligence towards IoT is discussed by Patel et al. [15]. A similar line of research has been carried out by Plageras et al. [16] towards performing the analytical operation in medical data. Discussion on Ambient intelligence towards emphasizing the performance in IoT was carried out by Ricciardi et al. [17]. Study towards emphasizing collaborative network in IoT was seen in the work of Sharma and Wang [18] and Silva et al. [19]. The implication of fog computing is carried out by Yang [20]. A significant application of mining approaches towards sensory data was discussed by Zhu et al. [21] towards studying various forms of events. A similar form of approaches towards analyzing event was seen in the work of Bhuiyan and Wu [22]. Hwang et al. [23] have presented a study that focuses on the stream of structured data generation along with an effective recommendation system on IoT data. Kumarage et al. [24] have emphasized on security aspect on the analytical data in IoT application. Sun et al. [25] have presented a discrete discussion on IoT from the viewpoint of analytical operation on big data approach with more insight on mobile sensing. Adoption of cognitive-based modeling towards mining operation in IoT was presented by Mishra et al. [26] over big data. Mishra et al. [27] have presented a model based on knowledge reengineering process. Ganz et al. [28] have discussed the process of handing various data as well as signifying the abstraction methods in IoT. Mikusz et al. [29] have presented a theoretical discussion on IoT whereas Zhu et al. [30] have adopted a Bayesian framework for performing the analytical operation in massive IoT data using predictive approach. Hence, there are various research approaches highlighted towards improving the performance of the IoT-based application in multiple fields. The next section outlines the research problems.

3 Problem Description

A closer look at the existing approaches in prior section exhibits that there is less number of standard prototyping or the framework toward leveraging the analytical performance in IoT. We also observe that there are no standard models and the benchmark test-bed being ever adopted for analyzing the performance of the IoT based application. Although, there is various existing research claiming to offer enhancement in the analytical approach using big data, however, they miss the validation in the form of the benchmarked model. Hence, the problem statement can be stated as *"To develop a prototype framework that could offer highly structurization of the massive sensory data for facilitating real-time data analysis using big data approach."* The next section outlines the proposed methodology to address such issues.

4 Proposed Methodology

The proposed study is an extension of our prior study where sensory data as a service has been presented. The proposed study considers an experimental research methodology where a prototype model has been constructed that directs the aggregated data for real-time analysis on the cloud. The schematic diagram of the proposed prototype is shown in Fig. 1.

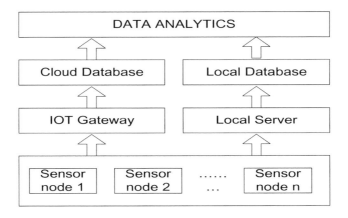

Fig. 1. Proposed schematic architecture

The frame work shows the flow from data collection process to the data analysis process. The sensor nodes are employed for collecting the data, which may be deployed at different geographical locations and connected by a network. The data collected from the nodes are updated to the local and the cloud databases through a local server and an IoT gateway device. The data collected in the databases are then used for performing the analysis. The next section discusses the algorithms being implemented.

5   Algorithm Implementation

The proposed system is designed for aggregating the data from the sensors (or smart connected devices) to perform real-time analytics operation on the top of it. However, to develop such concept, we classify operations into two parts, i.e., (i) data transmission among the IoT devices and (ii) performing analytical operations on a real-time basis. These section further briefs the algorithms as follows:

(i) Algorithm for Data Transmission

This algorithm is responsible for establishing the connectivity between the IoT devices and the cloud. The algorithm considers that there are various forms of heterogeneous IoT devices connected using a local server. The prime task in this regard is to aggregate the data and forward the data. The algorithm takes the input of m_{conf} (message) and n (number of IoT devices) that after processing yields successful delivery of m_{conf}. The steps included in the algorithm are as follows:

> **Algorithm for Data Transmission**
> **Input**: m_{conf} (message), n (number of IoT devices)
> **Output**: delivery of m_{conf}
> **Start**
> 1. init m_{conf}
> 2. **For** i=1:n
> 3.   **If** ($m_{conf}!=0$)
> 4.   forward $m_{ACK(msg\_ID)}$
> 5.   **Else**
> 6.   forward m_{REJC}
> 7. compute time
> $T_1 \rightarrow t_{ACK}*((2*retrans_{MAX})-1)*\alpha$
> $T_2 \rightarrow t_{ACK}*((2*retrans_{MAX})+1))-1)*\alpha$
> $T_3 \rightarrow (2*del_{max})+del_{proc}.$
> $T_4 \rightarrow T_1+(2*del_{max})+ del_{proc}$
> 8. Optimize, $g(x) \rightarrow T$
> 9. Forward m_{conf}
> 10. **End**
> **End**

For ensuring that the proposed algorithm always forward reliable sensory traffic to the local server, we tag the message as confirmable in its header, i.e., m_{conf} (Line-1). For all the IoT devices (Line-2), the algorithm checks there is a valid confirmable message (Line-3). As we are dealing here with the sensory network as the IoT environment, so receiver node always forwards an acknowledgment if it has successfully received m_{conf} (Line-4) or else it rejects m_{conf} (Line-6). The process of retransmission takes place by the sender IoT node until the data reaches the local server. The complete process of retransmission is governed by a counter that keeps track of number of retransmission and timeout of it. Initially, we configure the time-out factor to be random seconds if it has received new m_{conf} and counter for retransmission is configured to 0. Interestingly, the proposed algorithm offers good control of retransmission time to ensure effective forwarding data rate as well as ensure forwarding only non-redundant data packets to the local server. In case a received IoT device receives same mconf different number of time, it forwards the m_{ACK} with same message ID and hence easier to filter such messages. Hence, the phenomenon of message deduplication is ensured by the proposed algorithm. This process of message deduplication and data transmission is potentially affected by the time-factor. The first time factor is T_1 that represents highest time limit retrans$_{MAX}$from first to last instance of message retransmission using the random duration of acknowledgment i.e., α. The second time factor is T_2 that represent highest time limit retransmit from the first forwarding instance of message till the time when the transmitting node resets. The third time factor is highest round trip time, i.e., T_3 which depends on highest possible delay, i.e., del$_{max}$ and delay caused by processing, i.e., del$_{proc}$. The final time factor is T_4 that is the time from the initial point of message forwarding till the time when the message has been successfully forwarded, and there is no dependency on any forms of acknowledgment. Its dependable factors are T_1, i.e., highest duration of the transmission, del $_{max}$ maximum delay, and del$_{proc}$ delay caused by processing. The algorithm considers the complete network of IoT connected the device and applies a function g(x) to optimize the transmission time T. Finally the confirmable message is forwarded to the local server.

(ii) Algorithm for Real-Time Analytics

This algorithm is responsible for making the aggregated data suitable for performing real-time analytics. The algorithm takes the input of n (number of IoT Devices), p (types of IoT devices), and m (received the message) that after processing yields an output of m_2 (analyzed real-time data). The steps of the algorithm are as follows:

Algorithm for Real-Time Analytics
Input: n (number of IoT Devices), p (types of IoT devices), m (received message)
Output: m_2 (analyzed real-time data)
Start
1. init n_p
2. **For all** m
3.   **If** m(n_p) $\subseteq$ n$_{scope}$
4.   **Apply**f(x)$\rightarrow$m& obtain m_1 //$f$$\rightarrow$fusion, m_1—processed data
5.   **For** m_1=1:m(n_p)
6. $m_1$$\rightarrow$server$_{local}$
7.   **End**
8. $m_2$$\rightarrow$TS($m_1$)   //Express m_1 in increasing time scale
9.   update m_1 after x-sec
10.   **Else**
11.   Reject m
12. **End**
End

The proposed study considers that there are different forms of IoT devices that can capture different forms of physical attributes. We consider that different forms of the sensory data are being captured by the IoT devices where we consider that there is finite scope of the IoT devices, i.e., n$_{scope}$ (Line-3). For example, consider that n_1, n_2, n_3...

represent a set of IoT devices with three different specific forms. Hence, $n_{scope} = \{n_1, n_2, n_3\}$ represents maximum domains of a sensor network that are equally connected with the local server $_{local}$ (Line-6). We develop a prototype of the local server which performs explicit processing of the message m by applying a specific message conversion function $f(x)$ (Line-4), and this mechanism results in processed message m_1. Hence, m_1 represents processed data that is finally the outcome of the data fusion technique by the IoT device. After all the types of the IoT devices forwards their fused data to the local server, the server then further processes the data user-friendly to the user-interface that uses cloud application and can be accessed by the authorized users only. The aggregated data processed by the local server (Line-6) is then reposited in a specific data structure in the form of the increasing time scale TS (Line-8). The processed data m_1 is then analyzed concerning increasing time-scale (Line-9). However, if the incoming data doesn't map within any defined scope than it is aborted (Line-11).

6   Results Discussion

The design of the proposed prototype consists of mainly three types of hardware viz. (i) sensor nodes with the capability to trace light, (ii) sensor nodes with the capability to trace humidity and temperature, and (iii) Raspberry Pi 3. The study considers the local server to be implemented in Raspberry Pi 3 where the programming is carried out on IoT platform using Think Speak communication protocol [31]. The analysis of the numerical outcome is carried out considering $T_1 = 45$ s, $T_2 = 93$ s, $del_{max} = 100$ s, $del_{proc} = 2$ s, $T_4 = 247$ s, $\alpha = 1.5$ s.

Figure 2 exhibits the real-time analysis of the two physical attributes, e.g., light and voltage over present time scale in the x-axis. Similarly, Fig. 3 showcases the real-time analysis of other three different physical attributes, e.g., heat, temperature, and humidity.

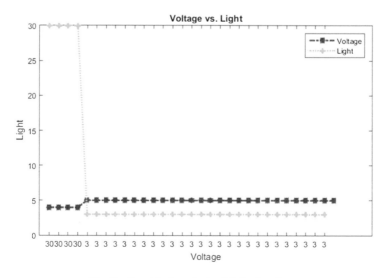

Fig. 2. Graphical analysis of light & voltage

The complete coding towards customizing the outcome of sensory data is carried out using MATLAB. The outcome exhibits the lag of few seconds that hardly affects the real-time analysis. The proposed system performs communication with the sensor node that uses Arduino microcontroller. The prototype enables the sensor node to acquire both the types of data, i.e., digital and analog. Using I2C, it performs full-fledge accessibility over the peripheral device. Usage of Think Speak allows a customized user interface to analyze the real-time data captured from the IoT device. The complete communication can be carried out via IEEE 802.11 standard as well as using the USB cable.

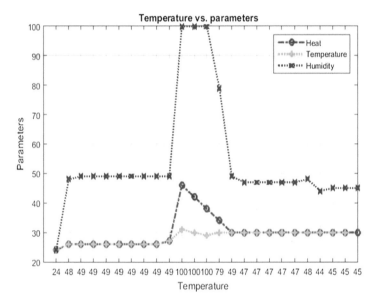

Fig. 3. Graphical analysis of temperature, humidity, and heat

Figure 4 showcases the real-time analysis of IoT data concerning increasing time scale. The data is accumulated in a structured database maintained by the Raspberry Pi that is ready for applying any forms of analytical operations.

The complete algorithm execution takes less than 0.3589 s in Raspberry Pi 3 which offers significantly less computational time. Usage of proposed prototype allows an effective conversion, integration, and computation of new form of data in the form of physical attributes. It also offers scheduled calculations to be executed at a required interval of time. The most interesting part of the prototype design is the visual interpretation of the obtained environmental data. It also offers a combination of the data from different forms of the communication channel.

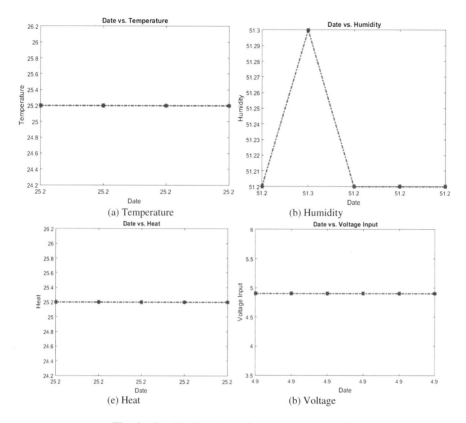

Fig. 4. Graphical analysis of concerning time-scale

7 Conclusion

There are various problems associated with analysis of dynamic and large scale data in IoT, e.g., (i) it is quite challenging to understand the actual pattern of the IoT data, and hence it is computationally challenging to identify the simplicity score of IoT data, (ii) evaluation of the appropriate frequency of the data captured from the sensors is another bigger problem that results in error-prone data, (iii) a correct detection of the complex data pattern over a time-factor is yet a np-hard problem in IoT, (iv) identifying and addressing the error prone or redundant data from sensors, etc. Existing studies are found to show that they are yet in its nascent stage and hence there is a scope of more research in this area. Hence, our prototype offers a simple model where various IoT applications can be tested over real-time analytics. Our framework is not only simpler to use but also offers highly customization towards any forms of advance analytical operations.

References

1. Geng, H.: Internet of Things and Data Analytics Handbook. Wiley, Hoboken (2017)
2. Greengard, S.: The Internet of Things. MIT Press, Cambridge (2015)
3. Kamila, N.K.: Handbook of Research on Wireless Sensor Network Trends, Technologies, and Applications. IGI Global (2016)
4. Mohanan, V., Budiarto, R., Aldmour, I.: Powering the Internet of Things With 5G Networks. IGI Global (2017)
5. Mukhopadhyay, S.C.: Internet of Things: Challenges and Opportunities. Springer, Heidelberg (2014)
6. Acharjya, D.P., Kalaiselvi Geetha, M.: Internet of Things: Novel Advances and Envisioned Applications. Springer, Cham (2017)
7. Tripathy, B.K., Anuradha, J.: Internet of Things (IoT): Technologies, Applications, Challenges and Solutions. CRC Press, Boca Raton (2017)
8. Tayeb, S., Latifi, S., Kim, Y.: A survey on IoT communication and computation frameworks: An industrial perspective. In: 2017 IEEE 7th Annual Computing and Communication Workshop and Conference (CCWC), Las Vegas, NV, pp. 1–6 (2017)
9. Sterbenz, J.P.G.: Smart city and IoT resilience, survivability, and disruption tolerance: Challenges, modelling, and a survey of research opportunities. In: 2017 9th International Workshop on Resilient Networks Design and Modeling (RNDM), Alghero, Italy, pp. 1–6 (2017)
10. Manujakshi, B.C., Ramesh, K.B.: SDaaS: framework of sensor data as a service for leveraging services in Internet of Things. In: International Conference on Emerging Research in Computing, Information, Communication, and Applications, pp. 351–363 (2017)
11. Akbar, A., Khan, A., Carrez, F., Moessner, K.: Predictive analytics for complex IoT data streams. IEEE Internet Things J. **4**(5), 1571–1582 (2017)
12. Cao, N., Nasir, S.B., Sen, S., Raychowdhury, A.: Self-optimizing IoT wireless video sensor node with in-situ data analytics and context-driven energy-aware real-time adaptation. IEEE Trans. Circ. Syst. I Regul. Pap. **64**(9), 2470–2480 (2017)
13. Conti, F., et al.: An IoT endpoint system-on-chip for secure and energy-efficient near-sensor analytics. IEEE Trans. Circ. Syst. I Regul. Pap. **64**(9), 2481–2494 (2017)
14. Marjani, M., et al.: Big IoT data analytics: architecture, opportunities, and open research challenges. IEEE Access **5**, 5247–5261 (2017)
15. Patel, P., Intizar Ali, M., Sheth, A.: On using the intelligent edge for IoT analytics. IEEE Intell. Syst. **32**(5), 64–69 (2017)
16. Plageras, A.P., et al.: Efficient large-scale medical data (eHealth Big Data) analytics in Internet of Things. In: 2017 IEEE 19th Conference on Business Informatics (CBI), Thessaloniki, pp. 21–27 (2017)
17. Ricciardi, S., Amazonas, J.R., Palmieri, F., Bermudez-Edo, M.: Ambient intelligence in the Internet of Things. Mob. Inf. Syst. (2017)
18. Sharma, S.K., Wang, X.: Live data analytics with collaborative edge and cloud processing in wireless IoT networks. IEEE Access **5**, 4621–4635 (2017)
19. Silva, B.N., Khan, M., Han, K.: Big data analytics embedded smart city architecture for performance enhancement through real-time data processing and decision-making. Wirel. Commun. Mob. Comput. (2017)
20. Yang, S.: IoT stream processing and analytics in the Fog. IEEE Commun. Mag. **55**(8), 21–27 (2017)

21. Zhu, M., Liu, C., Wang, J., Su, S., Han, Y.: Service hyperlink: modeling and reusing partial process knowledge by mining event dependencies among sensor data services. In: 2017 IEEE International Conference on Web Services (ICWS), Honolulu, HI, pp. 902–905 (2017)

22. Bhuiyan, M.Z.A., Wu, J.: Event detection through differential pattern mining in Internet of Things. In: 2016 IEEE 13th International Conference on Mobile Ad Hoc and Sensor Systems (MASS), Brasilia, pp. 109–117 (2016)

23. Hwang, I., Kim, M., Ahn, H.J.: Data pipeline for generation and recommendation of the IoT rules based on open text data. In: 2016 30th International Conference on Advanced Information Networking and Applications Workshops (WAINA), Crans-Montana, pp. 238–242 (2016)

24. Kumarage, H., Khalil, I., Alabdulatif, A., Tari, Z., Yi, X.: Secure data analytics for cloud-integrated Internet of Things applications. IEEE Cloud Comput. **3**(2), 46–56 (2016)

25. Sun, Y., Song, H., Jara, A.J., Bie, R.: Internet of Things and big data analytics for smart and connected communities. IEEE Access **4**, 766–773 (2016)

26. Mishra, N., Lin, C.-C., Chang, H.-T.: A cognitive adopted framework for IoT big-data management and knowledge discovery prospective. Int. J. Distrib. Sens. Netw. (2015)

27. Mishra, N., Chang, H.-T., Lin, C.-C.: An IoT knowledge reengineering framework for semantic knowledge analytics for BI-services. Math. Prob. Eng. (2015)

28. Ganz, F., Puschmann, D., Barnaghi, P., Carrez, F.: A practical evaluation of information processing and abstraction techniques for the Internet of Things. IEEE Internet Things J. **2**(4), 340–354 (2015)

29. Mikusz, M., Clinch, S., Jones, R., Harding, M., Winstanley, C., Davies, N.: Repurposing web analytics to support the IoT. Computer **48**(9), 42–49 (2015)

30. Zhu, X., Kui, F., Wang, Y.: Predictive analytics by using bayesian model averaging for large-scale Internet of Things. Int. J. Distrib. Sens. Netw. (2013)

31. ThinkSpeak. https://thingspeak.com/pages/learn_more. Accessed 06 Dec 2017

FAN: Framework for Authentication of Nodes in Mobile Adhoc Environment of Internet-of-Things

G. N. Anil[✉]

Department of Computer Science and Engineering, BMSIT, Bengaluru, India
gramaanil@gmail.com

Abstract. The Mobile Adhoc Network (MANET) has undergone significant improvement in the form of routing capabilities but still it lacks security potentials owing to its dynamic topology problems. Adoption of MANET in ubiquitous environment e.g. Internet-of-Things (IoT) could significant increase the communication capability but it could introduce significant level of threats too at same time. Therefore, the proposed manuscript introduces a Framework for Authentication (FAN) of mobile nodes where the technique first offers light-weight ciphering technique to initial stage of node communication followed by providing a significant access permission to authorized mobile nodes only to participate in data dissemination process. This analytical technique is also proven to offer communication efficiency apart from security against major potential threats on IoT environment.

Keywords: Internet-of-Things · Mobile adhoc network · Security
Access control · Secure permission · Ubiquitous

1 Introduction

There are various research works to report that security problems in Mobile Adhoc Network is yet unsolved [1–5]. Adoption of MANET is currently being experimented in ubiquitous environment like Internet-of-Things (IoT) in order to provide more mobility [6–10]. At the same time, the IoT is basically an integration of sensor network with cloud which also bridges the communication between them [11]. However, this also makes IoT environment more vulnerable to exponential attacks right from internet as well as from wireless environment of sensor nodes [12]. At present, there are less standard secure routing schemes in sensor network that claimed to provide 100% security against lethal attacks [13, 14]. Moreover they were never being testified that how a mobile node (infrastructure-independent) could possibly communicate with a sensor node (infrastructure dependent). With presence of massive amount of lethal threats on cloud and sensor network, IoT offers most vulnerable security challenges to network security [15, 16]. Therefore, there is a need of evolving up with a research approach to offer identification of lethal threats followed by mitigation policy for them. Therefore, this paper presents a framework to offer a robust authentication with successful and

© Springer International Publishing AG, part of Springer Nature 2019
R. Silhavy (Ed.): CSOC 2018, AISC 763, pp. 56–65, 2019.
https://doi.org/10.1007/978-3-319-91186-1_7

precise access control mechanism for mobile nodes communicating in IoT environment. Existing solutions towards such problems has been briefed in Sect. 2 and its associated pitfalls are outlined in Sect. 3 as problem description. Section 4 outlined the proposed methodology followed by illustration of solution in the form of algorithms. Finally, Sects. 5 and 6 discusses about results being obtained and conclusion of the paper respectively.

2   Related Work

This section discusses about the existing approaches of device security in IoT. At present, there are various security schemes in IoT towards its application as seen in the work of Nia and Kha [17]. Xu et al. [18] have used semantic-based approach towards network security. Studies towards different problems in privacy in IoT are discussed by Yang et al. [19] and Lin et al. [20]. Hu et al. [21] have discussed about privacy problems associated with IoT using fog computing. Studies towards security policies associated with software-defined network are discussed by Villari et al. [22]. Investigation towards a lethal attack in IoT has been discussed by Perazzo et al. [23] which is claimed to have highly adverse effect on routing. Lyu et al. [24] have presented a mechanism using stability function for optimal scheduling in order to resist illegitimate entries in IoT. Study in similar direction of introducing scheduling was also carried out by Qiu et al. [25] which offers both security as well as controls congestion. Chen and Zhu [26] have presented a technique of threat identification using pricing scheme. Study towards access control has been presented by Huang et al. [27] using attribute-based encryption. Wang et al. [28] have used compressed sensing in order to offer security to physical layer in IoT. Yang et al. [29] has theoretically discussed about the IoT security and Occuhiuzzi et al. [30] have presented a special design of identifying sensing in IoT that can be used for tracking the node behavior. Therefore, there is not much work being carried out toward IoT security most recently. The existing research works are also associated with issues as briefed in next section.

3   Problem Description

The identified problems are (i) existing research approaches do not have any consideration of involving adhoc network and its associated security risk in IoT, (ii) there is no standard protocols in IoT that offers significant security to IoT devices against connecting with external interfaces, (iii) the security protocols used in IoT network devices are mainly conventional cryptography e.g. SHA, AES, RSA, DES, etc. which has already reports of being ineffective against various potential threats on internet and other wireless networks, (iv) Existing security techniques doesn't claim to offer robust access mechanism even after authentication is offered in less effective way. Hence, problem statement is to design a novel computational framework that could offer integrated services of authenticated followed by highly resilient access control technique when MANET is implemented in IoT environment.

4   Proposed Methodology

Our prior study has presented solution towards security of MANET [31, 32]. This part of the study further extends the security features of MANET when the mobile nodes are considered to be operating using IoT environment. Designed using analytical research approach, the proposed study constructs a novel authentication provision using access rights as shown in Fig. 1.

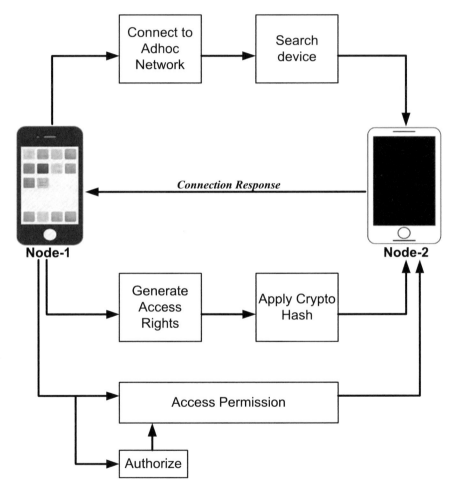

Fig. 1. Proposed schema of FAN

Figure 1 highlights the process of establishing secure communication between two mobile nodes in IoT environment, which is essentially divided into two phases viz. (i) the first phase consist of constructing a crypto message for validating the communicating nodes and (ii) second phase consist of offering the access permission to only legitimate mobile nodes. The next section outlines the algorithm involved in the process of design implementation.

5 Algorithm Implementation

The proposed algorithm is mainly responsible for ensuring higher degree of confidentiality as well as data integrity for the message being communicated by the mobile nodes in IoT environment. For this purpose, the node n_1 and n_2 computes (C_1, D_1) and (C_2, D_2), which is basically a generated key pairs, $(C_1, C2)$ are public keys while (D_1, D_2) are secret keys of node 1 and node-2 as shown in Fig. 2.

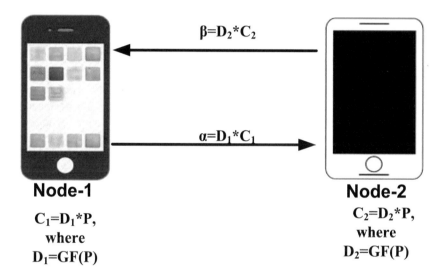

Fig. 2. Preliminary key sharing process

A novel algorithm is designed that performs a secure communication among the mobile nodes within IoT environment. The steps of the algorithms are as follows:

Primary Algorithm for forwarding ciphered data among IoT Nodes
Input:a,ts, θ, α
Output:$data_{ciph}$
Start
1. $n_1 \rightarrow a$, ts
2. Compute $k_{sec} = \theta (\alpha, ts)$
3. Perform encryption, $A = C_R (a)$, $ts_1 \rightarrow C\alpha (ts)$
4. $VC_1 \rightarrow VC (\alpha, A \| ds_1)$
5. Forward $data_{ciph} (A, ts_1, VS_1)$
End

The algorithm takes the input of a (prime number), ts (time stamp), θ (hash function), and α (common secret key) that after processing yields data$_{\text{ciph}}$ (ciphered date). The first step of the algorithm is that the node-1 will apply Galois Field randomly in order to generate a prime number a (Line-1) along with generation of time stamp (ts). The second stage of the algorithm is to apply a standard hash function θ of a number obtained by XORing α and ts (Line-2). The next step is to perform a ciphering operation where the prime number a is ciphered using a secret key that is obtained after applying ciphering function using a secret key k_{sec} on prime number a (Line-3). The algorithm also performs encryption of timestamp by α in order to obtain encrypted timestamp ts_1 (Line-3) using ciphered function C. The next art of the study construct a control message in mobile adhoc network that can perform internal verification of the node identity using digital signature and further performs concatenation of R and data structure ds (Line-4). This mechanism offers performing authentication for the IoT device. Finally, a combined controlled message as shown in Line-5 is forwarded by node-1 to node-2 (Line-5). A new timestamp is generated by the node-2 as well as the same node will decrypt ts_1 in order to obtain original ts value. The system then compares its current timestamp with the ts and is considered as valid node communication if current time stamp is found greater than ts. The secret key k_{sec} is generated by node-2 and performs deciphering in order to get a. The computation of verifier code VC_1 is also carried out by node-2 and is matched with node-1. Upon successful matching, the node-1 is considered to be a legitimate node to perform communication with node-2. The next stage of the algorithm involves reverse direction authentication. In this process, VC2 is generated by applying verifier code function on concatenated value of a and data-structure of node-2 i.e. ds$_2$. This leads to $VC_2 = VC$ (a∥ds$_2$). The ciphering of the prime number a is carried out by β very similar to Line-3 i.e. $k_{\text{sec2}} = C \beta$ (r). This secret key k_{sec2} is forwarded by node-2 along with VC_2. Both the prime numbers are now matched and upon successful match, the positive authentication is confirmed between node-2 to node-1.

After the algorithm performs successful authentication check between two communicating nodes in IoT, the next stage is to perform its structure granted rights, which is empirically represented as G = [node$_{\text{ID}}$, H, σ], where nodeID represented identity of a node, H is list of granted permission specific to node ID, and σ represents an arbitrary function in order to resist any packet tampering events. The computation of σ can be carried out by applying any standard cryptographic hash function on node$_{\text{ID}}$ and H. In this case, all the IoT devices are interconnected in order to check their device rights. It is required that all the perspective communication among the mobile nodes are required to be validated using the same access rights. Only the nodes that are validated in this process will be able to participate in data forwarding process. For this purpose, a discrete set of control message is being exchanged in highly encrypted form using any standard cryptographic hash function. More closely looking in the access policies in the proposed system will show that it offers permission to either forward the message as it is or it let the control message to perform certain changes in order to embed information related to dynamic topology of mobile nodes.

In simpler manner, the proposed algorithm displays a distributed characteristics where both the communicating nodes perform two different forms of authentication process for each other prior to communication. This process is further strengthened by offering comprehensive access rights to all the communicating nodes. It will mean that although the similar protocol is running in all the nodes, only the nodes found to be legitimate will be able to perform communication. Thereby a good balance between security as well as communication performance can be seen in the proposed system. The next section outlines the results being obtained.

6   Results and Discussion

The implementation of the proposed study is scripted using MATLAB considering 1000 mobile nodes with 10 meters of transmission range and initialized with 10 joules of energy. The nodes were dispersed in simulation area of 1700×1300 m^2. The complete implementation scenario also consists of 4 gateway nodes using OpenFlow technology in order to represent the IoT environment. The study outcomes were compared with conventional security protocols of SHA and AES that is frequently adopted in securing IoT environment. The study outcomes are assessed using performance parameters of energy consumption, overall delay, packet delivery ratio, and algorithm processing time.

The outcome clearly indicates that the proposed system offer significantly better performance in almost every performance scale. Proposed system uses non-recursive scheme of authentication which consistently keeps on updating and hence more number of nodes are aware of security information leading to less energy consumption (Fig. 3). Owing to inclusion faster authentication mechanism using data structure of node leads to lower delay (Fig. 4) and reduced allocation of resources offers significant productivity towards packet delivery ratio (Fig. 5). The study outcome also shows that FAN is computationally less complex compared to existing algorithms (Fig. 6).

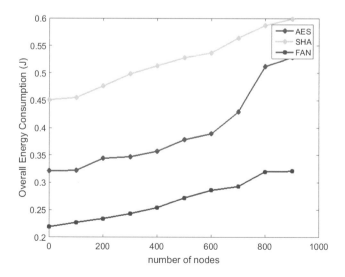

Fig. 3. Comparative analysis of energy consumption

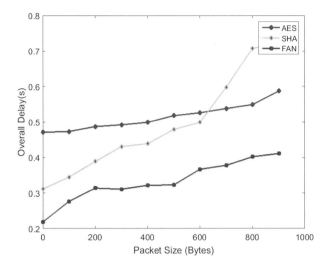

Fig. 4. Comparative analysis of overall delay

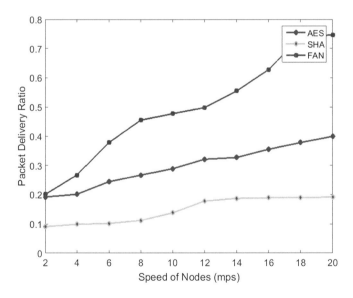

Fig. 5. Comparative analysis of packet delivery ratio

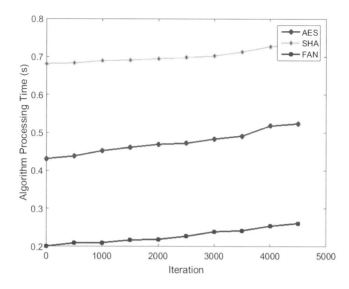

Fig. 6. Comparative analysis of algorithm processing time

7   Conclusion

Existence of mobile nodes in dynamic topology offers a serious treat in MANET environment. However, the threats doubles up when the MANET nodes are allowed to performed communication in an IoT environment due to multiple of reasons viz. (i) burden of increasing heterogeneous routing protocol in IoT when integrated with MANET, (ii) no authentication protocol exists without significant delay in MANET which further increase when added to MANET environment. As the proposed FAN computes and compensates the delay with each data being ciphered, it is never feasible for any man-in-middle attack to initiate any form of attacks owing to multiple level of anonymous dependencies of security parameter. Hence, bit-level security is quite high using smallest size of encryption key. Therefore, the proposed system i.e. FAN is the first attempt where such issues are identified in order to solve the security problem existing in communication between mobile nodes in IoT environment. The study outcome of proposed system shows that there is a least computational burden of FAN as well as it also offer better and lightweight security features in contrast to existing security protocols. It is also feasible for upgrading the proposed system by incorporating an enhanced version of Elliptical Curve Cryptography by generating only the prime keys for better balance between key generation process and authentication. As the backbone architecture of proposed FAN is based on Galois Field structure; hence; it will be also compatible for any future design using Elliptical Curve Cryptography.

References

1. Movahedi, Z., Hosseini, Z., Bayan, F., Pujolle, G.: Trust-distortion resistant trust management frameworks on mobile Ad Hoc networks: a survey. IEEE Commun. Surv. Tutorials **18**(2), 1287–1309 (2016)
2. Lima, M.N., dos Santos, A.L., Pujolle, G.: A survey of survivability in mobile Ad Hoc networks. IEEE Commun. Surv. Tutorials **11**(1), 66–77 (2009)
3. Govindan, K., Mohapatra, P.: Trust computations and trust dynamics in mobile Ad Hoc networks: a survey. IEEE Commun. Surv. Tutorials **14**(2), 279–298 (2012)
4. Surendran, S., Prakash, S.: An ACO look-ahead approach to QOS enabled fault- tolerant routing in MANETs. China Commun. **12**(8), 93–110 (2015)
5. Zhao, S., Aggarwal, A., Frost, R., Bai, X.: A survey of applications of identity-based cryptography in mobile Ad-Hoc networks. IEEE Commun. Surv. Tutorials **14**(2), 380–400 (2012)
6. Alam, F., Mehmood, R., Katib, I., Albogami, N.N., Albeshri, A.: Data fusion and IoT for smart ubiquitous environments: a survey. IEEE Access **5**, 9533–9554 (2017)
7. Verma, S., Kawamoto, Y., Fadlullah, Z.M., Nishiyama, H., Kato, N.: A survey on network methodologies for real-time analytics of massive IoT data and open research issues. IEEE Commun. Surv. Tutorials **19**(3), 1457–1477 (2017)
8. Anagnostopoulos, T., et al.: Challenges and opportunities of waste management in IoT-enabled smart cities: a survey. IEEE Trans. Sustain. Comput. **2**(3), 275–289 (2017)
9. Xu, L., Collier, R., O'Hare, G.M.P.: A Survey of clustering techniques in WSNs and consideration of the challenges of applying such to 5G IoT scenarios. IEEE Internet Things J. **4**(5), 1229–1249 (2017)
10. Yang, K., Blaauw, D., Sylvester, D.: Hardware designs for security in Ultra-Low-Power IoT systems: an overview and survey. IEEE Micro **37**(6), 72–89 (2017)
11. Chauvenet, C., Etheve, G., Sedjai, M., Sharma, M.: G3-PLC based IoT sensor networks for SmartGrid. In: 2017 IEEE International Symposium on Power Line Communications and Its Applications (ISPLC), Madrid, pp. 1–6 (2017)
12. Sen, A., Madria, S.: Risk assessment in a sensor cloud framework using attack graphs. IEEE Trans. Serv. Comput. **10**(6), 942–955 (2017)
13. Peter, S., Westhoff, D., Castelluccia, C.: A survey on the encryption of convergecast traffic with in-network processing. IEEE Trans. Dependable Secure Comput. **7**(1), 20–34 (2010)
14. Han, G., Jiang, J., Sun, N., Shu, L.: Secure communication for underwater acoustic sensor networks. IEEE Commun. Mag. **53**(8), 54–60 (2015)
15. Deogirikar, J., Vidhate, A.: Security attacks in IoT: a survey. In: 2017 International Conference on I-SMAC (IoT in Social, Mobile, Analytics and Cloud) (I-SMAC), Palladam, pp. 32–37 (2017)
16. Pongle, P., Chavan, G.: A survey: attacks on RPL and 6LoWPAN in IoT. In: 2015 International Conference on Pervasive Computing (ICPC), Pune, pp. 1–6 (2015)
17. Mosenia, A., Jha, N.K.: A comprehensive study of security of Internet-of-Things. IEEE Trans. Emerg. Topics Comput. **5**(4), 586–602 (2017)
18. Xu, G., Cao, Y., Ren, Y., Li, X., Feng, Z.: Network security situation awareness based on semantic ontology and user-defined rules for Internet of Things. IEEE Access **5**, 21046–21056 (2017)
19. Yang, Y., Wu, L., Yin, G., Li, L., Zhao, H.: A survey on security and privacy issues in Internet-of-Things. IEEE Internet Things J. **4**(5), 1250–1258 (2017)

20. Lin, J., Yu, W., Zhang, N., Yang, X., Zhang, H., Zhao, W.: A survey on Internet of Things: architecture, enabling technologies, security and privacy, and applications. IEEE Internet Things J. **4**(5), 1125–1142 (2017)
21. Hu, P., Ning, H., Qiu, T., Song, H., Wang, Y., Yao, X.: Security and privacy preservation scheme of face identification and resolution framework using fog computing in Internet of Things. IEEE Internet Things J. **4**(5), 1143–1155 (2017)
22. Villari, M., Fazio, M., Dustdar, S., Rana, O., Chen, L., Ranjan, R.: Software defined membrane: policy-driven edge and Internet of Things security. IEEE Cloud Comput. **4**(4), 92–99 (2017)
23. Perazzo, P., Vallati, C., Anastasi, G., Dini, G.: DIO Suppression attack against routing in the Internet of Things. IEEE Commun. Lett. **21**(11), 2524–2527 (2017)
24. Lyu, X., et al.: Optimal schedule of mobile edge computing for Internet of Things using partial information. IEEE J. Sel. Areas Commun. **35**(11), 2606–2615 (2017)
25. Qiu, T., Qiao, R., Wu, D.O.: EABS: an event-aware backpressure scheduling scheme for emergency Internet of Things. IEEE Trans. Mob. Comput. **17**(1), 72–84 (2018)
26. Chen, J., Zhu, Q.: Security as a service for cloud-enabled internet of controlled things under advanced persistent threats: a contract design approach. IEEE Trans. Inf. Forensics Secur. **12**(11), 2736–2750 (2017)
27. Huang, Q., Yang, Y., Wang, L.: Secure data access control with ciphertext update and computation outsourcing in fog computing for Internet of Things. IEEE Access **5**, 12941–12950 (2017)
28. Wang, N., Jiang, T., Li, W., Lv, S.: Physical-layer security in Internet of Things based on compressed sensing and frequency selection. IET Commun. **11**(9), 1431–1437 (2017)
29. Yang, Y., Peng, H., Li, L., Niu, X.: General theory of security and a study case in Internet of Things. IEEE Internet Things J. **4**(2), 592–600 (2017)
30. Occhiuzzi, C., Amendola, S., Manzari, S., Caizzone, S., Marrocco, G.: Configurable radiofrequency identification sensing breadboard for industrial Internet of Things. Electron. Lett. **53**(3), 129–130 (2017)
31. Anil, G.N., Venugopal Reddy, A.: Semantic probabilistic modelling of novel routing protocol with implication of cumulative routing attack in mobile adhoc network. Int. J. Comput. Sci. Issues **9**(1) (2012)
32. Anil, G.N., Venugopal Reddy, A.: Strategical modelling with virtual competition for analyzing behavior of malicious node in mobile Adhoc network to prevent decamping. Int. J. Comput. Sci. Issues **8**(6) (2011)

Towards a Conceptual Model of Intelligent Information System for Smart Tourism Destinations

Tomáš Gajdošík[(✉)]

Faculty of Economics, Matej Bel University, Tajovského 10, 975 90 Banská Bystrica, Slovakia
tomas.gajdosik@umb.sk

Abstract. Smart tourism destinations represent a new concept of application of information technologies and information sharing in destinations, leading to the higher competitiveness and satisfaction of all relevant stakeholders, including tourists as tourism product co-creators and co-promoters of a destination. Although the use of information technologies in tourism destinations is well analyzed, the requirements on information system integrating all relevant technologies have not been conceptualized so far. Therefore the aim of the paper is to propose a conceptual model of intelligent information system for smart tourism destinations based on the state-of-the art technologies applied in tourism destinations. The selected smart destinations are analyzed together with the use of information technologies during the travel behavior. The proposed model consists of three layers – data collection, processing and exchange. The article contributes to the new paradigm research – the management of smart destinations and use of big data in tourism development.

Keywords: Big data · Information technology · Intelligent system
Smart tourism destination

1 Introduction

The concept of smart tourism destinations is nowadays a well discussed topic in the scientific literature [1], leading to fruitful discussions not only in the field of tourism, but also in the computer science and data processing. This concept is mainly connected with the rise of new information technologies, leading to the change in the buying behavior of tourists and thus forcing destination management organizations (DMOs) to adopt new information systems in order to properly govern the destination. The challenge is to provide real-time personal services to visitors and interconnect all stakeholders in order to share information and knowledge.

The idea of smart tourism destination is derived from the concept of smart city, where smartness is incorporated in mobility, living, people, governance, economy and environment [2]. It is a city with knowledge center that manage information, technology and innovation, trying to reach efficient management, sustainable development and better quality of life for residents [3]. Since the inception in the urban environment, the smart

© Springer International Publishing AG, part of Springer Nature 2019
R. Silhavy (Ed.): CSOC 2018, AISC 763, pp. 66–74, 2019.
https://doi.org/10.1007/978-3-319-91186-1_8

approach has been applied also in tourism and the term smart tourism destination has arisen [4]. Implementing the smart concept in a tourism destination has been crucial since the connected, better informed and engaged tourist is dynamically interacting with the destination, leading to the need of co-creating tourism product and adding value for all tourism stakeholders [5].

In smart tourism destinations, it is important to stress the network system of stakeholders delivering services to tourists, complemented by a technological infrastructure aimed at creating a digital environment which supports cooperation, knowledge sharing and open innovation [6].

2   The Role of Information Technologies in Tourism Destinations

Due to the fragmented nature of tourism supply, the penetration of information technologies to tourism destinations was, comparing with other areas, very slow. The first sophisticated information system started to be used in the 1980s, when first destination management system (DMS) was introduced [7] as a database system allowing distribution of tourism products and information for visitors. Later, the geographic information systems (GIS) started to be used in destinations to support the overall sustainability of resources [8]. With the emergence of the Internet and web-based systems in 1990s, there has been a qualitative as well as quantitative boost in the use of information technologies in tourism destinations [9].

Nowadays tourism destinations can profit from a wide variety of information technologies that are applicable for their management and for co-creation of tourism experiences. From the management point of view, information technologies should allow destination management organizations and stakeholders to decide and act on the basis of collected and processed data. Stakeholders should be dynamically linked with information technology which enable them to create, collect and exchange real-time information in order to meet customer needs.

For managers there is a challenge for the implementation of open data, the application of big data analysis techniques in real-time databases, incorporating recommender and context aware systems, as well as supporting decision-making and the exchange of information between the stakeholders [4].

From the tourist point of view, information technologies should enhance experience by giving all the related real-time information about the destination and its services in the planning phase, enhance access to real-time information to assist tourists in exploring the destinations during the trip and prolong the engagement to relive the experience by providing the descent feedback after the trip [10].

In order to stimulate information sharing, automation, control and connectivity, destinations should take advantage of three technological components: cloud computing, Internet of things and end-user internet service systems [11]. Cloud computing allows the data to be accessible and ready to use anytime and anywhere with the use of the Internet. Internet of things allows the connection of everyday objects, and to collect, process and share data with minimum human intervention. End-user internet service systems comprise all the application and hardware that enable to use these other two technological components. They can include destination apps, augmented reality, sensors, NFC, QR codes or Wi-Fi. Interoperability and ubiquitous computing should ensure that everybody and everything is interconnected and processes are integrated towards generating value, through dynamic co-creation, sustainable resources and dynamic personalization.

Therefore there is a need for a technological platform on which information relating to tourism activities could be exchanged instantly and which will dynamically interconnect destination stakeholders [10]. This kind of information system should be able to obtain information from physical and digital sources and with the combination of advanced technologies be capable of transforming the data into experiences and business value propositions focused on efficiency, sustainability and experience enrichment [12]. It should also facilitate the touch points with the tourists by allowing the connection through a wide range of end-user devices supporting tourism experiences [13].

Despite the recognized need for an information system for smart tourism destination, there has been a little attempt to conceptualize the requirements into one model. Therefore the new horizon for the conceptual model of intelligent information system for smart destinations arises.

3   Materials and Methods

The aim of the paper is to propose a conceptual model of intelligent information system for smart tourism destination based on the state-of-the art technologies applied in tourism destinations.

The paper deals with secondary data sources of using the information technologies in selected smart tourism destinations. The research sample of best practices include destinations as Barcelona, Amsterdam, Helsinki, Singapore and Salzburg. The data are enriched with the study Intégration des technologies smart par les organisations touristiques en Suisse [14] and author's own research [15] on information technologies used in Slovak destinations. Based on the used technologies a conceptual model is proposed. The proposed model of intelligent information system is created using MS Power Pivot software.

4   Results

4.1   Information Technologies Used in Smart Tourism Destinations

There are several tourism destinations that are claiming themselves as "smart". The best case studies can be found in Europe and Asia, where smartness is oriented towards technological infrastructure and end-users application leading to the better satisfying the needs of tourists and more sophisticated destination management (Table 1).

Besides these state-of-art technologies, it is also important to consider technologies used during all phases of tourist buying behavior. In searching for information about attractions in the destination, accommodation, gastronomic and ancillary services, and when deciding to participate in tourism, tourists often use national travel portals, official

Table 1. The use of state-of-art information technologies in selected smart tourism destinations

Destination	Example of information technologies
Barcelona	- sensors to monitor traffic
	- smart streetlight
	- air quality monitoring
	- electronic complaints system
	- Wi-Fi in public places
	- digital bus stops
	- car and bike sharing systems
Amsterdam	- sensors for density of crowds
	- beacons based on BLE for translating signs
Helsinki	- real-time databases for energy monitoring
	- open data initiative
	- mobile apps with active use of geolocation
Singapore	- sensors and cameras for cleanliness, density of crowds and movement of vehicles
	- big data and analytics to improve planning
	- road sensors and smart parking
	- mobile app monitoring behavior
Salzburg	- traffic sensors
	- energy monitoring
	- tourist smart card
	- mobile apps

webpages of destinations or search using search engines (e.g. Google) and meta search engines (e.g. Trivago). An important source of data when deciding to participate on tourism are also review sites, where tourists can find reviews from other tourists.

Booking is most often done through the internet distribution systems (e.g. Booking.com), online travel agencies (e.g. Expedia) and the official web page of a destination using DMS. The advantage of these electronic booking systems is the possibility of booking from home 24/7. Due to the expansion of smart phones, information technologies are frequently used during the stay in the destination. Orientation and clarity of all the important information about the destination and its attractiveness is facilitated by mobile applications, based on geolocation. The integration of maps to mobile apps, along with the use of location-based services have created useful electronic tour guides. During and after travel, visitors are happy to share their experiences. They use social media to a large extent. They most frequently express their views on the review sites (e.g. TripAdvisor) or social networks (e.g. Facebook). A large increase has recently been reported by sites with the possibility of sharing media content where visitors publish their photos (such as Instagram, Pinterest) or videos (e.g. YouTube).

Based on the review of using of information technologies in selected smart tourism destinations and the available knowledge, the use of information technologies applicable in tourism destination concerning the frequency of use and phases of buying behavior is graphically designed (Fig. 1).

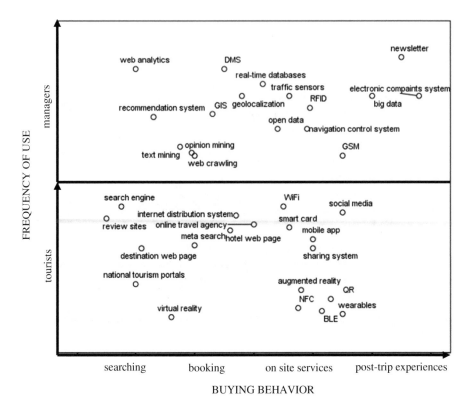

Fig. 1. The use and importance of information technologies in tourism destination

The *x* axis represents the stages of buying behavior of tourists, starting with information search, booking, using on site services and post-trip experiences sharing. The *y* axis represents the frequency of use of information technology, indicated separately for tourists and destination managers. The data on frequency of use of information technologies from the tourist point of view were obtained from Eurostat Preferences of Europeans Towards Tourism, while the sources of data from the view of destination managers are from the study Intégration des technologies smart par les organisations touristiques en Suisse [14] and author's own research [15]. These data are baseline for the conceptual model of intelligent system for smart tourism destinations.

4.2 Conceptual Model of Intelligent Information System for Smart Tourism Destinations

The conceptual model is based on state-of-art technologies focusing on their connection and contribution to smart destination concept. The intelligent information system for smart tourism destinations should contain three layers focusing on collecting data, their processing and exchange (Fig. 2).

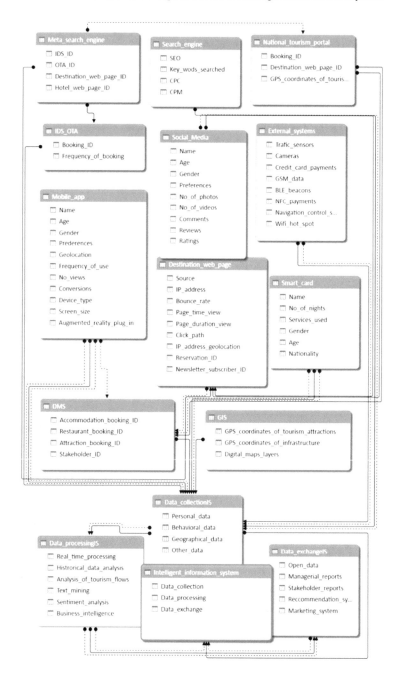

Fig. 2. Conceptual model of intelligent information system for smart tourism destination

The *data collection* layer of intelligent information system is composed of personal, behavioral, geographical and other data. The personal data contain name, age and gender of tourists obtained from social media (social networks, blogs, media-related pages), mobile apps of a destination and destination smart card. Moreover, it comprises booking ID (or newsletter subscriber ID) of a tourist from internet distribution systems, online travel agencies, destination web page and DMS. This personal database leads to better targeting of marketing management of a destination.

The behavioral data are of quantitative and qualitative matter. The source of quantitative data is web page and mobile app analytics or external systems. These can include bounce rate, page time view, page duration view and click path from destination web page, frequency of use as well as number of users of mobile app. The traffic sensors and cameras can provide information on number of vehicles or overcrowding in tourist attractions. Information about tourists' consumption in a destination can be derived from credit card payments or payments via NFC. Social media provide mainly the qualitative information of preferences, comments and reviews made by tourists. These data can be obtained by web scraping and crawling. These data are valuable in smart destination concept as they allow the personalization of destination product.

Geographic data give information on tourist origin, such as IP address geolocation, or tourist movement thanks to active location based services in a smartphone launched by a mobile app, or passive mobile localization via GSM data obtained from mobile operators. Moreover Bluetooth low energy (BLE) used in beacons, used mainly inside the tourist attractions, can provide useful information about tourists flows inside a building. GPS coordinates of tourism attractions and infrastructure, as well as digital map layers from desktop GIS are also important sources of geographic data for further analysis. The other collected data may include information on CPC and CPM search engine advertising, mobile device and its screen size and other useful data for management of smart tourism destinations.

The *data processing* layer is based on real-time processing of all collected data. The size, heterogeneous nature and complexity of these big data put a pressure on the automatic treatment and analysis. Therefore the traditional methods should be enriched with new ones to contribute to intelligent system. Except of actual data, the database of intelligent information system contains also historical data in order to apply revenue management concept and forecast demand, as well as apply dynamic pricing. The geo-localized data allow the identification of tourists' flows and therefore to identify the zones dedicated to tourism and density of tourists. The textual data from social media are analyzed by text and opinion mining as well as sentiment analysis to find out the emotions, opinions and attitudes of tourists. The cross-referencing of all data provides an automatic classification of destination zones, tourist typologies and customer relationship management leading to provision of highly customized destination product and thus better satisfying of tourist's need. The application of business intelligence (BI) allows not only data preparation and modelling, but also their storage in data warehouse (DWH) and the application of online analytical processing (OLAP) and data mining procedures.

The *data exchange* layer is composed of open data, in order to stimulate information sharing among tourism stakeholders. All collected and processed data are available through extranet and support the "soft smartness" of a destination. Intelligent information system provides interactive dashboards for managers and stakeholders leading to provision of real-time reports. These reports enable more effective and efficient destination and visitor management as well as they contribute to management of stake-holders' businesses. Recommendation system allows to perform right decisions, while marketing system automatically sends newsletters and push notifications to tourists and uses channel manager to adjust availability and prices of bookings.

5   Discussion and Conclusion

The proposed model of intelligent information system is a conceptual one, not taking into account the difficulties in obtaining, processing and sharing the data. The data acquisition may face certain problems such as ethical questions of involuntary tracking the tourists or time and money dedicated to collect data (e.g. fees to mobile operators for aggregated mobile positioning GMS data). However as [16] claim, one of the greatest difficulties in the analysis of digital trails is the incomplete character of the data. There is no access to complete picture of tourist's behavior but, rather, a collection of fragments that can be combined to produce valuable information.

Moreover another challenge is the interoperability and compatibility of technologies using various data formats and communication protocols. The willingness of stake-holders to share the data is also very low, as the competition is strong and fear from the misusing the data for own business purposes is very high.

Some destinations, e.g. Swedish Åre [17] have implemented information systems that have the features of an intelligent system. However the application of real-time intelligence, knowledge sharing and open innovation have been rather limited so far. Nevertheless, the added value of the intelligent information system for smart tourism destinations is in the provision of useful information on behavior, movement, timetable and visitation of tourist attractions; real-time processing of all collected data and offering a platform for open data exchange and decision making. Its application in the praxis contributes to the new paradigm – the management of smart destinations and use of big data in tourism development.

Acknowledgments. The research was supported by the research project VEGA 1/0809/17 Reengineering of destination management organizations and good destination governance conformed to principles of sustainable development.

References

1. Xiang, Z., Tussyadiah, I., Buhalis, D.: Smart destinations: foundations, analytics, and applications. J. Destination Mark. Manag. **4**(3), 143–144 (2015). https://doi.org/10.1016/j.jdmm.2015.07.001
2. Giffinger, R., Fertner, C., Kramar, H., Kalasek, R., Pichler-Milanovic, N., Meijers, E.: Smart Cities. Ranking of European Medium-Sized Cities, Vienna (2007)
3. Caragliu, A., Del Bo, C., Nijkamp, P.: Smart cities in Europe. J. Urban Technol. **18**(2), 65–82 (2011). https://doi.org/10.1080/10630732.2011.601117

 4. Ivars-Baidal, J.A., Celdrán-Bernabeu, M.A., Mazón, J.N., Perles-Ivars, Á.F.: Smart destinations and the evolution of ICTs: a new scenario for destination management? Curr. Issues Tourism **3500**(October), 1–20 (2017). https://doi.org/10.1080/13683500.2017.1388771

 5. Neuhofer, B., Buhalis, D., Ladkin, A.: Conceptualising technology enhanced destination experiences. J. Destination Mark. Manag. **1**(1–2), 36–46 (2012). https://doi.org/10.1016/j.jdmm.2012.08.001

 6. Del Chiappa, G., Baggio, R.: Knowledge transfer in smart tourism destinations: analyzing the effects of a network structure. J. Destination Mark. Manag. **4**(3), 145–150 (2015). https://doi.org/10.1016/j.jdmm.2015.02.001

 7. Benckendorff, P., Sheldon, P., Fesenmaier, D.: Tourism information technology. CABI International, Oxfordshire (2014)

 8. Bahaire, T., Elliott-White, M.: The application of Geographical Information Systems (GIS) in sustainable tourism planning: a review. J. Sustain. Tour. **7**(2), 159–174 (1999). https://doi.org/10.1080/09669589908667333

 9. Buhalis, D.: eTourism: Information Technology for Strategic Tourism Management. Pearson Education Limited, Edinburgh (2003)

10. Buhalis, D., Amaranggana, A.: Smart tourism destinations enhancing tourism experience through personalisation of services. In: Information and Communication Technologies in Tourism, pp. 377–389. Springer, Cham (2015). http://doi.org/10.1007/978-3-319-14343-9_28

11. Zhang, L., Li, N., Liu, M.: On the basic concept of smarter tourism and its theoretical system. Tour. Tribune **27**(5), 66–73 (2012)

12. Gretzel, U., Sigala, M., Xiang, Z., Koo, C.: Smart tourism: foundations and developments. Electr. Mark. **25**(3), 179–188 (2015). https://doi.org/10.1007/s12525-015-0196-8

13. Buonincontri, P., Micera, R.: The experience co-creation in smart tourism destinations: a multiple case analysis of European destinations. Inf. Technol. Tour. **16**(3), 285–315 (2016). https://doi.org/10.1007/s40558-016-0060-5

14. Schegg, R., Hébert, E.: Intégration des technologies smart par les organisations touristiques en Suisse. Sierre (2016)

15. Gajdošík, T.: Smart tourism destination? The case of Slovakia. In: Borseková, K., Vaňová, A., Vitálišová, K. (eds.) 6th Central European Conference in Regional Science Engines of Urban and Regional Development Banská Bystrica, Slovak Republic, 20–22 September 2017, pp. 217–225. Faculty of Economics, Matej Bel University in Banská Bystrica, Banská Bystrica (2017)

16. Cousin, S., Chareyron, G., Jacquot, S.: Big data and tourism. In: Lowry, L. (ed.) The SAGE International Encyclopedia of Travel and Tourism, pp. 151–155 (2017)

17. Fuchs, M., Höpken, W., Lexhagen, M.: Big data analytics for knowledge generation in tourism destinations – a case from Sweden. J. Destination Mark. Manag. **3**(4), 198–209 (2014). https://doi.org/10.1016/J.JDMM.2014.08.002

An Integrated Schema for Efficient Face Recognition in Social Networking Platforms

Ramesh Shahabadkar[✉] and S. Sai Satyanarayana Reddy

Vardhaman College of Engineering, Kacharam, Shamshabad, Hyderabad 501218, Telangana, India
ramesh.shahabadkar@gmail.com

Abstract. The conceptual background of face recognition (FR) evolved witnessing various contributions in the past two decades which has been extended towards a wide area of applications including commercial and law enforcement security solutions both. However, it has become a foundation of several breakthroughs on various research aspects associated with cloud computing (CC) driven big data analytics and machine learning platforms. The extended research track in this specific domain claimed to transform the conventional view of solving the problems associated with analytics based FR in social media platforms. The study also aimed to explore various scope of integrating conventional social media (SM) based big data analytics (BD) technology on FR considering an approach of machine learning (ML). Thereby it has formulated a novel framework well capable of face detection considering a machine learning approach on a cloud operated SN platforms. The study formulated analytical approach namely computationally efficient face recognition (CE-FR) schema for face tagging on big data driven SN platforms. The effectiveness of the study further evaluated to validate the performance of the proposed FR system.

Keywords: Cloud computing · Machine learning · Social networks
Face recognition

1 Introduction

The rapid advancement in the field of social media technology enhanced the foundation of conceptualization behind different facial recognition software applications possibly exercised to perform detection of different facial patterns of human beings [1, 2]. However, it is a matter of fact that a human being has always had the inborn potential ability to recognize and distinguish between facial patterns of one compared to another, yet computers also more likely found to perform the same ability owing to some instructions embedded with a set of conditions and rules. The timeline of the research track on FR claimed that scientist initiated developing computer programs to recognize human facial patterns in the year of 1960, since then the research track started to evolve to a large extent. Facial recognition software is driven with specific instructions well capable of distinguishing a face by computing various distinct features of that face [3]. There exist certain distinguishable features associated with every face which includes

© Springer International Publishing AG, part of Springer Nature 2019
R. Silhavy (Ed.): CSOC 2018, AISC 763, pp. 75–83, 2019.
https://doi.org/10.1007/978-3-319-91186-1_9

numerous landmarks, different peaks and valleys that allow facial features to construct a face. These distinguishable landmarks are often referred as nodal points [4, 5]. Usually a human face consists of approximately 80 nodal points where some these are distinguishable by software driven computational frameworks in terms of distance between the eyes, width of the nose and depth of the eye sockets etc. however, along with all these various distinct features of faces precisely nodal points can be represented by creating numerical codes which also represent a facial values within a database. The wide advancement and potential integration of FR immensely produced a range of applications in both commercial and law-enforcement settings at an incredible scale. FR has subsequently produced a large number of trailblazing innovations including Samsung Smart TVs, where built-in cameras are installed to provide authorized access control with respect to different aspects. However, owing to the potential advancement of FRtechnologies, the facial features are transformed into an imperative smart object processed by software/cloud driven computing to perform several active multitude of authentication paradigms including facial recognition supported by biometric data extractions. FR now a day becomes an integral part of *Internet of Things* (IoT) by enabling its consequent potential over CC, BD, SN and ML [6]. Despite of having effectiveness the conventional studies pertaining to FR mechanisms considering ML approach lack effectiveness in terms computational efficiency, large set of images i.e. big data volume from contemporary operational aspects. The following Fig. 1 exhibits a conventional approach of FR using ML for social media networks.

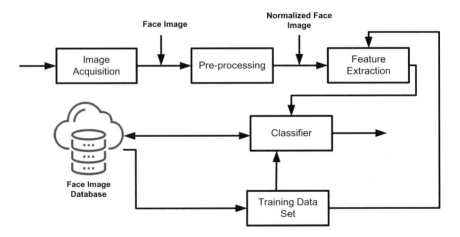

Fig. 1. Conventional ML based FR

The study thereby narrowed the discussion of FR in a correlation with different aspects of BD, CC, SM platforms and ML. Further it extended its contribution towards presenting an analytical design aspect of a novel FR mechanism computationally efficient to tag facial images on big data and cloud driven social media platforms. The analytical paradigms are numerically designed and simulated in a computing framework where the outcomes obtained exhibits that the proposed (CE-FR) schema achieves better accuracy and negligible computational complexity as compared to the existing system.

The manuscript is organized with a four-fold pattern where Sect. 2 represents the conventional FR approaches being exercised on different aspects such as BD, SN, ML and CC along with a brief discussion on related works. Along with this Sect. 3 explicitly presents the design aspects of the conceptual CE-FR modeling integrated with both cloud based FR engine and database. The experimental outcomes obtained by simulating the proposed CE-FR, further illustrated in Sect. 4, which also demonstrates a comparative analysis followed by conclusion of the overall study in Sect. 5.

2   Related Work

This section exclusively discusses about the FR techniques adopted into different field of studies with the aim of enhancing the accuracy of face tagging with optimal operational cost.

2.1   Studies Carried Out in Face Recognition Considering Big Data Analytics

Face tagging and big data, these two terms hardly have something in common but the recent advancements in the field of FR has led to a scenario where there is a scope still existing conveying that SM data (i.e. facial images) can be utilized to be emerged in impactful ways to aid criminal detection using FR and also on other potential face tagging applications.

The study of Chen et al. [7] extensively illustrated how big data drives face recognition paradigm to a significant level which converges image processing and facial detection both operated on a massive amount complex big data generated from several internet sources e.g. several face tagging within a specific image in a social media platform like *Face book* makes criminal detection task more crucial due to the complex features and their correlation presence in human faces. It also studied the several operational aspects of criminal apprehension through mug shot matching and its significant contribution towards finding a missing person. They study also calmed that big data analytics (BDA) can leads to improve the accuracy of a FR system as it processes a massive number of images for a specific task to execute. It also exhibits that BDA leads to better accuracy as it relies on sample sizes of input facial images and perform correlation on a set of facial attributes and extracted coefficients.

The study of Thorat et al. [8] conceptualized a scalable and flexible solution approach well capable of recognizing facial attributes from different CCTV footage operating on big data and also supported by distributed Hadoop clusters. The study incorporated software defined network which usually deployed an increasing number of CCTV for the security purpose. However, this ideology is claimed, that it has found better investigational records while processing a vast data generated by each CCTV, which has further led to better outcomes. It has utilized CCTV footage approximately in a size of 100 GB. A Hadoop and Map Reduce tool are deployed across a wide span of region to perform distributed parallel computation and video processing. The study has claimed to achieve better facial recognition on different models and distributed configuration of

Hadoop batch processing. The experimental outcomes also exhibited the effectiveness of Hadoop architecture on facial recognition in terms of complexity aspects.

2.2   Studies Carried Out in Face Recognition Considering Social Networking Platforms

Currently social networking platforms have gained a lot of attention from both personal and professional aspects. An insight into social networking platforms reveals the matter of fact which is inclined towards inherent behavior associated with the users connected to the SM word wide. It also subjected their various ways of expressing different emotions associated with their daily lives. May be it can include few fundamental activities including uploading and sharing personal facial images on timeline etc. FR has been implemented in this area in a large extent where its variety of characteristic feature includes tag suggestions to conform the appropriate face tags.

In the study of Indrawan et al. [9] introduced a novel face recognition system which integrates cloud services with user mobile devices. The study developed an application which has been implicitly deployed on mobile devices. It utilizes various facial images arranged and organized in a distributed web application namely Test.com API. The experimental prototyping further exhibited that the proposed model achieves considerable outcomes with respect to recognize face samples with 85% accuracy by obtaining very les processing time. The augmented reality translation time is found approximately 1.03 s while fetching someone's information.

2.3   Studies Carried Out in Face Recognition Considering Machine Learning Algorithms

Machine learning also witness a remarkable growth in the field of facial expression and face recognition applications both from theoretical and experimental perspectives. ML algorithms in FR are defined with certain abstract level of computation which involves distinguishing complex facial patterns and learning new patterns adaptively for better predictive outcomes of recognition systems. However, one of the most emerging trends associated with ML algorithms is Artificial Neural Networks (ANN) and it was having a broader scope into tacking pattern recognition problems in the past. However, despite of having these all significant attributes its success become limited to some specific FR applications which has been highlighted in the study of [10]. Therefore, further the scope of the study has been integrated with the advent of Deep Learning (DL) [11]. It is extensively studied that DL has a potential capability to transform and reuse ANN which thereby increases efficiency of a network towards transmitting more data. DN basically operates well while it fed with more of data patterns which are quite a similar scenario like providing more experience to a person. The concept of DNN added more advantages to the conventional classifiers such as support vector machines (SVM) and linear discriminant analysis (LDA) mechanisms. The potential advancement of DL mechanisms has been found in the study of Yi et al. [11] where they have utilized DNN to model high level complex data for the purpose of facial authentication mechanism. The study has also claimed that using the potential benefits of DNN on Deep Face traces with 100 million connections along with a substantial database, Face book achieves an accuracy of 97.35% on face recognition. Therefore DNN become one of the most powerful techniques to extract significant features from a substantial set of variables. It also has a capability to formulate solution with relatively small datasets by incorporating more data.

2.4 Studies Carried Out in Face Recognition Considering Cloud Computing Paradigm

Cloud computing has emerged as a potential model which proffers reliable services to its users residing in a remote location. The concept of FR has been implemented on cloud since long time where the FR engine is configured in a virtualized server located in a remote location with different computational capability. The ubiquitous cloud network also enabled remote processing of FR attributes necessary to process facial images. Various studies [12–14] have claimed that cloud-based FR systems can operate on a pool of resources ensuring better processing on real-time facial image entities. Several commercial as well as law-enforcement applications are witnessed to deploy privacy-preserving security protocols to ensure prompt and authentic connection between client-server models. The following segment of the study presents the system model of proposed CE-FR schema.

3 Proposed Methodology

The study adopted an analytical methodology to support the design essentials subjected to present the proposed CE-FR mechanism. The proposed system exclusively designed on the basis of a cloud based modeling where the both FR engine and database are located inside a virtualized cloud model. The aforementioned modules operates on a large scale database consisting facial images which is more often referred here as BD. The design principals of initiated CE-FR methodology designed with an aim of detecting faces in the social media generated facial images which are stored inside the cloud repositories deployed by Face book, Google etc. The design principles of the proposed architecture is defined with a set of procedures which mainly includes (1) enrollment of query or new faces to be tagged through a mobile or web application panel. (2) Secondly, in order to carry out an effective communication the system establishes a connectivity with the respective server through a cloud based web API running under https protocol stack. The cloud based paradigm contains a middle tier program called representational state transfer module operates on a FR engine and massive database containing a huge set of facial images generated by social networking sites like Face book, twitter etc. (i.e. Big data repositories).

The application operates on the newly enrolled facial images by performing encoding. After this, the image containing significant layer of features are transferred to the FR engine through a cloud based API. Once, the FR engine receives the encoded image, it thereby performs there different operational stages such as (1) face detection, (2) extracting feature vectors followed by (3) face matching. Along with these steps the pre-defined FR algorithm also performs additional stages to make the process of FR more efficient. The query image which is basically the user face image provided as input is processed by the cloud based FR engine to extract the feature vector and compared against the set of images residing inside a particular database (e.g. in social networking platform there exist a set of images with tagged attributes). Once the system finds a uninterrupted match of the individual facial vector with an image residing into the database, then it classify the query image by determining the fact that to whom this face is supposed to belong. Further the system tag the respective face and the queried outcome responded back to the interface of web application. Although the framework has been conceptualized specifically for face tagging or face detection but it can also be used to define efficient access-control mechanism by incorporating face authentication. The modular design of the proposed CE-FR is illustrated in the following Fig. 2

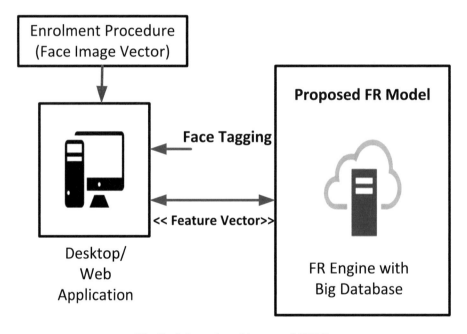

Fig. 2. Schematic architecture of CE-FR

The above highlighted Fig. 2 depicts the design methodology of the proposed system exclusively designed for privacy preservation and face detection in social media platforms driven by big data. The study also incorporated a multilevel security solution between the client-server models where the traffic flow has been controlled and the privacy has been extended by incorporating the following attributes:

- Integration of CE-FR with https protocol for secure data transfer.
- Incorporation of SSL protocols to utilize certificates for user authentication.
- Encryption of input face image.
- Restriction provided to secure the cloud service model using password based authentication.

The study also adopted a learning mechanism namely Decision based machine learning found to exhibit the conventional Deep Face technique in terms of computational complexity. In order to optimize the operational constraints initiated by Deep Face mechanism the study incorporated decision based ML which is preferred owing to its cost effective and light weight features. The proposed decision based ML approach found to overcome the slower learning speed problems associated with traditional optimization methods. The proposed ML approach achieves efficient performance while training a NN. The distinct properties of the proposed learning method typically include (a) the capability of attaining smallest training error and also have (b) lesser optimization constraints as compared to SVM. The proposed system basically incorporates the decision based learning approach to train the NN with the extracted features where the input weights and hidden layer biases are randomly selected. The proposed system also incorporates a differentiable activated mechanism which considers random assignment of input weights and hidden layer biases to determine the efficient predictive based decision. Further the proposed model exclusively simulated on a numerical computing platform with different optimization aspects. The system also minimized the error vector in order to train the proposed ML approach.

4   Results Discussion

The following are the outcomes obtained after simulating the proposed system in a numerical computing platform. The system specification includes 64-bit operating system along with 8 GB internal memory and Intel core i5 processor. The training time and the accuracy of face detection rate have been considered as performance parameters to evaluate the effectiveness of the proposed CE-FR system. The numerical outcomes are represented as follows: The following outcomes in Table 1 thereby represent the processing time accumulated during the numerical computation of learning phase of the cloud driven proposed CE-FR approach.

Table 1.   Processing time analysis for training phase

Simulation round	Processing time for training (sec) Proposed CE-FR	Processing time for training (sec) DeepFace
0	1.11	2.21
200	2.11	3.31
400	1.01	4.01
600	1.03	4.03
800	1.05	3.05
1000	1.08	2.08

The above mentioned Table 1 shows that the proposed system exhibits negligible computational complexity as it achieves very lesser processing time during the training phase of classifier in comparison with the conventional Deep face [9] learning mechanism. Thereby the adoption of multi-layer based decision oriented ML approach CE-FR achieves higher computational efficiency. The study performed another comparative performance analysis with respect to accuracy of FR is shown in below Fig. 3.

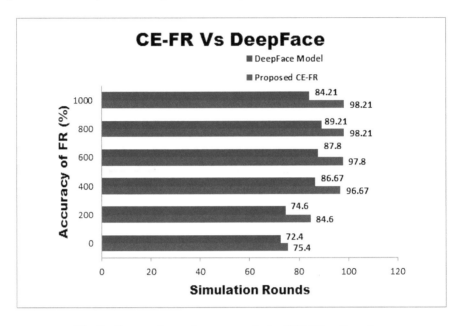

Fig. 3.   Comparative performance analysis of CE-FR vs Deep face

The above highlighted comparative graph clearly depicts that the proposed decision based ML approach achieves higher accuracy in terms of FR as compared to the conventional Deep face based learning mechanism the system found to excel superior outcomes at the simulation round 800 and 1000 where the proposed classifier has exhibited an accuracy of 98.21% in detecting the actual face.

5   Conclusion

The study initially performed a comprehensive discussion on FR and its significant impact on different aspects. It also formulated a background which converge and correlate FR with different recent advances such as big data analytics, machine learning etc. Further, the study has conceptualized a novel FR technique namely CE-FR, which operates on cloud environment, based social media platforms driven by big data. It also demonstrated the operational environment defined for face detection in the proposed cloud environment considering a novel decision based learning mechanism. The performance analysis further shows that the proposed CE-FR exhibits fast convergence speed in terms of learning and also attain better accuracy by means of completing FR tasks with maximum possible access control and effectiveness.

References

1. Ramanathan, N., Chellappa, R., R-Chowdhury, A.K.: Facial similarity across age, disguise, illumination and pose. In: International Conference Image Processing, ICIP 2004, vol. 3, pp. 1999–2002 (2004)

2. Keywebmetrcs article: How Big Data drives facial recognition. http://www.keywebmetrics.com/2013/08/big-data-drives-social-graph/. Accessed 28 Jan 2015

3. Animetrics article: Cloud face recognition services. http://animetrics.com/cloud-face-recognition-services/. Accessed Jan 2015

4. Computervisiontalks.com article: Image processing and cloud computing architecture overview. http://computer-vision-talks.com/articles/2011-04-13-image-processing-cloud-computing-architecture-overview/. Accessed Jan 2015

5. Metalife article: Face time. http://metalifestream.com/wordpress/?p=6498. Accessed 28 Jan 2015

6. Smartdatacollective.com article: Is Facebook taking Big Data analytics too far?. http://smartdatacollective.com/bernardmarr/121876/facebook-taking-big-data-analytics-too-far. Accessed 30 Jan 2015

7. Chen, M., Mao, S., Zhang, Y., Leung, V.C.M.: Big Data analysis. In: Big Data, pp. 51–58. Springer, Heidelberg (2014)

8. Thoughtworks.com article: New beginnings in facial recognition. http://www.thoughtworks.com/insights/blog/new-beginnings-facial-recognition. Accessed 28 Jan 2015

9. Indrawan, P., Budiyatno, S., Ridho, N.M., Sari, R.F.: Face recognition for social media with mobile cloud computing. Int. J. Cloud Comput. Serv. Archit. 3(1), 23–35 (2013)

10. Bishop, C.M.: Pattern Recognition and Machine Learning, vol. 4, no. 4. Springer, New York (2006)

11. Sun, Y., Chen, Y., Wang, X., Tang, X.: Deep learning face representation by joint identification-verification. In: Advances in Neural Information Processing Systems, pp. 1988–1996 (2014)

12. Bhatt, G.B., Shah, Z.H.: Face feature extraction techniques: a survey. National Conference on Recent Trends in Engineering & Technology, 13–14 May 2011

13. Peer, P., Bule, J., Gros, J., Štruc, V.: Building cloud-based biometric services. Informatica Int. J. Comput. Inf. 37(1), 115–122 (2013)

14. Chi, H., Chi, L., Fang, M., Wu, J.: Facial expression recognition based on cloud model. In: International Archives on the Photogrammetry, Remote Sensing and Spatial Information Sciences, vol. 38, Part II. Accessed 18 Dec 2017

Functional Specification to Support Security Risk Assessment of Large Systems

Neila Rjaibi$^{(\boxtimes)}$ and Latifa Ben Arfa Rabai

Laboratoire SMART, Institut Supérieur de Gestion,
Université de Tunis, Tunis, Tunisia
Rjaibi_neila@yahoo.fr, latifa.rabai@gmail.com

Abstract. Measuring the security of organizations is needed to obtain security evidence. We believe that common security identification and quantification related to system's functionalities can be extended to be used in other systems. Security measurements are common at the business process layer. This paper supports the development of security metrics according to each function of a related system. An elementary metric quantify risk by system's function. This leads to improve security risk analysis and communication for decision making.

Keywords: Cyber security metrics · E-Learning · Mean failure cost
Risk management · Security measures · Functional specification

1 Introduction

Metrics are defined as quantifiable measurements of some aspects of a system. It is compared to a scale or benchmark to produce a meaningful result. The term metrics describes a broad category of tools used by decision makers to evaluate data in many different areas of an organization. Thereafter, corrected actions are done according to the observed measurements [1]. Security metrics denote the maturity level of the security of system. And denote the extent to which security characteristic is present in a system. Performance and accountability are improved using collection and analysis of data. According to Jansen [1] main merits of security metrics are:

- Strategic support: decision making is supported by the assessment of security features. For example, product, service selection and resource allocation.
- Quality assurance: the software development cycle is supported by security metrics, in order to eliminate vulnerabilities. Examples fall into different perspectives such as during code production, during the identification of vulnerabilities and tracking security flaws.
- Tactical oversight: the monitoring and report of security status, the evaluation of the effectiveness of security measures and the management of risk. For example, we provide a practical area improvement.

When we intend to measure the overall security of a given network, a crucial issue is to correctly compose the measure of individual functional components. This leads to correct assessment and efficient result [2]. Our need is to develop a security risk

© Springer International Publishing AG, part of Springer Nature 2019
R. Silhavy (Ed.): CSOC 2018, AISC 763, pp. 84–89, 2019.
https://doi.org/10.1007/978-3-319-91186-1_10

analysis for each system's function. All the above metrics could be considered as exhaustive as they try to analyze possible threats on a system's assets. We intend to measure security metrics for reusable software functional components.

In the field of information security related risk management, there was a necessity to place a link between organizations' security information in order to exchange privileged information [3]. Unfortunately, in the open literature vulnerabilities and threat-sources do not have standards for defining, assigning, combining impact and probability ratings, and managing the impact of controls. Different methodologies, tools and approaches are used. The risk is difficult to be identified and assessed for one system and between different member's risk management processes. Organizations widely recognize the importance of information security related common risk management.

And this point of view is encouraged by recent standards and recommendations on the management of information systems [2].

Measuring security metrics for reusable software functions are justified by:

- Ensuring a common language or common terminology in the risk management community
- Ensuring a common functional risk management framework for other systems
- Incremental use of previous risk management study
- Helping interoperability between risk analysis performance using different methods
- Helping security analysis to a set of additional functionalities that have been identified
- Using a common objective security repository.

Our model should be as usual quantitative instead of qualitative and reusable. Our intention is a possible reuse of previous security risk analysis on a system, sub system or components and to include these results in a new analysis of the same system or in another system. Our model may support exchanging requirements and security analysis for systems interconnections if two systems are required to be interconnected to exchange information. Also, two system managers can exchange and understand the results of risk analysis related to common functions [2]. Our model intends to define and share a common threat and vulnerability repository by functions. This leads to produce common profiles for system's assets, threats, security requirements, and vulnerabilities. The importance is to facilitate the sharing of information and results. The creation of a common and shared threat and vulnerability repository is crucial for large systems sharing same functions.

2 Characteristics of the Functional Security Model

For one system we expect more computing functional security metrics (FS) which are equal to the number of the system's functions. For each function of the related large system, we focus on its respective stakeholders, security requirements, architectural components and security threats in order to develop a functional security metric. For examples, an e-learning functional specification includes assessment, communication tools, registration, certification, course management and course administration as functions

Mathematically, the Functional Security model (FS) can be formulated as follows: Let fc be the set of elements in one function, FC is the set of elements of the other system's functions. ST is the set of stakeholders, R is a set of requirements, and T is a set of threats. For each function fc, we define the functional security risk management FS (fc, FC) of element fc as follow:

$$FS_{(fc;FC)} = S\ TR_{(fc;FC)} \circ RFC_{(fc;FC)} \circ FCT_{(fc;FC)} \circ T_{(fc;FC)} \tag{1}$$

1. We denote by $\circ$ the matrix multiplication operation.
2. STR_{fc} is a matrix of size ($|S\ T|$; $|R|$) that each entry (i; j) represents the value of the stake that stakeholder ST_i has in meeting a requirement R_j. we denote by $|S\ T|$ (resp:$|R|$) the size of the set ST (resp. $|R|$).
3. RFC_{fc} is a matrix of size ($|R|$; $|FC|$) that each entry (i; j) represents the probability of failing requirements R_i due to a failure originating from elements FC_j.
4. FCT_{fc} is a matrix of size ($|FC|$; $|T|$) that each entry (i; j) represents the probability of failing FC_i once the threat T_j has materialized.
5. T_{fc} is a column vector of size $|T|$ that each entry i represents the probability that threat Ti has materialized during unitary period of time.

3   Applying the Functional Security Model in E-Learning Systems

E-learning systems development increase with the widespread of internet and the need of online courses. According to a survey conducted by Campus Computing (campuscomputing.net) and WCET (wcet.info) showed that almost 88% of the surveyed institutions have adopted a learning management system (LMS) as their medium for offering online courses [4]. Every element in the e-learning systems is considered as an educational asset; it can be a potential target of hacking or attacks [4]. It is crucial to discover the critical assets and to identify the limitations in current systems and to underline factors that affect the quality toward safe e-learning. To mitigate the security risks inherent in online learning, security policies and mature security measures needed to be developed.

The online learning management systems or e-learning systems are adopted but without a thorough understanding of the security aspects [5]. Also, bloggers have not discussed security in online learning with great frequency. In addition, online learning providers and practitioners have not considered security as a top priority [6, 7].

3.1   The FS Metric for the E-Learning Registration Function

We illustrate the usage of the proposed FS model using the examples to estimate the security of systems' registration function in terms of loss incurred by each stakeholder. For each function of the considered system, we define the lists of stakeholders, security requirements, threats and architectural components.

To compute FS metric for each system's function we use 4 steps:

- We keep the set of primary stakeholders that are administrator, teacher, student, technician which are applicable to any e-learning function. For example, Table 1 illustrates the STR registration.

Table 1. The STR registration

Usability	Consistent APTs	20	20	10	10
Physical protection	**Physical protection**	20	10	0	10
Access control	**Authorization**	10	30	5	5
	Identification	10	30	5	5
	Authentication	10	30	5	5
Availability	**Resource allocation**	22.5	22.5	1.5	7.5
	Expiration	22.5	22.5	1.5	3.75
	Response time	15	15	0.75	3.75
Non-repudiation	**Non-repudiation**	10	20	0	5
Integrity	**Software integrity**	7.5	4.44	0.38	1.47
	Personal integrity	10	6.6	1.66	2.1
	Hardware integrity	5	4.44	1.66	2.1
	Data integrity	7.5	4.44	0.83	1.05
Privacy	**Confidentiality**	40	20	0	10
Security requirements	**Security requirements sub factor/**	**Administrator**	**Teacher**	**Student**	**Technician**
		Stakeholders			

- We define the security requirements related to each system's function. For example Table 1 illustrates the main security requirements for each stakeholder. Thus at this matrix loss of users vary according to each function.
- From the mapping of security threats and the architectural components we infer also values from previous empirical investigations [6, 7] and their corresponding probabilities (Tables 2, 3, 4).
- We derive an original quantitative functional metric of current e-learning system's functions. The Mean Failure Cost with all the stakeholders for the function registration is presented in Table 5 (the metric FS registration)

The proposed metric leads to identify, analyse, and prioritize information security risks and improve vulnerability management activities by viewing them in a functional context. Managing risk is important to managing enterprise risk. The strength of this study is in developing risk response strategies appropriate for the organization's business requirements. Management is suitable to specific security goals. The gathered

Table 2. The RFC registration

Security requirements		Browser	Web server	Application Server	DB server	Firewall server	Mail server	No failure
Usability	Consistent APTs	5 10-4	5 10-4	5 10-4	5 10-4	0	5 10-4	9.97 10-1
Physical protection	Physical protection	0	0.7 10-3	0.7 10-3	0.7 10-3	0.7 10-3	0.7 10-3	9.965 10-1
	Authorization	0	4.2 10-3	4.2 10-3	4.2 10-3	4.2 10-3	4.2 10-3	9.79 10-1
Access control	Identification	0	4.2 10-3	4.2 10-3	4.2 10-3	4.2 10-3	4.2 10-3	9.79 10-1
	Authentication	0	4.2 10-3	4.2 10-3	4.2 10-3	4.2 10-3	4.2 10-3	9.79 10-1
	Resource allocation	0	3.3 10-3	3.3 10-3	3.3 10-3	0	3.3 10-3	9.868 10-1
Availability	Expiration	3.3 10-3	3.3 10-3	3.3 10-3	3.3 10-3	3.3 10-3	3.3 10-3	9.802 10-1
	Response time	3.3 10-3	3.3 10-3	3.3 10-3	3.3 10-3	3.3 10-3	3.3 10-3	9.802 10-1
Non-repudiation	Non-repudiation	2 10-2	3.3 10-2	3.3 10-2	0	1 10-2	3.3 10-2	8.71 10-1
	Software Integrity	7 10-3	7 10-3	7 10-3	7 10-3	7 10-3	7 10-3	9.58 10-1
Integrity	Personal Integrity	0	0	0	0	0	0	1
	Hardware Integrity	0	7 10-3	7 10-3	7 10-3	7 10-3	7 10-3	9.65 10-1
	Data Integrity	0	7 10-3	7 10-3	7 10-3	0	7 10-3	9.72 10-1
Privacy	Confidentiality	2 10-2	3.33 10-2	3.33 10-2	5 10-2	1 10-1	3.33 10-2	7.3 10-1

Table 3. The FCT registration

Threats	BroA	InsC	DoS	CryptS	DOR	InfL	Buff	CSRF	CSS	FURL	InjecF	MFile	No Threats
Components													
Browser	0,477	0,119	0,006	0	0	0	0	0	0	0,397	0	0	0
Web server	0,273	0,137	0,001	0	0	0	0,342	0,007	0,014	0,227	0	0	0
Application server	0,271	0,135	0,007	0	0	0	0,338	0,007	0	0,225	0,014	0,003	0
DB server	0,187	0,094	0,005	0,155	0,155	0,155	0,234	0,005	0	0	0,009	0,002	0
Firewall server	0,143	0,143	0,714	0	0	0	0	0	0	0	0	0	0
Mail server	0,375	0,187	0,009	0,028	0,028	0,028	0	0,009	0	0,312	0,019	0,005	0
No Failure	0,523	0,813	0,286	0,845	0,845	0,845	0,658	0,991	0,986	0,603	0,981	0,995	1

Table 4. The Vector T registration

Threats	Probability
Broken authentication and session management (BroA)	4.20 10-3
Insecure communication (InsC)	3.00 10-3
Denial of service (Dos)	3.08 10-3
Insecure cryptographic storage (CrypS)	7.00 10-4
Insecure direct object reference (DOR)	7.00 10-4
Information leakage and improper error handling (InfL)	7.00 10-4
Buffer overflow (Buff)	1.00 10-4
Cross Site Request Forgery (CSRF)	4.20 10-4
Cross Site Scripting (CSS)	1.80 10-4
Failure to restrict URL access (FURL)	9.80 10-3
Injection flaws (InjecF)	2.17 10-3
Malicious file execution (MFile)	5.04 10-4
No Threats	974.44 10-3

information can be mapped to any arbitrary application or technical specification. Since each e-Learning system is composed of a set of functions and our model quantifies the security risks by function, our FS model can be applied to any E-Learning system totally (if both systems present the same functions) or partially (if the two systems do not have the same list of functions). However, the FS model will be useful and usable for each E-learning system to which a risk management policy is intended.

Table 5. The metric FS registration

Stakeholders	FS registration
System administrator	193.178
Teacher	225.827
Student	32.39
Technician	66.8

4 Conclusions

The new approach resides on describing security requirements, components, threats and stakeholders at the business process layer. A complete and consistent security analysis model is built according to the different functions or applications of such a system. In practice, this section states the idea to develop a security quantification model related to the system's educational asset in order to cope with openness and future empirical reuse problem. This contribution is beneficial to further reuse quantification for other large systems. While taking account of the same system's function. Empirical data and risk information can be gathered and organized via interviews, documentation reviews, and technical analysis. Risk assessment is more strategic, objective and depends on one functional context.

References

1. Jansen, W.A.: NIST IR 7564: Directions in security metrics research. National Institute of Standards and Technology, US Department of Commerce, Gaithersburg (2009)
2. Wang, L., Singhal, A., Jajodia, S.: Toward measuring network security using attack graphs. In: Proceedings of the 2007 ACM Workshop on Quality of Protection, pp. 49–54. ACM (2007)
3. Final Report of Task Group IST-049, Improving Common Security Risk Analysis (2008)
4. Chen, Y., He, W.: Security risks and protection in online learning: a survey. Int. Rev. Res. Open Distrib. Learn. **14**(5) (2013)
5. MohdAlwi, N.H., Fan, I.S.: Threats analysis for e-learning. Int. J. Technol. Enhanced Learn. **2**(4), 358–371 (2010)
6. Rjaibi, N., Rabai, L.B.A.: Expansion and practical implementation of the MFC cybersecurity model via a novel security requirements taxonomy. Int. J. Secure Softw. Eng. (IJSSE) **6**(4), 32–51 (2015)
7. Rjaibi, N., Rabai, L.B.A.: Maximizing security management performance and decision with the MFC cyber-security risk management model. EAI Endorsed Trans. e-Learn. **4**(15) (2017)

The Way of Quality Management of the Decision Making Software Systems Development

O. N. Dolinina[1]([✉]) ⓘ, V. A. Kushnikov[1] ⓘ, V. V. Pechenkin[1] ⓘ, and A. F. Rezchikov[2] ⓘ

[1] Yury Gagarin State Technical University of Saratov, Saratov, Russia
odolinina09@gmail.com
[2] Institute of Precision Mechanics and Control of Russian Academy of Sciences,
Saratov, Russia

Abstract. The different characteristics of the decision making system's software quality are analyzed. In spite of a lot of research comprehensive criterion of the software quality management still exists only on an informal level. There are described the differences between Russian GOST R standard and ISO. It is shown that the quality of the software is a manageable indicator can be represented by an acyclic connected graph G, in which the upper level is represented by the following characteristics according to the standard ISO. The task of the providing of the planned quality level is formalized as the optimization one taking into consideration the vectors of the control activities and environment states. Special attention is given to the quality characteristics of the intellectual systems. Plan of the activities is validated by the Boolean functions, for this aim graph of the causal relationships is built and transferred to the logic scheme. The plan can be built at any stage of the software life cycle.

Keywords: Integrated criteria of the software quality · Optimization task
Vectors of the control activities · Graph of the causal relationships
Discrete logic scheme

1 Introduction

Methods of quality management at all stages of the life cycle of the software development are determined by a large number of factors. The main reason for this situation is the increasing complexity of modern systems in industry, economy, and social sphere. A major role in the need to improve the quality and reliability of software plays the integration of the decision-making systems, the use of the software in the areas that have high requirements for reliability of decision making. For example, this circumstance is important in the control systems for objects in the aerospace field, the management of rail transport, to ensure the uninterrupted operation of equipment in the energy sector, decision-making in medicine. The quality of decision-making software according ISO/ IEC 25000: 2014 is defined as the ability of the software product under specified conditions to meet specified or expected requirements. Software quality is a complex indicator, determined by the normative documents of different levels: international, national, industry standards, and enterprise standards as well as software projects. In general, it can be asserted that the quality of the software is a manageable indicator and can be represented by an acyclic connected graph G, in which the upper level is represented by the following characteristics according to the standard ISO/IEC 25010–2015:

- functionality;
- performance level;
- compatibility;
- ease of use;
- reliability;
- security;
- maintainable;
- portability (mobility).

It should be noted that each of these characteristics has associated metrics. For example, functionality is defined as interoperability, usability, standards compliance, security, accuracy (ISO 9126). An important indicator of quality is the reliability index, defined by ISO 9126, as maturity, ability to recover, compliance with standards, resistance to failures. This indicator directly affects the quality of decisions made by the software system. It should be noted that in the current literature there are differences in the meaningful definitions of quality characteristics. A very indicative example can be such a measure as dependability, defined by Russian Federal standard (GOST), as reliability, while international sources [1] define this term as dependability in which reliability is included as an integral part. In the ITU E800 and IEC 60300 standards, the warranty for the composition is narrower than reliability. According to the standards of IAEA, ECSS, properties such as confidentiality and integrity are considered as related to dependability. Because most quality indicators are qualitative expert methods are used to define them.

Among the studies on the development of software quality assurance methods it is necessary to mention the works of Lozinsky [2], Clarke, Gramberg [3], Markov, Pope, Hill, Hetzel, Davis which deal with methods of providing different quality SW indicators. The most developed are the testing methods described by Myers [4], Lipaev [5], Kaner [6], Crispin, Gregory [7], Beizer [8], Culbertson [9] which allows to detect errors in the software source code. However, the problem of quality management software as a comprehensive criterion still exists only on an informal level.

2 Method for Achieving Specified Software Quality Level

In general, the integrated quality measure of software for complex decision making systems contains a large number of quantitative and qualitative characteristics [10] that can be defined by an expert way. From a system analysis perspective the goal of achieving the required software quality level can be presented as a problem of optimizing of the complex criterion.

$$K = f(P_Q(\vec{x}(t), \vec{u}(t))) \tag{1}$$

where $\vec{u}(t) \in \left\{ \vec{U}(t) \right\}$ – control vector, $\vec{x}(t) \in \left\{ \vec{X}(t) \right\}$ – environmental state vector. The subintegral expression of the effectiveness criterion is the function

$$P_Q(\vec{x}(t), \vec{u}(t)) \tag{2}$$

which characterizes the probability of maintaining, at a given time interval $[t_b, t_e]$, the required level of software quality in an unfavorable combination of circumstances.

An unfavorable coincidence of circumstances is a combination of events in the process of the software system functioning, each of which does not significantly affect the quality of the software, but in the aggregate leads to a significant decrease in its quality. The formal definition of the software quality optimization problem is as follows

$$K_1 = \int_{t_H}^{t_k} P_Q(\vec{x}(t), \vec{u}(t))dt \rightarrow \max \tag{3}$$

constraints have the following form

$$F_i(\vec{x}(t), \vec{u}(t)) < 0, \ i = \overline{1, n_2} \tag{4}$$

$$F_i(\vec{x}(t), \vec{u}(t)) \geq 0, \ i = \overline{1, n_1} \tag{5}$$

boundary conditions are

$$F_i^{(t_b)}(\vec{x}(t), \ \vec{u}(t)) = 0, \ i = \overline{n_2 + 1, n_3} \tag{6}$$

$$F_i^{(t_e)}(\vec{x}(t), \ \vec{u}(t)) = 0, \ i = \overline{n_3 + 1, n_4} \tag{7}$$

$n_1 - n_4$ – known constants;
$\left\{ \vec{X}(t) \right\}$ and $\left\{ \vec{U}(t) \right\}$ – the range of admissible values for the vectors $\vec{x}(t)$ and $\vec{u}(t)$, respectively.

Boundary conditions are defined by specific of the object management.

The described problem belongs to the class of variational calculus tasks on a conditional extremum. It is characterized by high dimensionality, uncertainty of the number of scalar components for the state vector of the environment $\vec{x}(t) \in \left\{ \vec{X}(t) \right\}$ at time intervals $[t_b, t_e]$ of considerable length. Both the X and U vectors have quantitative and qualitative variables which are also of fuzzy nature. For synthesis $\vec{u}(t)^*$ it is necessary to solve in real time complex system of nonlinear differential equations of high order.

To solve the task of synthesizing the vector of control actions $\vec{u}(t)^*$, we use a proven statement in accordance with which it is sufficient to develop and implement a detailed comprehensive plan $Pl(t)$ to improve the quality of software functioning in order to solve the task of maximizing the criterion $K_1 = f(P_Q(\vec{x}(t), \vec{u}(t)))$ [11, 12]. If in time moment t the plan can be implemented this means that the software quality is improved to the maximum. Otherwise, it is necessary to identify the reasons that prevent the implementation of this plan, as well as to develop and implement a set of actions to eliminate them.

If it is impossible to ensure the feasibility of the action plan $Pl(t)$ under the existing conditions, the task (3) does not have a solution and the optimization of its criterion will require the development of a different solution way.

On the basis of the above discussion, the following heuristic procedure for constructing an operation plan is proposed, based on the requirements of the GOST ISO/IEC 9126-93, GOST/IEC 25010-2015 and consisting of the following steps:

1. Development a graph G_1 of the hierarchical causality between the elements of the decision system software for which the task is being solved; definition of the major software modules of the system $N_i, i = \overline{1, m}$ that affect the quality of its operations.
2. For each program module in accordance with GOST R ISO/IEC 25010-2015, characteristics and properties (indicators) of software quality functioning are selected. As a result, we form a graph G_2 of hierarchical cause-effect relationships between the characteristics and performance indicators of the software modules.
3. The graph G_3 characterizes a complex of cause-effect relationships between the quality indicators of program modules and the most common types of software errors in the decision-making system that affect the quality of its functioning.

The most common types of software errors, as well as their impact on the quality of software modules are discussed in detail in the literature [2–4]. For specialized decision-making systems, intelligent systems, new classes of errors inherent in knowledge bases can be added [13–16]: incompleteness, inconsistency, redundancy.

The graph of an event tree G_{ET} is defined as the union of graphs

$$G_{ET} = G_1 \cup G_2 \cup G_3 \tag{8}$$

Let's consider an example for a fragment of the graph describing the upper levels of the event tree used in analyzing the software quality of the GAZDETECT intelligent decision-making system that determines the causes of malfunctions of gas-pumping aggregates [11] (Fig. 1). The edges of this graph are conjunctive edges. In Fig. 1 used the following symbols for software quality indicators:

u_3 - usability;
u_{16} - understandability;
u_{17} - learnability;
u_{19} - attractiveness;
u_{20} - consistency of the modules (compliance);
u_{77} - deficiencies identified in the examination of documentation;
u_{78} - difficulties in understanding input and output data;
u_{79} - disadvantages when using the help system;
u_{80} - contradictions in the implementation of the user interface;
u_{81} - difficulties in mastering software according to documentation;
u_{82} - difficulties in mastering the software on a test case;
u_{99} - contradictions in the implementation of the user interface;
u_{100} - contradictions in the diagnosis of the system;
and so on.

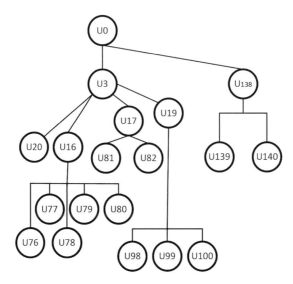

Fig. 1. An example of the upper levels G_{ET} used in analyzing the quality of intelligent systems software.

Vertices u_{138}, u_{139}, and u_{140} correspond to special requirements for knowledge bases (KB) in case of the intellectual systems: u_{138} - errors in the KB of the intelligent decision-making subsystem; u_{139} - absence of structural errors in the KB and u_{140} - absence of errors in the KB such as "forgetting about exclusion" and "critical combination of the facts" [13–16].

One must mention that the u_{139} for the rule-based systems can be satisfied by the development of the special verifier of the knowledge base graph [13, 14]. The most complicated is satisfying the u_{140} as it is connected with the errors in the field of study.

To verify the action plan, we use the Boolean functions apparatus and transform the graph G_{ET} to the form of the discrete device circuit D_u (Fig. 2) described by a logical function:

$$
\begin{aligned}
\varphi(&u_3, u_{16}, u_{17}, u_{18}, u_{19}, u_{20}, u_{76}, u_{77}, u_{78}, u_{79}, u_{80}, u_{81}, u_{82}, \\
&u_{98}, u_{99}, u_{100}, u_{138}, u_{139}, u_{140}, e_1, e_2, \ldots, e_y) = \\
&u_3 \wedge u_{16} \wedge u_{17} \wedge u_{18} \wedge u_{19} \wedge u_{20} \wedge u_{76} \wedge u_{77} \wedge u_{78} \wedge u_{79} \wedge \\
&\wedge u_{80} \wedge u_{81} \wedge u_{82} \wedge u_{98} \wedge u_{99} \wedge u_{100} \wedge u_{138} \wedge u_{139} \wedge \\
&\wedge u_{140})
\end{aligned}
\tag{9}
$$

where u_i are the vertices of the graph, e_j – the arcs of the graph, respectively. When testing, we assume that if all events preceding the event u_i occur, then this event will occur also.

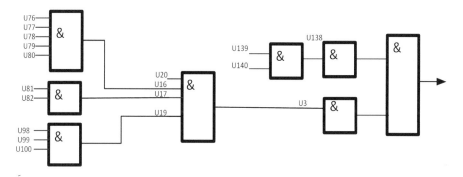

Fig. 2. The circuit of a discrete logic device D_u used in verifying the feasibility of the plan $Pl(t)$ at the time t.

The values of the input variables for D_u are defined (0 or 1) depending on whether performed or not certain operations of the plan $Pl(t)$, characterizing the corresponding u_i. The expression (10) allows to determine the possibility of implementation of the plan at any time moment $t \in [t_b, t_e]$.

$$\varphi(u_3, u_{16}, \ldots, u_{139}, u_{140}) = \begin{cases} 1, & \text{if } Pl(t) \text{ can be executed} \\ 0, & \text{if } Pl(t) \text{ can not be executed} \end{cases} \quad (10)$$

If at the moment of time $t \in [t_H, t_k]$ the value of a Boolean function is equal to 1

$$\varphi(u_3, u_{16}, \ldots, u_{139}, u_{140}) = 1,$$

this means that the problem (3) is solved and the scalar components of the vector $\vec{u}(t)$ are determined and they correspond to those actions that must be performed to satisfy the condition $u_3 = 1, \ u_{16} = 1, \ldots, \ u_{140} = 1$.

If $\varphi(u_3, u_{16}, \ldots, u_{138}, u_{139}, u_{140}) = 0$ the plan $Pl(t)$ at the time $t \in [t_H, t_k]$ can not be executed, and the value of the software quality at a given time is not optimal.

In the process of analyzing the spread of the zero signal, it is possible to determine the reasons, which do not allow to carry out the plan in time t. In [11, 12], it is shown that the implementation of each plan operation $Pl(t)$ affected by the conditions associated with the peculiarities of the decision making software system that depend on the environment. These conditions U for the operation of the plan $O_i \in Pl(t)$, $i = 1, l$ corresponding to the vertex $u_i \in G_{ET}$ are formalized in this way:

IF

$$< Y_1(\vec{x}, \vec{u}, t) \ R_1 \ Y_2(\vec{x}, \vec{u}, t) \ R_2 \ldots \ R_{k-1} \ Y_l(\vec{x}, \vec{u}, t) >$$

THEN PLAN OPERATION

$$O_i \in Pl(t), \ i = \overline{1, l}$$

WILL BE PERFORMED/NOT PERFORMED

$$R_i \in \{AND, OR, NOT, AND - NOT, OR - NOT; \leq; \geq; =\}, \ i = \overline{1, k-1} \qquad (11)$$

For example, u_i
IF

$$< \xi_i \leq \varepsilon_i >$$

THEN *SOFTWARE*
DOES NOT HAVE AN INDICATOR OF QUALITY REQUIRED VALUE
(ε_i – coefficient of significance [12])

If the number of conditions is one or two, the rules (11) have the following form
$Y_1(\vec{x}, \vec{u}, t)$ or $Y_1(\vec{x}, \vec{u}, t) \ R_1 \ Y_2(\vec{x}, \vec{u}, t)$.

Taking into account these conditions, we can evaluate the feasibility of the action plan $Pl(t)$ at each time point $t \in [t_b, t_e]$, as the described rules contains the experts' knowledge about the specific details of plan's operations affected by different factors. Often the rules reflect the standards established at the enterprise or established specifically for this software project. In view of the above, we use the design of discrete device D_u (Fig. 2) as a basis for the feasibility check of the action plan $Pl(t)$. For each discrete element this design is completed with inputs f_i, $i = \overline{1, c}$ which meet the requirements (11). It is possible to analyze the errors propagation during the fulfillment of the plan.

One must mention that the example shown at the Figs. 1 and 2 contains the conjunction nodes only but in general the graph G_{ET} can contain disjunction nodes because the software developers even could define their own quality characteristics and decide which of them it is necessary to satisfy mostly and that's why use OR operation.

The introduction of expert rules requires the change of respective design of discrete device D'_u, used for feasibility check of the action plan $Pl(t)$ at time point t, see the modified design on Fig. 3. The check is made according to the procedure described above. f_i are the control actions aimed on the fulfillment of the plan.

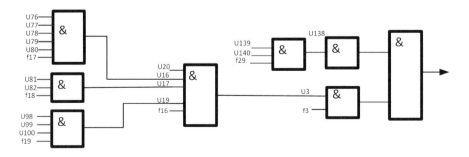

Fig. 3. The circuit of a discrete logic device D'_u used in verifying the feasibility of the plan $Pl(t)$ at the time interval $t \in [t_b, t_e]$.

The developed methods of action plan $Pl(t)$ feasibility check during time interval $t \in \left[t_b, t_e\right]$ help to build a feasible action plan for quality assurance of decision-making software. The plan $Pl(t)$ can be built at any stage of the software life cycle. The suggested method of the software quality planning was successfully used for the development of the expert system GAZDETECT for the detection of the gas turbines faults.

3   Conclusion

It is shown that the increasing of the software quality problem for the decision making systems is formalized as the task of the vector optimization with the limitations by equations and inequations. This task can be solved as the task of the synthesis of the activities complex plan to increase of the software quality and by the verification of it with the various control actions and states of the environment. There is suggested the method of the plan verification by means of the Boolean functions and expert rules which allows to present the plan with the set of the discrete devices where the "1" output shows the possibility of the plan's realization. The described methodology was successfully used for the development of the expert system GAZDETECT which detects the gas turbines faults.

References

1. Avizienis, A., Laprie, J.-C., Randell, B., Landwehr, C.: Basic concepts and taxonomy of dependable and secure computing. IEEE Trans. Dependable Secure Comput. **1**, 1–33 (2004)
2. Lozinsky, A., Shubinsky, I.: Defining requirements to software (in Russian). http://www.ibtrans.ru/Requirements.pdf. Accessed 06 May 2016
3. Klark, E., Gramberg, O., Peled, D.: Verification of Program Models: Model Checking. MTSNMO, Moscow (2002). (in Russian)
4. Mayers, G.: Quality of software. Mir, Moscow (1980). (in Russian)
5. Lipayev, V.: Methods of quality assurance for large-scale software (In Russian). Sinteg, Moscow (2003). (in Russian)
6. Kaner, K., Folk, D., Nguyen, E.: Software testing. Fundamentals of business application management. DiaSoft, Kyiv (2001). (in Russian)
7. Crispin, L., Gregory, J.: Agile Testing: A Practical Guide for Testers and Agile Teams. Vilyams, Moscow (2010). (in Russian)
8. Beyzer, B.: Testing the Black Box. Technologies for Functional Testing of Software and Systems. Piter, St. Petersburg (2004). (in Russian)
9. Culbertson, R., Brown, C., Cobb, G.: Rapid Testing. Vilyams, Moscow (2002)
10. ISO/IEC 25010:2011 Systems and software engineering – Systems and software Quality Requirements and Evaluation (SQuaRE) – System and software quality models. https://www.icc-iso.ru/toclients/standard
11. Shlychkov, E., Pokhaznikov, M., Kushnikov, V., Kalashnikova, O.: Feasibility analysis of the action plans in operating control of the machine engineering factory. Vestnik Saratovskogo gosudarstvennogo tekhnicheskogo universiteta **1**(1–21), 88–95 (2007). (in Russian)

12. Sklyomin, A., Kushnikov, V., Rezchikov, A.: Models and algorithms of feasibility analysis of the action plans in industry enterprise management. Vestnik Saratovskogo gosudarstvennogo tekhnicheskogo universiteta **3**(1–67), 145–152 (2012). (In Russian)
13. Dolinina, O.: Method of the debugging of the knowledge bases of intellectual decision making systems. In: Automation Control Theory Perspectives in Intelligent Systems, Proc. of the 5th Computer Science Conference 2016 (CSOC 2016), vol. 3, pp. 307–315. Springer (2016)
14. Dolinina, O., Rezchikov, A., Suchkova, N.: Formal models of structural errors in intellectual systems' knowledge bases. Sovremennye naukoyomkie tekhnologii, vol. 3 (2017), http://www.top-technologies.ru/ru/article/view?id=36607. Accessed 13 Apr 2017
15. Ginsberg, A.: Knowledgebase reduction: a new approach to checking knowledge bases for inconsistency & redundancy. In: Proceedings of 7th National Conference on Artificial Intelligence (AAAI 1988) (St. Paul MN), vol. 2, pp. 585–589 (1998)
16. El-Korany, A., Shaalan, K., Baraka, H., Rafea, A.: An approach for automating the verification of KADS-based expert systems. New Rev. Appl. Expert Syst. **4**, 107–124 (1998)
17. Avramov, M., Antropov, P., Gubin, N., Dolinina, O., et al.: Expert system of fault diagnosis of gas pumping units at compressor stations. Intellektualnye sistemy v proizvodstve **15**(1), 20–25 (2017). (in Russian)

An IoT Approach to Positioning
of a Robotic Vehicle

Miroslav Dvorak and Petr Dolezel[✉]

Faculty of Electrical Engineering and Informatics, University of Pardubice,
Pardubice, Czech Republic
{miroslav.dvorak,petr.dolezel}@upce.cz
http://www.upce.cz/

Abstract. This paper presents and evaluates one approach to the problems of automatic control of a vehicle movement in a large outdoor area. The positioning of the vehicle in the area is provided by iBeacons, located at the edges of the given surface. The iBeacon is a small and low-power device which periodically transmits its UUID (Universally Unique Identifier) number through the interface of a Bluetooth 4.x. The vehicle should be able to calculate its position according to the power of the signal, considering the location of the iBeacons. To be more specific, the triangulation method is applied to determine the position. According to the set of experiments presented at the end of the paper, the position error of a robotic vehicle is mostly less then 1 m.

Keywords: iBeacon · Trilateration · LLS · RSSI · Bluetooth
Kalman estimator

1 Introduction

Positioning is an essential part navigation projects either for single vehicle or for a group of subjects. Despite of being a very well examined phenomenon [1,2], it is still useful to explore new possibilities of positioning, especially concerning modern electronic elements.

Generally speaking, localization means a process of setting the position of a given subject. Triangulation [3] belongs to the most-known methods, where having the knowledge of two solid points will define the position. The solution presented in this contribution uses trilateration, which includes three points and their locations. Furthermore, multilateration is a generalized term of trilateration, where four or more solid points are used and compared. The solid points are, in our case, iBeacons, i.e. transmitters based on the technology of Bluetooth Low Energy (BLE). When considering the measured power of a signal,

© Springer International Publishing AG, part of Springer Nature 2019
R. Silhavy (Ed.): CSOC 2018, AISC 763, pp. 99–108, 2019.
https://doi.org/10.1007/978-3-319-91186-1_12

called RSSI (Received Signal Strength Indicator) [4], a distance from the transmitter can be estimated. In addition, a more accurate value of distance can be determined using various filtration methods.

Although this approach has already been explored by several project teams, the literature sources have provided very heterogeneous results [5–9]. Thus, one of the aims of this contribution is to reproduce and either prove or disprove those findings.

2 iBeacon

The presented approach is based on iBeacons used as the main component. The iBeacon is a small and low cost device which uses a protocol developed by Apple [10]. The iBeacon periodically transmits its Universally Unique Identifier (UUID) through the interface of Bluetooth Low Energy. For the experiments presented at the end of the paper, a nRF51822 chip [11] and RPi (Raspberry Pi) version 3 were used. Both devices provided similar measurement results.

Apart from the fact that the iBeacon transmits its UUID regularly, a vehicle is also able to measure the iBeacon signal strength (referred to as Received Signal Strength Indicator (RSSI)). Using this information, the robotic vehicle should estimate the distance from the iBeacon.

3 2D Trilateration

As mentioned above, trilateration is examined in this contribution. Therefore, the group of "solid points" represented by iBeacons is distributed on the vertices and square edges. Although the issue of the exact position of solid points may seem attractive to solve, in fact, the different locations of individual iBeacons have very little effect on the accuracy of position determination, see experiment [12]. Then, the receiver inside the vehicle periodically scans the iBeacons around, and it also estimates the distances from each individual transmitter. The distance is estimated using the power of the signal [4]. The dependence of the signal power on the distance from the iBeacon is calculated by the following relation

$$RSSI = -20 log_{10} r + A, \tag{1}$$

where $RSSI$ means the received signal strength, r is the distance and A is the signal power at a distance of 1 m.

A set of circles is prepared from the gained data, as shown in Fig. 1. The localization of the vehicle - intersections - is determined from circle equation

$$(x - m)^2 + (y - n)^2 = r^2, \tag{2}$$

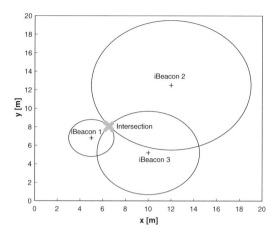

Fig. 1. Sample 2D trilateration.

having $S = [m, n]$ the circle center and r its radius. This equation can be transformed into a basic circle equation, as follows

$$x^2 - 2xm + y^2 - 2yn = r^2 - m^2 - n^2. \tag{3}$$

For i found iBeacons, a set of i equations is obtained. Under ideal conditions, this set of equations can be solved by a classic method for getting the circle intersection.

$$
\begin{aligned}
x^2 - 2xm_1 + y^2 - 2yn_1 &= r_1^2 - m_1^2 - n_1^2, \\
x^2 - 2xm_2 + y^2 - 2yn_2 &= r_2^2 - m_2^2 - n_2^2, \\
&\vdots \\
x^2 - 2xm_{i-1} + y^2 - 2yn_{i-1} &= r_{i-1}^2 - m_{i-1}^2 - n_{i-1}^2, \\
x^2 - 2xm_i + y^2 - 2yn_i &= r_i^2 - m_i^2 - n_i^2.
\end{aligned}
\tag{4}
$$

The equations are transformed into linear ones by subtraction of the second line from the first line in each corresponding pair. This operation provides a set of $i - 1$ equations

$$
\begin{aligned}
-2x(m_1 - m_2) - 2y(n_1 - n_2) &= r_1^2 - m_1^2 - n_1^2 - r_2^2 + m_2^2 + n_2^2, \\
-2x(m_2 - m_3) - 2y(n_2 - n_3) &= r_2^2 - m_2^2 - n_2^2 - r_3^2 + m_3^2 + n_3^2, \\
&\vdots \\
-2x(m_{i-1} - m_i) - 2y(n_{i-1} - n_i) &= r_{i-1}^2 - m_{i-1}^2 - n_{i-1}^2 - r_i^2 + m_i^2 + n_i^2.
\end{aligned}
\tag{5}
$$

The equations are then rewritten into a matrix form $\mathbf{Ax} = \mathbf{b}$ and solved by common operations of linear algebra, the matrix $\mathbf{x}$ indicates the coordinates of the intersection.

$$\mathbf{A} = -2 \begin{pmatrix} m_1 - m_2 & n_1 - n_2 \\ m_2 - m_3 & n_2 - n_3 \\ \vdots & \vdots \\ m_{i-1} + m_i & n_{i-1} - n_i \end{pmatrix},$$

$$\mathbf{x} = \begin{pmatrix} x \\ y \end{pmatrix},$$

$$\mathbf{b} = \begin{pmatrix} r_1^2 - m_1^2 - n_1^2 - r_2^2 + m_2^2 + n_2^2 \\ r_2^2 - m_2^2 - n_2^2 - r_3^2 + m_3^2 + n_3^2 \\ \vdots \\ r_{i-1}^2 - m_{i-1}^2 - n_{i-1}^2 - r_i^2 + m_i^2 + n_i^2 \end{pmatrix}.$$

Under real conditions, there is usually no exact intersection of circles due to imprecise $RSSI$ acquisition. Thus, the position of the robotic vehicle has to be calculated approximately using the least squares method.

4   Linear Least Squares Method

A linear least squares method (LLS) is used for estimation of the best-fitting position of the vehicle by minimizing the sum of the squares of the offsets ("the residuals") between the curves [13]. An overdetermined set of the equations $\mathbf{Ax} \approx \mathbf{b}$ is then solved. As one approach to solution, QR decomposition is recommended. Thus, the aim is to decompose the matrix $\mathbf{A}$ to the product of matrices $\mathbf{Q}$ and $\mathbf{R}$, where $\mathbf{Q}$ is the orthogonal and $\mathbf{R}$ is the upper triangle matrix.

$$\mathbf{A} = \mathbf{QR}, \tag{6}$$
$$\mathbf{Rx} = \mathbf{Q}^{\mathrm{T}}\mathbf{b}. \tag{7}$$

An estimated position of the vehicle is given by the solution. The precision of the provided solution should increase with the number of iBeacons.

$$\mathbf{x} = \mathbf{R}^{-1}\mathbf{Q}^{\mathrm{T}}\mathbf{b}. \tag{8}$$

The calculation of the position of the vehicle using LLS is illustrated in Fig. 2.

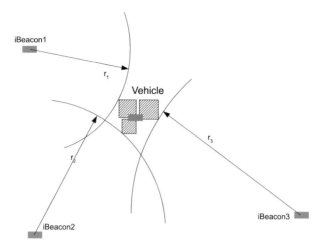

Fig. 2. Linear least squares method.

5   RSSI Values Filtration

Since the variance of the signal power received through time is significantly high, it is necessary to implement some type of filtering (refer to measured RSSI values at a distance of 1 m in Fig. 3). Three types of filtering are considered in this contribution: arithmetic mean [14], median [14], and Kalman estimator [15]. A comparison of experimental results provided by the selected filtering methods is presented in Fig. 4.

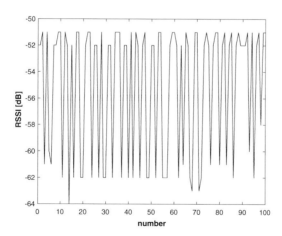

Fig. 3. Measured RSSI values at distance 1 m.

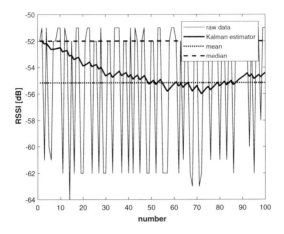

Fig. 4. Filtered measured of RSSI values at distance 1 m.

6 Experiment

A set of experiments is presented here to consider the possibility of RSSI measurement for position estimation. In the following paragraphs, the precision of the distance measurement related to the received RSSI is presented.

Data acquisition was performed within the range of distances 0.5 to 2 m with an increment of 15 cm. The experimental conditions are described in Table 1. The filtering methods described in Sect. 5 were applied to measured RSSI values, the results are shown in Table 2.

In addition, RSSI values received from iBeacons from different distances are shown in Fig. 5. These values should have a decreasing trend with an increasing distance from the iBeacon.

The aim of the experiment was to prove the position precision using BLE technology. Figure 6 captures these deviations within a given experiment range.

Table 1. Description of used equipment

	iBeacon	Measuring device
Device	Raspberry v3	Raspberry v3
OS	Raspbian Nov 2017	Raspbian Nov 2017
Kernel	4.9	4.9
Package	bluetooth, bluez	bluetooth, bluez
Using commands	hciconfig, hcitool	hciconfig, hcitool
Period transmit [ms]	100	-

Table 2. Comparison of experimental result for filters RSSI values

Real distance (m)	Type of filter	Error (m)		
		Min	Max	Average
0.50	Kalman estimator	0.00	0.22	0.07
	Mean	0.00	0.13	0.07
	Median	0.04	0.16	0.12
0.65	Kalman estimator	0.42	0.46	0.45
	Mean	0.43	0.46	0.44
	Median	0.40	0.47	0.44
0.80	Kalman estimator	0.01	0.58	0.53
	Mean	0.04	0.58	0.52
	Median	0.06	0.58	0.54
0.95	Kalman estimator	0.08	0.31	0.21
	Mean	0.09	0.27	0.19
	Median	0.05	0.21	0.15
1.10	Kalman estimator	0.02	0.46	0.24
	Mean	0.00	0.43	0.24
	Median	0.24	0.64	0.43
1.25	Kalman estimator	0.24	0.52	0.38
	Mean	0.31	0.53	0.41
	Median	0.39	0.62	0.51
1.40	Kalman estimator	0.01	1.05	0.26
	Mean	0.04	0.53	0.26
	Median	0.32	0.54	0.40
1.55	Kalman estimator	0.07	0.60	0.20
	Mean	0.00	0.41	0.15
	Median	0.19	0.55	0.47
1.70	Kalman estimator	0.48	0.93	0.81
	Mean	0.63	0.90	0.81
	Median	0.70	0.91	0.83
1.85	Kalman estimator	0.77	1.24	0.90
	Mean	0.73	1.39	0.93
	Median	0.81	1.66	1.20
2.00	Kalman estimator	0.15	1.69	0.84
	Mean	0.71	1.65	0.86
	Median	0.63	1.78	0.96

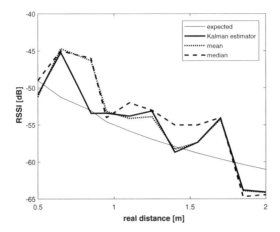

Fig. 5. Filtered measured of RSSI values dependence on the iBeacon distance.

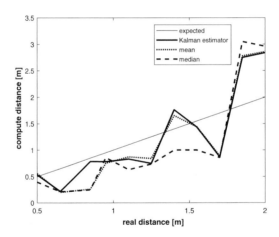

Fig. 6. Computed distance dependence on the iBeacon distance.

An example of trilateration and RSSI as tools for robotic vehicle positioning is shown in Fig. 7. Measured RSSI values for each iBeacon are represented as circles. The current position as well as the calculated position are shown as a ring and an asterisk, respectively. The position difference is indicated in the graph legend.

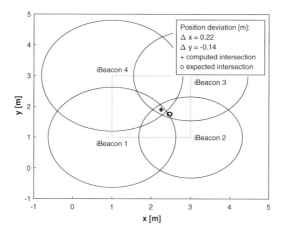

Fig. 7. Experiment with trilateration.

7   Conclusion

The possibility of robotic vehicle positioning using iBeacons is discussed in this contribution. The experiment deals with the determination of the precision of positioning. Selection of a suitable filter of measured RSSI values was also discussed. The experiments were performed in an outdoor environment and, as iBeacon device, RPI3 with 100 ms transmit period was used. The Kalman estimator and arithmetic mean appear to provide more appropriate filtering. Expectedly, the overall accuracy of the position of the robotic vehicle was approximately 0.2 m. Therefore, without proper improvement, this approach is not able to provide data for precise positioning. However, our future work within this project will consider these prelimiraly results and will include other possibilities of low energy positioning, where RSSI values will provide rather supplementary information.

Acknowledgment. The work has been supported by the Funds of University of Pardubice, Czech Republic. This support is very gratefully acknowledged.

References

1. Sharma, K.R., Honc, D., Dusek, F.: Sensor fusion for prediction of orientation and position from obstacle using multiple IR sensors an approach based on Kalman filter. In: 2014 International Conference on Applied Electronics, Pilsen, pp. 263–266 (2014). https://doi.org/10.1109/AE.2014.7011716
2. Sharma, K.R., Dusek, F., Honc, D.: Comparitive study of predictive controllers for trajectory tracking of non-holonomic mobile robot. In: 2017 21st International Conference on Process Control (PC), Strbske Pleso, pp. 197–203 (2017). https://doi.org/10.1109/PC.2017.7976213

3. Guan, Z., Zhang, B., Zhang, Y., Zhang, S., Wang, F.: Delaunay triangulation based localization scheme. In: 2017 29th Chinese Control And Decision Conference (CCDC), Chongqing, pp. 2627–2631 (2017). https://doi.org/10.1109/CCDC.2017.7978958

4. Mahapatra, R.K., Shet, N.S.V.: Experimental analysis of RSSI-based distance estimation for wireless sensor networks. In: 2016 IEEE Distributed Computing, VLSI, Electrical Circuits and Robotics (DISCOVER), Mangalore, pp. 211–215 (2016). https://doi.org/10.1109/DISCOVER.2016.7806221

5. Pradeep, B.V., Rahul, E.S., Bhavani, R.R.: Follow me robot using bluetooth-based position estimation. In: 2017 International Conference on Advances in Computing, Communications and Informatics (ICACCI), Udupi, pp. 584–589 (2017). https://doi.org/10.1109/ICACCI.2017.8125903

6. Schwiegelshohn, F., Wehner, P., Werner, F., Gohringer, D., Hubner, M.: Enabling indoor object localization through Bluetooth beacons on the RADIO robot platform. In: 2016 International Conference on Embedded Computer Systems: Architectures, Modeling and Simulation (SAMOS), Agios Konstantinos, pp. 328–333 (2016). https://doi.org/10.1109/SAMOS.2016.7818366

7. Zhou, M., Lin, J., Liang, S., Du, W., Cheng, L.: A UAV patrol system based on Bluetooth localization. In: 2017 2nd Asia-Pacific Conference on Intelligent Robot Systems (ACIRS), Wuhan, pp. 205–209 (2017). https://doi.org/10.1109/ACIRS.2017.7986094

8. Gorovyi, I., Roenko, A., Pitertsev, A., Chervonyak, I., Vovk, V.: Real-time system for indoor user localization and navigation using bluetooth beacons. In: 2017 IEEE First Ukraine Conference on Electrical and Computer Engineering (UKRCON), Kiev, pp. 1025–1030 (2017). https://doi.org/10.1109/UKRCON.2017.8100406

9. Gu, Y., Ren, F.: Energy-efficient indoor localization of smart hand-held devices using bluetooth. IEEE Access **3**, 1450–1461 (2015). https://doi.org/10.1109/ACCESS.2015.2441694

10. Apple Developer, "iBeacon". https://developer.apple.com/ibeacon/. Accessed 03 Jan 2018

11. NORDIC SEMICONDUCTOR, "NRF51822 Bluetooth Smart Beacon Kit". https://www.nordicsemi.com/eng/Products/Bluetooth-low-energy/nRF51822-Bluetooth-Smart-Beacon-Kit. Accessed 03 Jan 2018

12. Aman, M.S., Jiang, H., Quint, C., Yelamarthi, K., Abdelgawad, A.: Reliability evaluation of iBeacon for micro -localization. In: 2016 IEEE 7th Annual Ubiquitous Computing, Electronics & Mobile Communication Conference (UEMCON), New York, NY, pp. 1–5 (2016). https://doi.org/10.1109/UEMCON.2016.7777904

13. Hereman, W., Murphy, W.S.: Determination of a Position in Three Dimensions Using Trilateration and Approximate Distances. https://inside.mines.edu/~whereman/papers/Murphy-Hereman-Trilateration-MCS-07-1995.pdf. Accessed 03 Jan 2018

14. Miao, Z., Jiang, X.: Further properties and a fast realization of the iterative truncated arithmetic mean filter. IEEE Trans. Circuits Syst. II Express Briefs **59**(11), 810–814 (2012). https://doi.org/10.1109/TC-SII.2012.2218473

15. Qi, W., Cong, S., Zhang, P.: Robust steady-state Kalman estimators for discrete-time system with uncertain noise variances. In: 2016 35th Chinese Control Conference (CCC), Chengdu, pp. 2961–2966 (2016). https://doi.org/10.1109/ChiCC.2016.7553814

Preparing Influence Analysis of Meteoparameters on Production Process

Jela Abasova[✉] and Pavol Tanuska

Faculty of Materials Science and Technology in Trnava,
Slovak University of Technology in Bratislava, Bratislava, Slovakia
{jela.abasova,pavol.tanuska}@stuba.sk

Abstract. The aim of the paper was to describe analyzing influence of meteoparameters on production process. We compiled proposal for collecting data and recommendations for researchers solving similar problems in the future. The first part defines terms related to the topic (Big Data, Knowledge Discovery in Databases, and Data Mining) and analyze the chosen topic via opinions of experts. Weather influences production process mostly via human factor, so this part describes influence that meteoparameters have on human health, behavior and job performance. The second part deals with two types of input data – meteoparameters (parameters of weather) and data from production process. It describes the data, their cleaning, integration and selection, and generating of an additional dataset. The third part focuses on acquiring knowledge from the data via several data mining methods. It describes statistical analysis and consequent corrections of the data, building of data mining model, and compares individual methods. That results in proposal for collecting data and recommendations, both of which are based on problems that had arisen in the process of the analysis. The fourth, final part concludes with summarizing sequence of steps of the process.

Keywords: Big data · Knowledge discovery in databases · Data mining
Weather influence · Production data

1 Introduction

Today, companies collect and store big amounts of data, what costs them a lot of time and money. But they are often not able to use full potential of data. Data are stored for months without any use, then removed to clear space for another data. Companies are not aware of various data analysis options, available nowadays. Aim of the paper is to show, how useful can be information extracted from data, on a model problem.

Knowledge discovery in databases (KDD) is a process of discovering potentially useful information from big data sets, finding patterns and concepts that are not noticeable at first glance. The term is almost interchangeable with term Big Data. Big Data is a term used for large datasets and also for working with these datasets [1].

The process of KDD/Big Data cannot be fully automatized. Human inclusion is required, as only human can select appropriate data, choose suitable method(s) of analysis and interpret the results [2].

© Springer International Publishing AG, part of Springer Nature 2019
R. Silhavy (Ed.): CSOC 2018, AISC 763, pp. 109–120, 2019.
https://doi.org/10.1007/978-3-319-91186-1_13

Process of KDD includes, according to the book Data Mining: Concepts and Techniques, seven following steps: cleaning data, integrating data, selecting data, transformating data, data mining (DM), evaluating patterns, presenting discovered knowledge [3].

KDD deals with two basic types of problems: description and prediction. There is a wide range of topics solvable via methods of KDD. It is usable in many fields, as biology, biomedicine, medicine, social sciences, economics, trade etc. KDD is used in industry to the large extent, mostly for optimization of production, for making production process more effective and for security and safety reasons [4].

The paper deals with analyzing influence of meteoparameters on production process. Meteoparameters (attributes of weather) have confessed influence on health and life of living organisms, including human.

According to studies by Kalkstein and Valimont, the most destructive impact on human health have following meteoparameters: temperature extremes, low humidity (especially in winter months), snowing, rapid weather changes. Impact depends on sex, age and race, but the opinions of experts differ in deciding which group is the most endangered one [5].

Other researchers describe influence of weather on job performance. Sarah Parfitt summarize weather impacts on job performance: sunlight affects job performance positively; cloudiness lowers motivation; cold weather increases productivity; rain can lower stress of people working inside, but can also cause sleepiness; high temperature lowers speed of performance; lack of natural lighting (and artificial lighting as well) has negative impact on performance [6].

According to study by United States Environmental Protection Agency (UEPA), level of stress can be increased even by reading forecasts. Weather extremes affects not only health of individual persons, but also health of society [7].

Research from Tokyo investigated influence of three meteoparameters (temperature, amount of rainfall and speed of wind) disclosed, that weather influences human health and activities differently in different parts of a day [8].

Other studies introduce slightly different results, but they all are united in two points: weather definitely has impact on job performance; and interior workers are less affected than exterior ones.

Majority of researches identifies temperature, humidity, rainfall, cloudiness (and connected lack of light) as the most critical parameters. Because of this we have to analyze every case individually, considering local differences.

2   Methods

Getting the right data is a prerequisite for correct performance of KDD. Analysis is only as good as data it is based on. Therefore we paid enough attention to the phases of obtaining and preparing data.

We used two types of data – meteoparameters and production process data. Both types include multiple datasets.

2.1 Meteodata

Meteodata include the second half of one calendar year, 2016. They consist of two datasets, the first one with hourly measurements, the other one with daily measurements.

The hourly dataset consists of eleven parameters: ID and name of measuring station, date + time ([dd:MM:yyyy h]), average, minimum and maximum of temperature ([°C]), relative air humidity ([%]), average air pressure ([hPa]), duration of sunlight ([min]), amount of rainfall ([mm]), average wind speed ([m/s]) (Table 1).

Table 1. Hourly meteodata

Date [dd.mmm yyyy h]	Temperature (avg) [°C]	Temperature (max) [°C]	Temperature (min) [°C]	Humidity (avg) [%]	Air pressure (avg) [hPa]	Sunlight [min]	Rainfall [mm]	Wind speed [m/s]
17.5.2016 7	14,2	14,6	14,1	60	988,8	0,0	0	3,5
17.5.2016 8	14,3	14,7	14,0	62	989,0	0,0	0	4,1
17.5.2016 9	15,2	17,2	14,5	61	989,0	10,7	0	2,1
17.5.2016 10	17,5	18,7	16,4	55	989,0	50,0	0	3,1
17.5.2016 11	18,3	19,3	17,4	53	988,8	37,8	0	2,8
17.5.2016 12	17,5	18,1	17,2	55	988,5	13,0	0	3,5
17.5.2016 13	18,3	19,4	17,6	52	988,0	46,2	0	3,7
17.5.2016 14	19,4	20,2	18,7	49	987,4	55,5	0	4,3
17.5.2016 15	19,4	20,0	18,9	49	987,0	55,3	0	4,2
17.5.2016 16	18,8	19,5	18,2	51	986,7	28,3	0	3,4

The daily dataset contains the same attributes as the previous one (with date format dd:MM:yyyy) plus cloudiness (scale 1–10) plus amount of snowfall ([cm]); but included data are less detailed. They are measured only three times a day – at 7 a.m., at 14 p.m., and at 21 p.m.

These data contain all six attributes that are, according to Parfitt, most critical for human health and job performance: sunlight, cloudiness and lack of (natural) lighting in attributes "duration of sunlight" and "cloudiness"; rain (effects of which the expert introduces as ambivalent) in attribute "amount of rainfall", cold weather and high temperature in attributes of min, max and average temperature. Datasets also contain information about relative (average) air humidity and rapid weather changes (which can be discovered via several parameters, e.g. difference between "max" and "min" temperature during one hour/day), which are parameters marked as critical by Kalkstein and Valimont.

Adverse extremes of each parameter and its occurrence in both datasets are:

Temperature

- optimal value: −10 °C to 30 °C
- measured values: −10.2 °C ("arctic day") to 39.9 °C ("tropical day") – both extremes potentially dangerous to human

Air Humidity

- optimal value: 40–60% [10]

- measured values: 14–100% (both dangerous extremes), average 68% (value slightly higher than recommended)

Air Pressure

- optimal value: 1013.25 hPa (normal barometric pressure/standard atmosphere) [11]
- extremes: depression (less than 1005 hPa) – can affect human health and performance; anticyclone (more than 1010 hPa) – has no negative effect on human
- measured values: 956.6 hPa (depression) to 1025.5 hPa (anticyclone), average 996.4 hPa, which means depression prevails

Duration of Sunlight

- measured values: 0–60 min, average 18 min, which may look like lack of sunlight, but night shift is included as well, what should be taken into account

Amount of Rainfall

- measured values: 0–17.4 mm, annual rainfall total 876.1 mm, average amount of rainfall 0.066, which means the majority of days were not rainy

Average Wind Speed

- Beaufort scale: numbers 0 to 12
- optimal value: numbers 0 (Calm; less than 0.3 m/s) to 5 (Fresh breeze; 8–10.7 m/s)
- extremes: numbers 6 (Strong breeze; 10.8–13.8 m/s) to 12 (Hurricane; 32.7 m/s or above); "wind chill factor" occurs – increase of heat loss under influence of wind [8]
- measured values: 0 m/s (Calm; number 0) to 16.6 m/s (High wind; number 7), average 3 m/s (Light breeze; number 2)

These data are complex and to large extent self-explanatory. They require only minor modifications. It is appropriate to remove attributes with no use (for us that means excluding ID and name of measuring station, as values of these parameters always remain the same).

2.2   Production Data

We had one set of production data, which contains 65,516 records of error states. Values were recorded for the second half of one calendar year, 2016.

There are eight parameters in the dataset: Localization, Duration, Start, End, Tag, Name, Text, and Compass.

We can divide these into two groups: description attributes and time attributes.

Description attributes, as the name suggests, describe the problem. "Name" goes for standardized title of an error; "Localization" describes where an error occurred; "Tag" provides further information about localization. "Description" is optional (and due to that not very useful) attribute with additional info about error; Compass is one-value attribute (and therefore not very useful as well).

Time attributes contain "Start", "End" and "Duration". We can reduce the three of them by removing "End", which is redundant in context of knowing the duration.

We did not get any documentation, plans or explanatory notes to datasets, making working with them more difficult; and, which is even more problematic, there are no information about production process itself – about non-error states. That means that results we would get from analysis could be misleading (as we are not able to tell ratio of error and non-error states etc.).

As we had no records of non-error data, we decided to design a simulation model of the real process, to get data we need (Fig. 1).

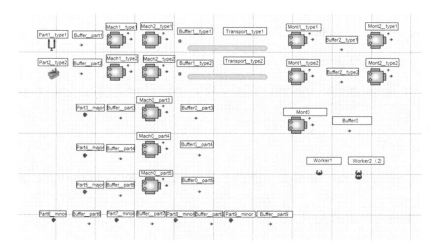

Fig. 1. Simulation model

We used Witness, simulation tool developed by British company Lanner. It is flexible technology capable of developing complex models and testing scenarios for different system settings without the risk associated with real-world testing [12].

Simulation takes into account working time divided into three shifts – morning, afternoon and night ones. Output are four attributes – "number of produced components", "WIP" (number of components that are actually in progress of being produced), "average time" of producing one component and "costs".

Our Witness model simulates real system, but it is still simulation; it would not consider weather changes unless it is ordered to. Our aim is to describe sequence of steps for process of analyzing influence meteoparameters have on production process. Therefore we need to simulate that influence so we can evaluate our process later.

While simulating influence, we used information gained from experts which we described in previous sections.

Our intervention looks as following:
High temperature (1):

$$high\_temp = IF(temperature > 30; (temperature - 30)/2; 0) \tag{1}$$

Low temperature (2):

$$low\_temp = IF(temperature < 10; (10 - temperature)/4; 0) \tag{2}$$

Depression (3):

$$depress = IF(pressure < 990; (1000 - pressure)/50; 0) \tag{3}$$

Big amount of rainfall (4):

$$rain = IF(rainfall > 1; (rainfall - 1); 0) \tag{4}$$

So, for example, time added to "average time" is (5):

$$added\_time = high\_temp + low\_temp + depress + rain \tag{5}$$

We did not consider influence of "relative humidity" and "average speed of wind" so we can evaluate process of our simulation later.

3   Results

After preparing data follows phase of DM, which is a key phase of KDD process.

3.1   Tools and Methods

For connecting and pre-processing data and for DM analysis we used software tool STATISTICA Data Miner. It is one of business systems of system STATISTICA. All of STATISTICA products include optimized management of data that enables processing big datasets and supplies multitasking [10].

STATISTICA Data Miner is a package of tools dedicated to DM. It offers a broad scale of integrated, easy modifiable DM methods, OLAP analysis, tools for cleaning and filtering data, general classifier, general predictor etc. [10].

3.2   Selection and Connection of Data

Selection of data for DM analyzes is an important step in KDD process, because using adverse data or data that analyst did not understand correctly may cause distortion of results.

We have four sets of data: hourly meteodata, daily meteodata, error messages from production process and production process data got via simulation. From two meteosets we decided to use the hourly one, for it is more detailed; the other set have more parameters (cloudiness and amount of snowfall in addition) but these parameters cannot be used as they are measured daily and adding them to the hourly set would be quite inaccurate. As the second dataset we chose data we gained from simulation of a real system.

In the previous section we cleaned datasets of inappropriate attributes and cases. We merged datasets in STATISTICA Data Miner by primary key date + time. Unmatched cases (for both files) were excluded rather than filled by median.

Result file consists of primary key (date + time), five meteoparameters (average temperature, average humidity, average air pressure, amount of rainfall, average speed of wind) and four columns of production data (number of produced components, WIP, average time, costs) (Table 2).

Table 2. Results of data simulation

Date [dd.mmm yyyy h]	Produced components	Components in progress	Average time [s]	Costs
1.5.2016 0	0	4,3	0,0	15,0
1.5.2016 1	3	5,3	28,8	141,8
1.5.2016 2	8	5,6	35,3	175,1
1.5.2016 3	9	5,8	39,0	160,0
1.5.2016 4	9	5,9	41,2	175,0
1.5.2016 5	8	6,0	41,4	156,0
1.5.2016 6	9	6,0	42,1	174,3
1.5.2016 7	8	6,0	42,0	6,2

At first glance we noticed rows with 0 produced components and 6 units of costs. These are data collected during the third shift (night shift), when no one is working. The amount of costs is due to using warehouses, which are active at night.

3.3 Descriptive Statistics

Before DM analysis itself we made descriptive statistics out of joined datasets (Table 3). Our aim was to evaluate whether an error occurred in data preparation process.

Table 3. Descriptive statistics

Variable	Descriptive statistics (Merge variables)				
	Valid N	Mean	Min	Max	Std. Dev.
Temperature (avg) [°C]	4416	17,98	0,40	37,30	6,98
Humidity (avg) [%]	4416	64,64	14,00	96,00	18,96
Air pressure (avg) [hPa]	4332	983,06	956,60	1004,50	5,48
Rainfall [mm]	4416	0,08	0,00	17,40	0,50
Wind speed [m/s]	4416	2,23	0,00	8,50	1,36
Produced components	4416	5,26	0,00	12,00	3,99
Components in progress	4416	6,15	0,00	6,95	0,36
Average time [s]	4416	71,72	0,00	137,26	8,05
Costs	4416	105,58	0,00	177,14	74,90

Number of cases in attribute "average air pressure" is lesser than required. Some cases can be absent due to short-term failure of measuring device or wrong transformation in pre-processing phase. As the amount of absent cases is relatively small (84, what is less than 2%), we can expect it not to affect results of analysis. We did not have to exclude the attribute, rather we replaced missing values by median.

Histograms of meteoparameters show nearly Gaussian distribution, what is consistent with data collected over a half year.

The greatest difference between minimum and maximum value has the attribute "relative humidity", followed by "air pressure" and "temperature". We can expect these meteoparameters to be leading in influence at production process, unlike "amount of rainfall" and "average speed of wind". (Normalization of meteoparameters should be considered.)

Out of production process attributes, costs have the biggest difference in minimum and maximum. As said before, that is due to the night shift, when no components are produces.

After minor adjustments based on descriptive statistics, we continued to DM analysis itself.

3.4   DM Model

We designed model in STATISTICA Data Miner that includes loading and joining two datasets, descriptive statistics, and analysis with several DM methods (Fig. 2).

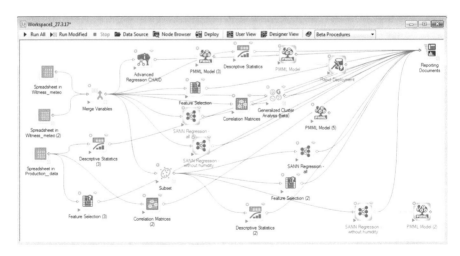

Fig. 2. DM model

Feature Selection. In feature selection method we chose all five meteoparameters as independent and four attributes of production process as dependent (Fig. 3). That means we included even meteoparameters, influence of which we did not simulate ("relative humidity" and "average speed of wind").

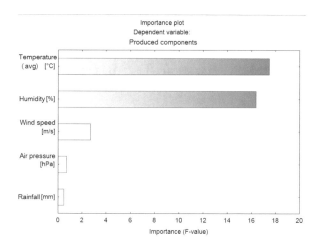

Fig. 3. Results of feature selection

For "Produced components" was "temperature" the most important factor; for "Components in progress" it was "air pressure". Attribute "average speed of wind" confirmed our assumptions, as it was not involved as important factor in any of the results.

Problems arose with input parameter "relative humidity", as it was the third most important factor for "Components in progress" and the second important factor for "Produced components".

We did not simulate influence of "relative humidity" on production process in previous phase. But it was possible for meteoparameters to correlate, and our next step was to verify the hypothesis.

Correlation Matrices. We use correlation matrices to reveal a possible correlation between meteoparameters (Table 4). Analysis revealed strong indirect correlation between "temperature" and "relative humidity". That is in line with researches we mentioned in one of the previous sections. Excluding one of the correlating parameters should be considered.

Table 4. Correlation of input parameters

Variable	Descriptive statistics (Merge variables)				
	Valid N	Mean	Min	Max	Std.Dev.
Temperature (avg) [°C]	4416	17,98	0,40	37,30	6,98
Humidity (avg) [%]	4416	64,64	14,00	96,00	18,96
Air pressure (avg) [hPa]	4332	983,06	956,60	1004,50	5,48
Rainfall [mm]	4416	0,08	0,00	17,40	0,50
Wind speed [m/s]	4416	2,23	0,00	8,50	1,36
Produced components	4416	5,26	0,00	12,00	3,99
Components in progress	4416	6,15	0,00	6,95	0,36
Average time [s]	4416	71,72	0,00	137,26	8,05
Costs	4416	105,58	0,00	177,14	74,90

Neural Nets – Regression. We used method of neural net regression (SANN regression) to search for influence of meteoparameters. We determined meteoparameters as continuous inputs and Number of produced components as continuous target. As strategy for creating predictive process we chose Automated network search (ANS). Results showed only a weak link between inputs and outputs – neural net was not able to predict correctly.

That is most likely due to existence of big amount of input and output combinations. In the process of learning and counting synaptic weights, only very weak connections are constructed. Neural nets are not very usable in process of solving our problem.

Use of Subset. We mentioned problem with third, night shift, when nothing is produced and which can therefore misrepresent results. We decided to design subset, where this third shift no longer figures. For this subset we repeated some of analysis and compared results.

Neural nets did not show better results, what confirms this method is not suitable for solving our problem. On the other hand, feature selection method showed slightly change results, when "Components in progress" is used as output. In case "Produced components" is an output, results stay the same as when we did not use subset. That means that using subset can (but not necessarily will) affect results and therefore should be considered.

3.5   Results

After going through all the phases of KDD process we have assembled proposal for collecting data used for similar problems:

- Collected data should be as detailed as possible.
- Data that have no clear reporting value have to be supplemented by plans and explanatory notes.
- While marking alarms and reports, severity level should be always mentioned.
- Parameters with only one value are not useful; discrete parameters with numerous values are not useful as well, but can get to use if they are split into additionally manufactured categories (e.g. via parsing or other methods).
- Error log should be always complemented by production process data, because using only error data can misrepresent results.

While working on the paper we have encountered some problems and dead ends. Accordingly, we have formulated following recommendations for future investigators of similar issues:

- It is appropriate to do research in the aim area before starting analysis, familiarize oneself with expert opinions, typical problems and possible solutions.
- Every case needs to be assessed individually, without overusing generalization, with taking individualities and specifics into account.
- It is appropriate to choose the most detailed data with plenty of attributes.
- It is necessary to understand every attribute used (and considered), so potentially useful attribute cannot be excluded.

- Joining data sets should be done in analytical tool, not before that, in case one (or more) data set(s) needs to be diversified.
- It is appropriate to remove potentially misleading cases (measured outside a workday, during the rise time etc.).
- Statistical analysis should be done before using DM tools, as basic statistics can early detect problems imperceptible at first sight, e.g. missing data.
- Strong correlation between input data should influence further analysis. Excluding one of the correlating parameters should be considered.
- Results interpretation needs to be done continuously, not just at the end of the KDD process.

These results and recommendations can be used while working on similar topic and to further investigation in the area.

4   Discussion

Aim of the paper was to describe sequence of steps for analyzing influence of meteo-parameters on production process. To conclude, we got through following steps in the process:

We identified the problem. Companies collect and store big amounts of data, which they do not use properly. We defined aim of the paper. We decided to design sequence of steps for analyzing data. As model topic we chose influence of meteoparameters on a production process.

Basic terms we dealt with were Big Data, DM and KDD. We introduced these methods areas where they are used. We prepared an overview of expert work, focusing on influence of weather on human health and job performance. We described software tools we used – STATISTICA Data Miner and Witness Horizon.

We described data chosen for the project: meteodata and data of production process. As production dataset was quite incomplete, we simulated real process via Witness Horizon. We described process of simulating influence of meteoparameters, necessary for validating results.

In the phase of pre-processing data we excluded useless attributes and modified format of some columns. We joined two datasets in STATISTICA Data Miner by date + time as primary key. We did basic statistics to better understanding of data. According to these partial results we adjusted the datasets.

We analyzed data, using multiple methods of DM. Feature selection very usable in solving our problem. In accordance with the results, we repeated tests for subset without useless cases, and then we compared results. We did correlation matrices for input data, what exposed strong indirect correlation between two meteoparameters, in line with expert findings.

On the basis of partial findings, we interpreted the results and used them for further DM analyzes, as described above. We came to a conclusion that this type of problem can really be solved via KDD, in case data are detailed and annotated, what help analyst in understanding process.

Finally, we compiled sequence of steps, proposal for collecting data and recommendation for future solvers of similar problems.

In the future we plan to use acquired knowledge in solving other Big Data problems. The analysis described in the paper allowed us to compile sequence of steps usable in wide range of topics. As we focus on production process data, we can also use recommendations we produced for improving quality of collecting and pre-processing data. We plan to extend our analysis via using other data mining tools and methods with usage of these results.

Acknowledgement. This publication is the result of implementation of the project: "Increase of Power Safety of the Slovak Republic" (ITMS: 26220220077) supported by the Research & Development Operational Programme funded by the ERDF and project VEGA 1/0272/18: "Holistic approach of knowledge discovery from production data in compliance with Industry 4.0 concept" supported by the VEGA.

References

1. Techopedia Inc Homepage. https://www.techopedia.com/definition/25827/knowledge-discovery-in-databases-kdd. Accessed 28 Apr 2017
2. TUKE Homepage. http://neuron.tuke.sk/zvada/statnice/II/08/index.html. Accessed 28 Apr 2017
3. Han, J., Kamber, M., Pei, J.: Data Mining: Concepts and Techniques, 3rd edn. Morgan Kaufmann, [s. l.] (2011)
4. Fayyad, et al.: The Primary Tasks of Data Mining. [s. n.], [s. l.] (1996)
5. Kalkstein, L., Valimont, K.: Climate effects on human health. In: Potential Effects of Future Climate Changes on Forests and Vegetation, Agriculture, Water Resources, and Human Health. EPA Science and Advisory Committee Monograph no. 25389, 122-52. U.S. Environmental Protection Agency, Washington, D.C. (1987)
6. Parfitt, S.: From Sunshine to Rain: How the Weather Affects Your Productivity (2017). https://www.business.com/articles/how-the-weather-affects-your-productivity/. Accessed 28 Apr 2017
7. United States Environmental Protection Agency. https://www.epa.gov/climate-impacts/climate-impacts-human-health. Accessed 23 Jan 2018
8. Horanont, T., et al.: Weather Effects on the Patterns of People's Everyday Activities: A Study Using GPS Traces of Mobile Phone Users. https://www.ncbi.nlm.nih.gov/pmc/articles/PMC3867318/. Accessed 2012
9. Klanova, K.: http://www.stop-vlhkosti.cz/vlhkost-v-bytech/. Accessed 23 Jan 2018
10. KSTST Homepage. http://www.kstst.sk/pages/vht/meteo/tlak.htm. Accessed 28 Apr 2017
11. Lanner Homepage. https://www.lanner.com/technology/witness-simulation-software.html. Accessed 28 Apr 2017
12. STATSOFT CR s. r. o.: Ovládání a základy statistiky v softwaru STATISTICA., p. 59. StatSoft CR s. r. o., Praha ([s. y.])

Data Model Design in Automatic Transit System (PRT) Simulation Software

Jakub Lorenc, Wiktor B. Daszczuk$^{(\boxtimes)}$ (ID), and Waldemar Grabski (ID)

Institute of Computer Science, Warsaw University of Technology,
Nowowiejska Street 15/19, 00-665 Warsaw, Poland
`jlorenc@mion.elka.pw.edu.pl`, `{wbd,wgr}@ii.pw.edu.pl`

Abstract. Simulation has become a very important factor in the field of Automated Transit Network – Personal Rapid Transit (ATN-PRT) design. Multiple traffic conditions, as well as model structure and movement parameters lead to increase in the number of simulation experiments which must be performed to evaluate ATN control algorithms. This article aims to show some guidelines for design of such simulation systems, with particular emphasis on data model design in object oriented programming (OOP) for massive simulations. These guidelines are presented in the context of Feniks Personal Rapid Transit (PRT) simulator development, but are also valid for other graph-based simulation software.

Keywords: Automated Transit Network · Personal Rapid Transit
ATN simulation · OOP · Parallel programming data structures · Software design

1 Introduction

Transport systems are subject for precise planning from traffic, social, economic and other perspectives. Simplest cases can be described analytically [1], but a system having complicated structure, many traffic and demand conditions and several parameters of fleet management algorithms require simulation for proper resolution.

In Warsaw University of Technology, the Feniks ATN-PRT (Automated Transport Network – Personal Rapid Transit) simulator [2] was elaborated under Eco-Mobility project [3]. First issues were built in typical way, having a network structure and other parameters as an input file, or supplied on-line using a graphical user interface. The results of simulations are displayed graphically or can be saved to output files. Several ATN-PRT simulators are constructed such way, for example Hermes [4], Beamways [5] or Netsimmod [6]. The purpose of those simulators is an analysis of an ATN system of a given structure, in given conditions. After preparing of a network structure and other conditions, a simulation is performed (typically allowing its observation in animation mode) and then output results are given to a designer. A simulation for changed structure and/or conditions can be executed using the same procedure from the start. Alternatively, a consecutive simulation for the same network structure and some changed parameters, can be performed without leaving the simulator, using the same topology.

© Springer International Publishing AG, part of Springer Nature 2019
R. Silhavy (Ed.): CSOC 2018, AISC 763, pp. 121–131, 2019.
https://doi.org/10.1007/978-3-319-91186-1_14

The contribution of this paper is organization of simulation data model to facilitate massive parallel simulations of ATN systems. The paper is structured as follows: Sect. 2 presents an ATN transport system, existing Feniks simulator and data structures which describe an ATN model. Data structures needed for performing multiple parallel simulations are discussed in Sects. 4 and 5, for static and dynamic data, respectively. Section 6 covers experimental evaluation of described solutions. Section 7 contains concluding remarks.

2   ATN System and Feniks Simulator

Personal Rapid Transit [7] is an individual (1–6 passengers in a group) mean of transport that uses autonomous vehicles traveling on separated (usually elevated) unidirectional track. The structure of a network is defined by a directed graph, illustrated in Fig. 1. The nodes are divided into stations, capacitors and intersections. The intersections are either of "fork" or "join" type and there are no track crossings. The edges are track segments connecting the nodes.

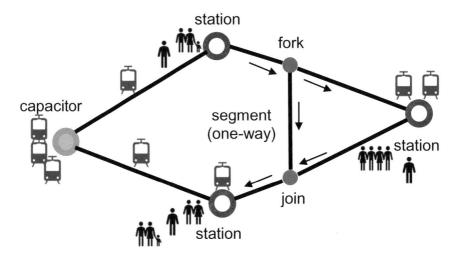

Fig. 1. The structure of ATN system

There are two levels of control in ATN systems: vehicle behavior on the track (keeping up, behavior on intersections, joining the traffic) and management algorithms.

The operation of ATN systems is investigated under Feniks simulation environment (Fig. 2). It is an event-driven simulator with network design facility.

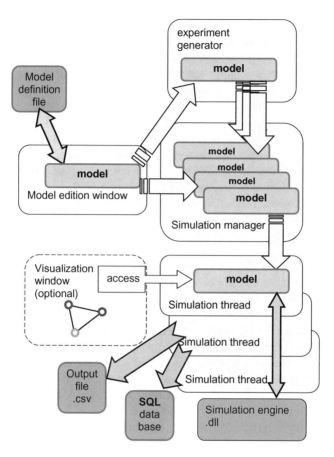

Fig. 2. The structure of Feniks simulator

A model simulated in Feniks consists of several data sets:

- *Network structure* defining location of nodes, length of edges and their parameters (number of berths in a station, maximum velocity of a segment, etc.).
- *Operation parameters*: number of vehicles, maximum velocity of individual vehicles, acceleration and deceleration, boarding and alighting time distribution, empty vehicles management factors and thresholds [8].
- *Demand parameters*: input distribution on individual stations, origin-destination matrix.
- *Dynamic data*: current positions and velocities of individual vehicles and passenger queues lengths.

The former three data sets are static for a given investigation, but they differ in variability from one simulation to another. A common example is a situation where network structure is static in a given research, operation parameters generate a large set of experiments while input mean value is different for every simulation.

If a series of simulations should be performed, the only way of simulation automation was initially running a simulator in a batch using its command-line version (available for example in PRTsim [9]). Tuning of network or algorithm parameters, which requires many simulations to run, was hardly possible in such simulation environment. Therefore, a new version of Feniks has been prepared [2], in which series of experiments can be planned and executed using simulation generator. Selected parameters can be assigned ranges of values with subrange steps, which form a Cartesian product of value vectors. Every experiment receives a set of common elements (for all experiments) of network structure, movement parameters, demand and other variables, and a set of specific parameter values. This principle allows for performing sets of several dozens of simulations to analyze a network performance in various conditions. The experiments are run in parallel, without animation, or with animation of selected experiments, achieving maximum simulation speed. Proper organization of simulator data structures allows the simulator to achieve maximum performance and facilitates parallel execution of multiple experiments.

Every simulation gives a set of output parameter values such as average waiting time, average trip distance, average trip delay, vehicle daily mileage etc. Retrieving the results from dozens of output files is highly error-prone if done manually (and it is awfully boring for a researcher). Therefore, in Feniks the results are stored in SQL database. Finding a dependence between input and output parameters is extremely easy because it may be achieved using a single SQL query.

Because the number of experiments to perform is so high there is no need for parallelization of a single simulation, although such solutions also exist and are being actively developed [10]. Instead, the goal is to minimize time needed to conduct all of the experiments using limited amount of hardware. This can be achieved by running multiple single thread simulations at once, avoiding time consuming network communication or synchronization as well as additional memory usage that are typical to solutions that parallelize single simulation experiments [10, 11]. However, these memory gains can only be apparent if concurrent experiments share large portions of their data, as opposed to working in completely independent environments.

3   Data Model

Data model creates a wireframe upon which application functionality is being build. It is often designed to mirror relations between objects in real world. In many cases it can also be changed according to technical needs to improve application efficiency and decrease memory usage, but if these changes have no support in the domain itself, they can often lead to more harm than good. In the case of ATN simulation software boundaries between real and technical worlds fade. On one level of description the model can use terminology such as stations, tracks and vehicles, while on the other simulation models, experiments, results and their aggregations. The second level belongs to both domains of simulation and implementation. Incorporating it into the data model can go a long way in improving simulation efficiency.

A typical simulation software is built upon how experiments are being conducted. Typically, there is a need for a basic simulation model. Such a model aims to describe simulated world at the beginning of simulation and can be used to conduct numerous experiments. These experiments aim to provide two important features. Firstly, they allow the simulated world to change slightly, giving researchers ability to conclude about how some parameter or graph change modifies the output parameter values. Secondly, they give the ability to conduct experiment on exactly the same world, but with a different random number generator seed, which provides aggregated results with statistical significance. The most straightforward way to think about such model is to build a data model based on each experiment. This way it is easy to achieve a situation where each experiment can be conducted without interference with others and with no dependence on architecture used. No matter whether the simulation software is based on a monolith or multiple runner services, this design gives certainty that one simulation will not block the other and that we can even change the architecture used during software evolution.

There are some questions that arise in connection with this design model. Do we really need to load the same world model before each execution of experiment on a particular runner instance? If the instance could run multiple experiments at once, would it improve the performance and reduce memory usage? If it would, then by how much? Is it worth leaving that simple model behind? We try to answer these questions below.

During software creation process it is often needed to choose between many advantageous traits of resulting program. To improve the basic data model, we assume the following priorities:

1. *Data access time minimization* - short data access time allows for increasing the pace of simulation, thus decreasing a time cost of conducting experiments. This is one of the most important features that distinguish good simulation software. It is worth noting that data model should account for the fact that some data may need a decrease in read time while the other in write time.
2. *Concurrency* - shared data will be accessed by multiple simulations and in case of visualization also graphical threads. Resulting data model should allow for quick way of accessing these data that does not lead to locking.
3. *Memory usage minimization* - less memory usage lets modern central processing units (CPU) perform advanced optimization of access time, for example using mechanisms such as CPU caches. These optimizations can lead to noticeable improvements in simulation time but their effect largely depends on the size and type of used cache [12] as well as the size of data model. That is why they have been put on the last position.

In the case of ATN simulation software, it is possible to describe some basic concepts that underline every program belonging to that group. First of all, the data model of each simulation will usually have a structure of directed cyclic graph. Furthermore, it can also be assumed that during the simulation the size of that graph is not going to change. Lastly, data stored within simulator instance can be divided into three groups:

- *frequently changing* - simulation state variables that potentially change in every simulation step and have different values in each concurrent experiment,

- *static model* - set only once when the model loads, it usually consist of graph structure and additional data used only for presentation purposes such as node names.
- *infrequently changing* - parameters that change only on the beginning of simulation, or at a specific, known moment in simulation time. They are meant to be used in order to check how same network operates in different conditions. In some simulation programs they can also be changed via lookup interface when the simulation is put on hold.

It is important to note that in case of static and infrequently changing data there is additional line of division between features that are object dependent such as node position and features that are shared between multiple objects forming a kind of group, with identical features, for example number of berths in a station or maximum speed and acceleration of vehicles.

4   Unification of Static Data

According to previous classification, each simulation has to use some data that is completely static during simulation. It is also very common to run multiple experiments using exactly the same initial model. This means it may be possible to keep the static, initial data model of simulation in memory only once per simulator instance.

4.1   Graph Structure

Graph structure is a part of static model that describes connections between nodes in simulated network, which means that data stored in that model is being used by simulation algorithm to access neighbors of current node in the graph. For the model to be useful for simulation purposes, at least some of the nodes also need to contain simulation dependent dynamic data that is being updated many times during calculation. This leads to the need of representing each node by, from experiment perspective, two different objects, where it is possible to access the dynamic part using a reference to the static one. Such effect can be achieved in two different ways:

- Static object with a collection of dynamic objects of size n, where n is a number of simulations run in parallel.
- Static object with index or identifier that is used in simulation-dependent collection to retrieve dynamic object.

Both of these concepts use a collection, which leads to prohibitive increase of access time in many situations. Even if the collection used is an array, it leads to at least 2 additional memory access operations, as the program has to retrieve index and use it to get an address of dynamic object from the array. In most modern objective languages such as Java or C# there is a need for one more memory access, because of arrays being separate objects [13, 14]. In effect, if the object discussed represents a node, using this technique may significantly increase time needed to perform as common operation as accessing a dynamic property of node's neighbor. In the simple model, such an operation takes only 2 reads, one for accessing neighboring node and the other for its property.

However, if both of these objects are split in the way described above, the program has to access the static object of the current node first, then it can access the static object of its neighbor, its dynamic object and finally a property. This leads to the need for performing as many as 6 reads, as shown in Fig. 3. Such unacceptable increase in access time has also been verified in benchmark tests. That is why we would not recommend using this method unless memory usage becomes too much of a problem.

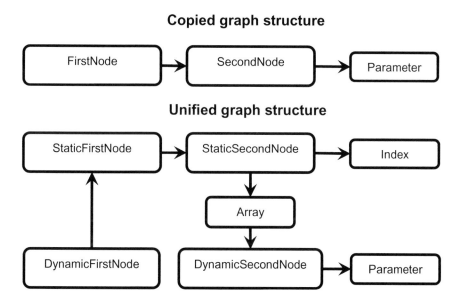

Fig. 3. Memory reads needed to perform a property access from neighboring node.

4.2 Graph-Independent Static Data

As explained in foregoing section, it is not possible to unify graph connections without negative effects on simulation performance. Still, there is other static data in the model that possibly can be unified. Such data consists mostly of node names, numbers, positions and other data that is not used in every step of simulation, but rather during starting and finishing phases. Such data is used for a very short time during simulation, therefore extracting it to a separate object will not change simulation time in any significant way. It may also have positive effects such as decreasing memory usage and size of objects used for most of simulation time effecting in cache miss reduction. It is important to realize that in case of this change the effect may vary depending on simulation algorithm and size of extracted objects so it is advised to execute performance tests before and after the change to confirm positive effect.

5   Template Objects

As discussed before, it may not be possible to perform a far-reaching unification of static model. Furthermore, dynamic data cannot be unified at all because it changes too often. This leaves only infrequently changing data to the discussion.

A distinctive characteristic of properties belonging this data type is that, from the standpoint of simulation algorithm, they do not change. At the same time the information they store is exactly what differs experiments between each other, as their effect on simulated reality is what user tries to examine.

Another interesting feature is that it is common for multiple objects in one experiment to have exactly same set of values for infrequently changing parameters, e.g. when studying the effect of using faster vehicle in the network on average travel time. In such situation each running simulation contains a set of objects that are identical to some template. It is of course possible that in described example there is more than one vehicle type, though it would be very rare for each vehicle to have distinct characteristics.

Taking this into consideration, it may be profitable to separate discussed parameters into a new object that can be shared among all objects belonging to a template. Such sharing can be performed not only between similar objects belonging to one experiment, but even between different experiments run in parallel. This causes the memory gain to quickly increase with the number of simulations and model size. The cost of this change is using additional reference when accessing parameters moved to template object. This cost can affect simulation performance but is somewhat reduced by CPU cache system, as the same object is going to be accessed multiple times in each simulation step. From the standpoint of additional memory usage, it is needed to keep one additional small collection for each template object class. The collection is needed to perform a search for an appropriate template when some infrequently changing parameter does eventually change. High cost of the operation is supposed to be mitigated by its low frequency, as it is assumed that such changes are possible only few times during the whole simulation. Additionally, the collection needs to keep reference counters to destroy a template when it is not used any more. These memory costs can be considered small, but it is important not to use this solution for classes with only a few parameters to avoid creating a structure bigger than the gain.

6   Performance Testing

To check how adding additional reference affects access performance in template object solution, a benchmark study was performed. The benchmark program has been written in C# for .NET 4.6.1 and was run on two different machines described in Table 1.

The benchmark measured access time in 3 different structures:

1. List of objects directly containing a value, further called *flat objects*.
2. List of objects with a reference to template object containing a value, named *template objects*.
3. List of values used for reference.

Table 1. Testing platforms specification

Platform identifier	P1	P2
CPU name	Intel Core i5 2500K	Intel Core i7 3520M
CPU microarchitecture	Sandy Bridge	Ivy Bridge
CPU frequency	4,2 GHz	3,5 GHz
Number of cores/threads	4/4	2/4
Cache size	6 MB	4 MB
RAM size	16 GB	8 GB
RAM frequency	1600 MHz	1600 MHz
Operating system	Microsoft Windows 8.1	Microsoft Windows 10

The latter structure was measured in order to receive the amount of time spent on iteration, so that it can be subtracted from the two former measurements. All the lists had same, constant length. The difference between experiments was the average number of objects belonging to the same template, which in this case means having same value of a property. Each experiment has been conducted a hundred times and the results have been aggregated using a mean.

Results in Fig. 4 show that even though additional reference pass has been added, access time difference is very small as long as number of objects is at least of the order of ten, which is normally always the case.

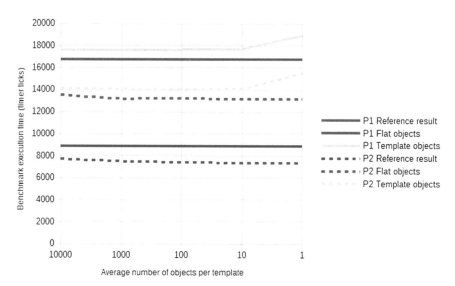

Fig. 4. Template object performance test results

7 Summary

Data model analysis and tests have shown that it is possible to significantly decrease memory usage of ATN simulation software without affecting simulation time or synchronization in a negative way. As a result two efficient methods have been developed:

1. Template object method, that can be used to reduce redundancy in simulation parameter storage.
2. Graph-independent data unification, which relies on choosing the parameters that are used limited number of times during simulation.

These two methods can be broadly used in ATN simulation software.

Additionally, a third method has been created, that is recommended to use only in situations when memory becomes the most important resource. The technique of graph structure unification provides large memory usage gains with the cost of data access time increase. That is why it should not be used in most situations.

Some research may require massive experiments to be performed. The empty vehicles management algorithm described in [8] was tested for four changing parameters, two values for each parameter, which gave 16 simulations (for every number of vehicles and every demand assumed). Various network structures were tested, so hundreds of experiments were run. Taking a set of four values of a parameter, instead of two values, would multiply the number of required simulations by two. As a result, enormous number of simulations might be necessary to analyze given management algorithm or to find optimal parameter values for a given network structure. This may be achieved using genetic algorithms, which require massive simulations to evaluate every parameter values set (treated as a chromosome) [15, 16].

Analogous situation concerns machine learning procedure [17], which can be used to cluster the sets of algorithm parameters for given network structure and traffic conditions. It requires hundreds or thousands of experiments to learn a neural network.

Massive simulations can be efficiently performed in a distributed environment. A distributed simulator is under development. Distribution of the simulator requires a simulation manager which prepares the experiments, spreads them over a computer network and collects the results. Every computer taking part in research should be equipped with local version of simulator manager and a local data structure following the principles described in this paper.

As was shown in [18] some programming constructs, nice and useful for a programmer, may significantly decrease the performance of an application, sometimes even 30 times. In future, ours algorithms can be refined, and maybe sped up, by rewriting some programming constructs according to the results presented in [18].

The presented methodology of building data structures for massive experiments may be applied in wide range of simulation environments, not only those concerning ATN or other transport systems.

References

1. Grabski, W., Daszczuk, W.B.: A study of urban transport means cooperation: PRT and light rail. Transp. Probl. **12**(4), 5–14 (2017). http://transportproblems.polsl.pl/pl/Archiwum/2017/zeszyt4/2017t12z4_01.pdf
2. Daszczuk, W.B.: Discrete event simulation of personal rapid transit (PRT) systems. Autobusy-TEST **17**(3), 1302–1310 (2016). ArXiv:1705.05237
3. Eco-Mobility. http://www.eco-mobilnosc.pw.edu.pl/?sLang=en
4. Hermes. http://students.ceid.upatras.gr/~xithalis/simulation_en.html
5. Future transit system for Uppsala. http://www.beamways.com/wp-content/uploads/uppsalaReportTranslatedBrief.pdf
6. NETSIMMOD. http://prtconsulting.com/simulation.html
7. McDonald, S.S.: Personal rapid transit (PRT) system and its development. In: Encyclopedia of Sustainability Science and Technology. pp. 7777–7797. Springer, New York (2012). https://doi.org/10.1007/978-1-4419-0851-3_671
8. Daszczuk, W.B., Mieścicki, J., Grabski, W.: Distributed algorithm for empty vehicles management in personal rapid transit (PRT) network. J. Adv. Transp. **50**(4), 608–629 (2016). https://doi.org/10.1002/atr.1365
9. PrtSim. https://users.soe.ucsc.edu/~elkaim/elkaim/PRT_Simulator.html
10. Fujimoto, R.M.: Parallel and distributed simulation. In: Yilmaz, L., Chan, W.K.V., Moon, I., Roeder, T.M.K., Macal, C., Rossett, M.D. (eds.) 2015 Winter Simulation Conference (WSC), Huntington Beach, CA, 6–9 December 2015, pp. 45–59. IEEE (2015). https://doi.org/10.1109/wsc.2015.7408152
11. Fujimoto, R.M.: Parallel and Distributed Simulation Systems, pp. 87–90, 142–144. Wiley Interscience, New York (2000). ISBN:0-741-18383-0
12. Ramasubramanian, N., Srnivas, V.V., Gounden, N.A.: Performance of cache memory subsystems for multicore architectures. Int. J. Comput. Sci. Eng. Appl. **1**, 59–71 (2011). arXiv:111.3056
13. Java Language Specification – Arrays. https://docs.oracle.com/javase/specs/jls/se7/html/jls-10.html
14. Microsoft Docs - Arrays as Objects (C# programming guide). https://docs.microsoft.com/en-us/dotnet/csharp/programming-guide/arrays/arrays-as-objects
15. Chebbi, O., Fatnassi, E., Chaouachi, J., Nouri, N.: Cellular genetic algorithm for solving a routing on-demand transit problem. In: Proceedings of the 2016 on Genetic and Evolutionary Computation Conference - GECCO, Denver, CO, 20–24 July 2016, pp. 301–308. ACM Press, New York (2016). https://doi.org/10.1145/2908812.2908921
16. Rezgui, D., Chaouachi-Siala, J., Aggoune-Mtalaa, W., Bouziri, H.: Application of a memetic algorithm to the fleet size and mix vehicle routing problem with electric modular vehicles. In: Genetic and Evolutionary Computation Conference Companion on - GECCO 2017, Berlin, Germany, 15–19 July 2017, pp. 301–302. ACM Press, New York (2017). https://doi.org/10.1145/3067695.3075608
17. Czejdo, B., Daszczuk, W.B., Baszun, M.: Using machine learning to enhance vehicles traffic in ATN (PRT) systems. Autobusy-TEST **18**(12), 1484–1489 (2017). ArXiv:1712.05990
18. Bluemke, I., Gawkowski, P., Grabski, W., Grochowski, K.: On the performance of some C# constructions. In: Zamojski, W., Mazurkiewicz, J., Sugier, J., Walkowiak, T., Kacprzyk, J. (eds.) DepCoS-RELCOMEX 2017. AISC, vol. 582, pp. 39–48. Springer, Cham (2018). https://doi.org/10.1007/978-3-319-59415-6_4

Patch-Based Denoising with K-Nearest Neighbor and SVD for Microarray Images

S. Elavaar Kuzhali[1(✉)] and D. S. Suresh[2]

[1] VTU Research Centre, CIT, Tumkur, Karnataka, India
kuzhalisubbiah@yahoo.com
[2] Channabasaveshwara Institute of Technology,
Gubbi, Tumkur, Karnataka, India
sureshtumkur@yahoo.co.in

Abstract. Irrespective of certain major advancement in filtering process in medical images, the denoising operation in microarray images are still considered to be unsolved and offers a large scope of research. Existing denoising principles are less investigated on such complex and massive dimensional microarray image that leads to the development of the proposed system. We present a method of performing simple denoising operation considering the presence of Gaussian noise in microarray image. From the target image denoising method, an improved version of patch-based denoising approach has been developed considering various forms of distance-based matching methods. The study outcome of the proposed system has been found to offer better peak signal-to-noise ratio and structural similarity index in contrast to existing filtering techniques.

Keywords: Filtering · Denoising · Microarray image · Patch-based
Euclidean distance · KNN · Patch matching

1 Introduction

Microarray technology provides the capability of monitoring gene expression profiles simultaneously in a cell under different experimentation conditions. cDNA microarray technology is an advanced process that can be used to understand and explore the functional aspects of life development, genetic causes of occurrence of diseases, evaluate activities of new drug etc. With the advancement of genomics, microarray technology is gaining momentum to a very large moment with its vast range of applicability in sequencing genes, genetic screening, and genetic diagnostic [1]. Microarrays are arrays of discrete DNA sequences printed by a robotic array on a glass slides, forming spots or probes in a circular shape of known diameter. These gene spots are then analysed with the aid of various computational techniques, for the understanding of underlying biological phenomena. Normally, features with lower intensity exist in microarray images that are quite challenging to be understood while visually analysing it [2]. The outcome of microarray experiment is usually a 16-bit image file. Unfortunately, the process of experimentation and image digitization process results in inclusion of potential noise although the technology offers high tolerances to noise [3]. There are

© Springer International Publishing AG, part of Springer Nature 2019
R. Silhavy (Ed.): CSOC 2018, AISC 763, pp. 132–147, 2019.
https://doi.org/10.1007/978-3-319-91186-1_15

various possibilities of sources from where the noise is originated in microarray image e.g. reflection from laser light, presence of dust in observation slide, photonic noise, etc. [4]. Presence of such noise results in higher degree of uncertainty acting as an impediment in the analysis of it e.g. classification problem [5, 6]. In order to deal such noise-related issue, various forms of denoising principles have been evolved that is essentially meant for addressing additive noise from such image. A closer look into existing research contribution shows that denoising principles have been less investigated on microarray images although there are various filtering approaches in bio-signals and images. Therefore, it is really a challenging task to truncate the adverse effect of noise from microarray image. Problems associated with multiplicative as well as additive noise towards gene arrays have been studied in [7–9]. With less number of observations due to high experimentation cost as well as inclusion of noise as particulars in microarray image will significantly lead to false positives. Hence, this paper presents a solution to address the noise problem in microarray images. Section 2 discusses about existing denoising techniques followed by briefing of problems in Sect. 3. The discussion of proposed methodology is briefed in Sect. 4, algorithm implementation in Sect. 5, result discussion in Sect. 6, and conclusion in Sect. 7.

2 Related Work

This section extends the discussion of our prior study [10] by briefing various denoising approaches. There has been good number of implementation of denoising algorithm in recent times considering biomedical signals and images. Most recently, wavelet-based approach was found to assist in denoising technique [11, 12] developing a good mathematical base of implementation. Adoption of adaptive approach of multi-resolution is another alternative for performing denoising operation in biomedical images [13]. Some other frequently used techniques are dictionary-based learning approach [14] to offer better denoised image. Adoption of adaptive edge detection approach along with wavelets is also reported to offer better filtering performance [15]. Singular value decomposition [16] as well as error-based estimation techniques [17] is also witnessed to offer better denoising performance on biomedical images. It has been also noticed that patch-based denoising approach is gaining an increasing momentum as a novel denoising principle as seen in the work of [18–26]. Different forms of patch-based denoising techniques evolved most recently as optical flow-based [18], geodesic path [19], quadratic programming [20], non-local means and dictionary based [21], Weiner filter [22, 23], joint usage of homogeneity and structural similarity [24], regression using non-parametric method [25], and iterative mechanism with weight factor [26]. All these techniques are mainly assessed with respect to signal quality and structural similarity factor and study outcomes are found to be in favour of good image quality accomplishment. However, irrespective of good number of literatures towards improving denoising performance, there are very less work carried out considering the complexities in microarray images. The method of employing image-processing techniques for cancer detection [27] proposes a filter based on wiener filtering and morphological operations for noise reduction in microarray images. Some of the significant work carried out towards denoising microarray images published before 2010 are [28–30] has used Gaussian scale mixtures, multi-resolution technique, and wavelet techniques respectively. Other denoising approaches considering microarray images found in literatures are found to use complex wavelets with bivariate estimation technique [31], Boolean orthonormalizer network [32], Global background noise correction and morphological based approach [33], and usage of Complex Gaussian scale mixture model in complex wavelet domain [34]. The next section briefs about significant problems being addressed in proposed system by reviewing the issues in existing literatures.

3   Problem Description

The identified research problems are (i) complexity associated with microarray images are quite large and different compared to other forms of images and hence existing denoising principles are yet to show its efficiency on it, (ii) patch-based denoising offers good methodology to minimize the presence of noise; however, it has never being experimented with microarray images, (iii) it is still undefined if the existing denoising principle could yield better signal quality of complex and massive dimensional of microarray image in reality, (iv) Most of the methods proposed have considered high SNR (signal-to-noise ratio) images, does not consider the image characteristics and (v) various parametric assumptions are made during the design of an image filter making them infeasible to understand the design concept. Noise in the microarray image degrades the performance of high level downstream processing and analysis. Thus, there is an urgent need to design a filter for microarray images based on the characteristics of images and with no or less parametric assumptions. These problems are addressed to be solved in proposed study that is briefed in next section.

4   Proposed Methodology

An analytical research methodology is considered in performing denoising operation in proposed system. In this paper, cDNA microarray image denoising filter to remove additive noise is introduced. In microarray image denoising, the goal is to restore a clean image from its noisy observation. The mixture of various types of noises is usually modelled as additive Gaussian noise. We consider the additive Gaussian noise model:

$$X = G + \varepsilon \tag{1}$$

Goal is to estimate $G \in \mathbb{R}^2$ from $X \in \mathbb{R}^2$, where X denotes the noisy image, G denotes ground truth image and $\varepsilon \in \mathbb{R}^2$ denotes additive Gaussian noise with zero mean and σ^2 variance. In this work, we are proposing patch based denoising technique. The technique is not a new technique but never been employed for denoising of microarray images. The method exploits the statistics from a dataset and considers the prior information for denoising. Dataset should be created with high quality clean images. Practically it is not possible to get high quality clean microarray images, we create dataset of real microarray images that are visually good from Stanford Microarray Database (SMD). We also create database of simulated microarray images that considers realistic characteristics [35] to validate the proposed methodology. Normally, microarray images consist of colour intensity of red, and green, where majority of the clinical investigation is validated using red and green or their combination of different levels of their intensity. Therefore, the proposed system emphasizes on this fact by transforming the coloured microarray image to grayscale. It is then corrupted by Gaussian noise and is then forwarded to the system executing the denoising algorithm in order to mitigate the presence of noise. The system is also connected with database of simulated clean microarray images, which is used for patch-based removal of noises from the given input image. The flow of the proposed scheme is highlighted in Fig. 1.

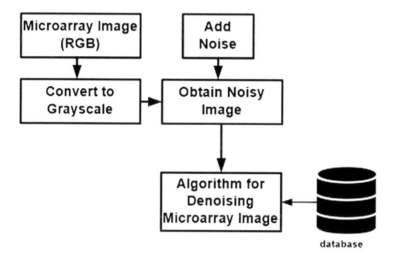

Fig. 1.   Proposed denoising scheme

The proposed system (Fig. 2) considers simulated microarray image of dimension 750×330. The algorithm partitions noisy input image into overlapping patches of size 8×8. Each noisy patch is considered for denoising utilizing the statistics of reference patches and the algorithm combines all the denoised patches to yield a final denoised image. Each noisy block of 8×8 is now checked for their pattern matching with the clean images on the database of microarray images (also called as search image). For simpler search operation, a smaller search window on the search image is only considered. In order to narrow down the searching process K-Nearest Neighbor search (KNN) with Euclidean distance measurement is used to obtain clear reference patches (i.e. N2) from the database for the given noisy patch. Since all the reference patches are from images similar to noisy image, not all the reference patches are required as prior information. The technique considers only the m-nearest reference patches (i.e. N3) resulting in dimensionality reduction of prior information. This process results in

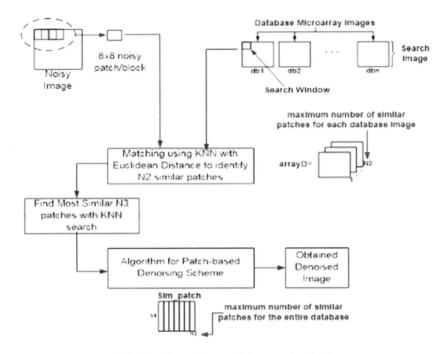

Fig. 2.   Flow diagram of the steps involved

generation of data matrix of prior learned from microarray image database. The obtained outcome is further subjected to patch-based denoising scheme using Singular Value Decomposition (SVD) in order to obtain two dimensional denoised image.

Therefore, the algorithm performs selection of specific number of similar patches to the patch that is inflicted with noise (i.e. input image) where distance-based approach is used for checking the similarity. The denoising principle uses singular value decomposition on the obtained reduced dataset and the noisy patch. This implementation is detailed in next section.

5   Algorithm Implementation

The algorithm mainly performs patch-based minimization of significant noises from the microarray images. The core principle of proposed algorithm is to perform partition of the noisy image in order to generate overlapping patches followed by minimizing of noises from the entire patch. A patch from noisy image is utilized in order to assist in this denoising process that further integrates the entire obtained denoised patches in order to obtain final denoised image. The basic steps of the core algorithm are as follows:

i) Algorithm for Denoising Microarray Image
Input: I, σ, d_{opt}, n
Output: I_{den}
Start
1. init σ, d_{opt},
2. $y_1 \rightarrow \alpha(I)$
3. $y \rightarrow \beta(y_1)$
4. $z \rightarrow y + \sigma$
5. **For** i=1: n
6.   data.db=[data.db, $\{\beta(\alpha(i))\}$]
7. **End**
8. $I_{den} \rightarrow Apply\ Algo\text{-}2(y, z, \sigma, data, d_{opt})$
End

The above algorithm represents the core algorithm that performs denoising of microarray image. The algorithm initially applies a method α for extracting the input microarray image from .mat file (Line-2). The image y_1 is then converted to grayscale (Line-3) using predefined function β (for converting RGB image to grayscale) for obtaining the grayscale image y. A noise is further added to the image in order to obtain the noisy image z (Line-4). The next phase of the algorithm initiates with the database creation. For this purpose, a data-structure *data* is created that considers all the simulated microarray images and starts converting the entire digitized RGB image into grayscale (Line-6). It is to be noted that proposed algorithm applies the algorithmic steps 1 to 4 for the test microarray image whereas step-5 to step-7 is applicable for database of simulated or realistic microarray images. Finally, the denoising algorithm is applied considering the input arguments of grayscale ground truth image y, noisy image z, noise variance σ, data-structure *data*, and options d_{opt} for specifying distance metric. The proposed system is capable of performing denoising using various distance-based approaches.

(ii) Algorithm for Patch-based Denoising Scheme with KNN and SVD
This algorithm applies non-parametric methods for performing search operation for most similar patches. The algorithm takes various forms of input arguments e.g. (i) Ground truth image free from noise y, (ii) image embedded with noise z, (iii) standard deviation of noise σ, and (iv) an explicit data structure consisting of external database of digitized simulated or realistic microarray images. The outcome of the algorithm is a final denoised image i.e. I_{den}. The core steps of the proposed algorithm are as follows:

Algorithm for patch-based denoising scheme with KNN
Input: y, z, σ, data.db, dopt
Output: I_{den}
Start
1. Init N_1, N_{step}, N_s, N_2, N_3.
2. Obtain (h, w)→size (z)
3. **For** size of noisy image z
4. Extract overlapped patch χ of size N_1xN_1 with N_{step}
5. **For** each image of db
6.   Initialize the dimension of search window N_s x N_s
7.   Obtain search window sw from each image of db
8.   arrayD_1→pm(χ, sw, N_2)
9.   Store arrayD_1 for further steps
10. **End**
11. sim_patch→KNNSEARCH(arrayD,χ,d_{opt})
12. rec_ref→Apply Algo-3(sim_patch, χ, N_1, σ)
13. I_{den}→(rec_ref, N_1, weight)
14. **End**
End

Line-1 in above steps of proposed algorithm is the initialization step, it provides the initial values for (i) patch size N_1, (ii) number of steps required to consider the next consecutive patch (i.e. N_{step}), (iii) a search window of dimension Ns for finding matching patches, (iv) the number of patches that are found similar in each external databases (i.e. N_2) and (v) the number of most similar patches from the complete database (i.e. N_3). The algorithm then extracts height h and width w from the noisy microarray image z. The algorithm partitions noisy image z into overlapping patches χ of size $N_1 \times N_1$ with step size N_{step}. For each clean or high quality image from database, a search window sw of dimension $N_s \times N_s$ is selected. To obtain clear reference patches from the database for the given noisy patch, patch matching function pm is designed using KNN with Euclidean distance that takes the input of noisy patch χ, SW, and N_2, in order to obtain a data matrix $arrayD\_1$. The data matrix $arrayD\_1$contains the reference patches rp1, rp2 …rpq that has small $\ell2$ distance from noisy patch χ. The measurement of similarity is carried out on the basis of $\ell2$ distance between the database $\{rp\}_{j=1}^{k}$ and noisy patch χ, where k > q, as

$$d(\chi, rp_j) = \left\| \chi - rp_j \right\|^2 \text{for} j = 0, 1 \cdots \cdots k \qquad (2)$$

Final data matrix of reference patches $arrayD$ is formed considering the data matrix $arrayD\_1$of each image from the database. Since all the reference patches are from images similar to noisy image, not all the reference patches are required as prior information. The technique considers only the m-nearest reference patches resulting in dimensionality reduction of prior information. K-nearest neighborhood search algorithm with distance-based approaches such as Chebychev, City block, Euclidean, and

Minkowski is applied on the obtained data matrix *arrayD* and noisy patch that results in construction of reduced dimension database with nearest *m* reference patches i.e. sim_patch ($\{rp\}_{j=1}^{m}$). Filtering procedure Algo-3 is used to denoise each patch. Average estimate of every pixel is considered for the final denoised image. The filtering procedure uses singular value decomposition to the sim_patch obtained, along with noisy patch χ, N_1, and σ. Singular value decomposition is used to preserve the principal image features and eliminate the trivial noise present in the image. The mechanism of denoising using singular value decomposition is to initially formulate weight vector as well as formulation of projection matrix A and sparse matrix S as follows:

Given $\{rp\}_{j=1}^{m}$, the algorithm computes the projection matrix A using singular value decomposition (SVD).

$$[A\,S] = eig(rp.rp^T) \tag{3}$$

Where A is the basis matrix and S is the diagonal matrix containing the spectral coefficients.

The reference patches obtained have deviations in terms of similarity with noisy patch. The data matrix sim_patch ($\{rp\}_{j=1}^{m}$) is upgraded by including a diagonal weight matrix W.

$$\mathbf{W} = \frac{1}{\mathbf{Z}}\,\mathbf{diag}\left[e^{-\|n-rp_1\|^2/h^2}\,\cdots\cdots\cdots,e^{-\|n-rp_1\|^2/h^2}\right] \tag{4}$$

h is the tuning parameter and Z is a normalization constant.

The equation (Eq. 3) becomes

$$[A\ S] = eig(rpWrp^T) \tag{5}$$

An estimation $\hat{O}(x,y)$ of original patch O(x,y) is provided by $\hat{O}(x,y) = ASA^T\chi$ such that the estimate has minimum mean squared error with respect to the ground truth O(x,y).

Thus Denoising filter is obtained by solving the optimization problem

$$(A\,S) = \underset{A\,S}{\arg\min}\ E\left[\left\|ASA^T\chi - O\right\|_2^2\right] \tag{6}$$

iii) Algorithm for denoising using Singular Value Decomposition
Input: sim_patch ($\{rp\}_{j=1}^{m}$), χ, σ
Output: rec_ref
Start
1. Formulate weight vector W

2. Compute Eigen vector and Eigen values $[A\ S] = svd(rpWrp^T)$

3. Compute $S=((S+\sigma^2 I))^{-1}(S)$

4. The patch is denoised by evaluating rec_ref$\rightarrow \hat{O} = ASA^T\chi$
End

Therefore, the denoising mechanism using singular value decomposition assists in obtaining *rec_ref*, which is further reshaped with respect to squared matrix of size N_1. Using this received reference, weight, N_1, row and column information, proposed system computes denoised image Iden. Average estimate of every pixel is considered for the final denoised image. In the entire process of algorithmic operation, the proposed system always ensures simpler block-based approach for minimization of noise. The algorithm results in superior image quality in terms of peak signal-to-noise ratio as well as structural similarity index. Notations used in the algorithm are also described in Table 1. The next section discusses the outcomes obtained after implementing the proposed algorithm of denoising the microarray image.

6 Result Discussion

The proposed study considers the database of microarray images [35, 36]. MATLAB is used for scripting the proposed denoising logic. The algorithm considers various amount of noise added to the image. Owing to larger dimension of such images, they are transformed to matrix form for easy accessibility in execution process. The default

Table 1. Notation Used

Notation	Meaning
y	Clean image
z	Noisy image
α	Predefined function for reading image
β	Predefined function for obtaining grayscale image
σ	Noise level
data	Data structure of external database
d_{opt}	Distance options for performing denoising
I_{den}	Denoised image
data. db	External database of simulated or realistic microarray images
N_1	Size of reference and noisy block
N_{step}	Step size for processing consecutive reference and noisy image block
N_s	Size of search window
N_2	Highest number of patches found similar in each image of the database
N_3	Highest number of patches found similar in complete database
χ	Noisy patch
pm	Patch matching
sw	Search window
W_vec	Non Local Means Weights
rec_ref	Received reference
O(x,y)	Original patch
$\hat{O}(x,y)$	Estimated patch

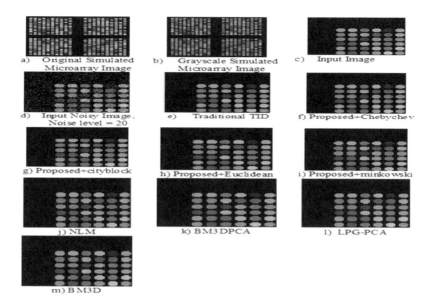

Fig. 3. Magnified visual outcomes of denoising results of simulated microarray images

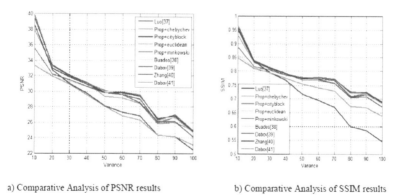

a) Comparative Analysis of PSNR results   b) Comparative Analysis of SSIM results

Fig. 4. Comparative Analysis of PSNR & SSIM results at different noise levels and by different methods for simulated microarray image

value of N_1, N_{step}, N_s, N_2, and N_3 are considered as 8, 1, 101, 20, and 40 respectively. Figure 3(a) highlights the original simulated microarray image that is further converted to grayscale image (Fig. 3(b)). The proposed system considers the input grayscale image (Fig. 3(c) which is further corrupted by Gaussian noise with noise level $\sigma = 20$ (Fig. 3(d)). Finally, the denoised images (Fig. 3(f–i)) is obtained by applying proposed KNN and singular value decomposition based denoising algorithm. The complete implementation is further testified for multiple forms of distance-based method integrated with KNN approach by evaluating the signal quality in terms of Peak Signal-To-Noise Ratio (PSNR) and Structural Similarity Index (SSIM) as the performance parameters. To validate the efficiency of the proposed system we tested our method on realistic microarray image taken from SMD database. Figure 5(a) shows the original realistic microarray image that is further converted to grayscale image (Fig. 5(b)). The proposed system considers the grayscale image which is corrupted by Gaussian noise with noise level $\sigma = 20$ (Fig. 5(c)). The methodology based on Chebychev distance in KNN search presented a sensible improvement in denoising quality (Fig. 5(d). The visual outcomes of the proposed system for simulated and realistic microarray images are as shown below:

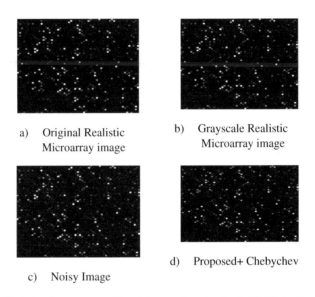

a)   Original Realistic   b)   Grayscale Realistic
       Microarray image   Microarray image

d)   Proposed+ Chebychev

c)   Noisy Image

Fig. 5.   Magnified visual outcomes of denoising results of realistic (SMD database) micro- array images

The investigation (Tables 2, 3, 4 and 5) is carried out considering 4 different distance-based approaches e.g. Chebychev, City block, Euclidean, and Minkowski. All these 4 approaches are jointly tested with respect to PSNR and SSIM. For an effective analysis, the study outcome of the proposed system is compared with similar category of work being carried out by certain existing researchers e.g. Luo et al. [37], Buades et al. [38], Dabov et al. [39], Zhang et al. [40], and Dabov et al. [41]. Visual outcome of the proposed system (Figs. 3 and 5) shows the denoised images with natural appearance without any spurious patterns. A closer look into the outcome (Fig. 4) shows that proposed system offers better PSNR (Figs. 4(a) and 6(a)) in contrast to all the existing system as well as better SSIM too (Figs. 4(b) and 6(b)). The prime reason behind this trend is that existing mechanisms of denoising was constructed by assuming that there could be similar forms of patches in an image at different positions. For an example, the works of Buades et al. [38] as well as Dabov et al. [41] have used approach for exploring same forms of patches. Hence, such approaches are not much applicable for larger and complex microarray images where the distance across diverse forms of

Table 2. PSNR results at different noise levels and by different methods for simulated microarray images

Noise levels Methods	10	20	30	40	50	60	70	80	90	100
'Luo [37]'	38.50	32.92	31.81	30.87	29.82	29.74	29.33	26.52	26.81	24.74
'Prop + cheby'	37.86	33.22	31.91	31.00	29.34	29.13	28.41	25.83	25.96	24.94
'Prop + citybloc'	39.8	33.33	32.06	31.00	30.06	29.55	28.99	25.91	26.68	24.86
'Prop + euclid'	39.54	33.33	32.07	31.01	29.98	29.84	29.58	26.25	26.63	24.67
'Prop + minko'	39.38	33.32	32.02	31.06	29.81	29.53	28.51	26.06	26.06	24.15
'Buades [38]'	34.04	31.89	31.22	30.56	29.76	29.90	29.46	26.27	26.97	24.98
'Dabov [39]'	35.53	32.20	31.46	30.62	29.78	29.86	29.45	26.34	26.95	24.94
'Zhang [40]'	38.53	32.71	30.99	29.72	28.11	27.30	26.79	24.32	24.10	22.44
'Dabov [41]'	33.27	31.90	30.85	29.49	28.06	26.79	26.30	24.33	24.14	23.10

Table 3. SSIM results at different noise levels and by different methods for simulated microarray images

Noise levels Methods	10	20	30	40	50	60	70	80	90	100
'Luo [37]'	0.947	0.832	0.809	0.789	0.775	0.775	0.769	0.725	0.721	0.685
'Prop + cheby'	0.929	0.831	0.810	0.789	0.768	0.766	0.755	0.707	0.715	0.688
'Prop + citybloc'	0.964	0.840	0.812	0.789	0.777	0.774	0.766	0.710	0.716	0.687
'Prop + euclid'	0.960	0.839	0.813	0.792	0.773	0.775	0.771	0.719	0.717	0.683
'Prop + minko'	0.958	0.836	0.811	0.792	0.772	0.770	0.757	0.705	0.708	0.671
'Buades [38]'	0.859	0.812	0.801	0.785	0.774	0.779	0.772	0.722	0.725	0.691
'Dabov [39]'	0.886	0.815	0.804	0.786	0.774	0.778	0.773	0.722	0.724	0.689
'Zhang [40]'	0.951	0.835	0.795	0.766	0.717	0.695	0.670	0.599	0.586	0.545
'Dabov [41]'	0.846	0.809	0.794	0.773	0.753	0.740	0.729	0.673	0.668	0.638

Table 4. PSNR results at different noise levels and by different methods for realistic microarray images

Noise levels Methods	10	20	30	40	50	60	70
'Luo [37]'	32.62	32.36	31.69	30.85	29.70	28.74	27.92
'Prop + cheby'	39.56	36.13	33.68	32.25	30.22	28.84	27.74
'Prop + citybloc'	39.65	36.12	33.81	32.04	30.46	29.31	27.99
'Prop + euclid'	39.67	36.16	33.69	31.96	30.26	29.35	28.03
'Prop + minko'	39.67	36.16	33.69	31.96	30.26	29.35	28.03
'Buades [38]'	31.65	31.14	30.13	28.97	28.45	28.34	27.28
'Dabov [39]'	34.39	32.80	31.39	30.43	29.21	28.39	27.27
'Zhang [40]'	34.61	32.47	30.33	28.78	27.04	25.94	24.59
'Dabov [41]'	36.56	33.40	31.99	30.61	29.75	28.88	27.83

microarray images should be considered. Such techniques are also witnessed with a problem where quantity of patches grows with the growth of the image numbers. Moreover, the proposed system also shows improvement in comparison to Luo et al. [37] work however the denoising principle is further enhanced using KNN search process in proposed system. Hence, better improvement of PSNR and SSIM is observed in the proposed system. This outcome shows that proposed system is highly capable of retaining signal quality of the denoised image where the performance could be further improved using different forms of distance-based approaches.

Table 5. SSIM results at different noise levels and by different methods for realistic microarray images

Noise levels	10	20	30	40	50	60	70
Methods							
'Luo [37]'	0.887	0.850	0.821	0.789	0.750	0.712	0.683
'Prop + cheby'	0.948	0.891	0.829	0.791	0.713	0.656	0.636
'Prop + citybloc'	0.948	0.892	0.839	0.779	0.723	0.683	0.640
'Prop + euclid'	0.949	0.893	0.831	0.779	0.709	0.666	0.641
'Prop + minko'	0.949	0.893	0.831	0.779	0.709	0.666	0.641
'Buades [38]'	0.875	0.844	0.811	0.769	0.746	0.719	0.687
'Dabov [39]'	0.895	0.858	0.816	0.783	0.743	0.699	0.668
'Zhang [40]'	0.885	0.788	0.667	0.583	0.498	0.423	0.351
'Dabov [41]'	0.911	0.856	0.834	0.793	0.765	0.722	0.656

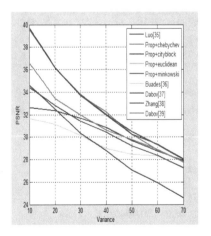

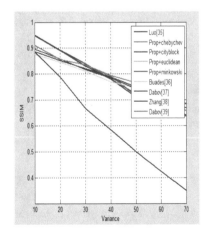

a) Comparative Analysis of PSNR results   b) Comparative Analysis of SSIM results

Fig. 6. Comparative analysis of PSNR & SSIM results at different noise levels and by different methods for realistic microarray image

7   Conclusion

From the review of literature, it is explored that there has been some good advancement observed in denoising principles in different forms of images except in microarray images. Microarray image is a result of a sophisticated procedure that is also accompanied certain fault leading to inclusion of noises. Experimental analyses of different existing denoising principles have not being extensively carried out on microarray images, which leads to a significant research gap. This paper presents a very simple denoising principle where the grayscale version of microarray image is taken as an input and it is further processed based on block-wise operation. The proposed algorithm considers 8×8 blocks of input image and starts checking for its match in its search window of search image in database. Using Euclidean distance, the process of matching of the query image with database image is carried out. The algorithm further implements K-nearest neighbor and singular value decomposition to facilitate the process of denoising. Our results show that multiple forms of distance-based methods-integrated with K-nearest neighbor serves better performance as compared with other existing approaches. Our outcomes are compared with existing denoising techniques to find that proposed system offers better PSNR and SSIM as compared to others. The research work will be continued in the direction of optimization for further enhancing the denoising performance and to reduce the computational complexity.

References

1. Kumar, A., Shaik, F., Abdul Rahim, B., Sravan Kumar, D.: Signal and Image Processing in Medical Applications. Springer (2016)
2. Srimani Shanthi Mahesh, P.K.: An effective automated method for detection of grids in DNA microarray. In: Springer-ICT and Critical Infrastructure: Proceedings of the 48th Annual Convention of Computer Society of India, vol. II, pp. 445–453 (2014)
3. Yakovlev, A.Y., Klebanov, L., Gaile, D.: Statistical Methods for Microarray Data Analysis: Methods and Protocols. Springer (2013)
4. Rueda, L.: Microarray Image and Data Analysis: Theory and Practice. CRC Press (2014)
5. Fraser, K., Wang, Z., Liu, X.: Microarray Image Analysis: An Algorithmic Approach. CRC Press (2010)
6. Scherer, A.: Batch Effects and Noise in Microarray Experiments: Sources and Solutions. Wiley (2009)
7. Gohlmann, H., Talloen, W.: Gene Expression Studies Using Affymetrix Microarrays. CRC Press (2009)
8. Do, K.-A., Müller, P., Vannucci, M.: Bayesian Inference for Gene Expression and Proteomics. Cambridge University Press (2006)
9. Balding, D.J., Bishop, M., Cannings, C.: Handbook of Statistical Genetics. Wiley (2008)
10. Kuzhali, E., Suresh, D.S.: A comprehensive study of enhancement and segmentation techniques on microarray images. i-manager's J. Pattern Recognit. **2**(2) (2015)
11. Pad, P., Alishahi, K., Unser, M.: Optimized wavelet denoising for self-similar α-stable processes. IEEE Trans. Inf. Theo. **63**(9), 5529–5543 (2017)

12. Boubchir, L., Boashash, B.: Wavelet denoising based on the MAP estimation using the BKF prior with application to images and EEG signals. IEEE Trans. Sig. Process. **61**(8), 1880–1894 (2013)
13. Lahmiri, S.: Denoising techniques in adaptive multi-resolution domains with applications to biomedical images. Healthc. Technol. Lett. **4**(1), 25–29 (2017)
14. Zeng, X., Bian, W., Liu, W., Shen, J., Tao, D.: Dictionary pair learning on grassmann manifolds for image denoising. IEEE Trans. Image Process. **24**(11), 4556–4569 (2015)
15. Wang, G., Wang, Z., Liu, J.: A new image denoising method based on adaptive multiscale morphological edge detection. Hindawi Math. Prob. Eng. (2017)
16. Guo, Q., Zhang, C., Zhang, Y., Liu, H.: An efficient SVD-based method for image denoising. IEEE Trans. Circuits Syst. Video Technol. **26**(5), 868–880 (2016)
17. Golshan, H.M., Hasanzadeh, R.P.R.: An optimized LMMSE based method for 3D MRI denoising. IEEE/ACM Trans. Comput. Biol. Bioinform **12**(4), 861–870 (2015)
18. Buades, A., Lisani, J.L., Miladinović, M.: Patch-based video denoising with optical flow estimation. IEEE Trans. Image Process. **25**(6), 2573–2586 (2016)
19. Chen, X., Kang, S.B., Yang, J., Yu, J.: Fast edge-aware denoising by approximated patch geodesic paths. IEEE Trans. Circuits Syst. Video Technol. **25**(6), 897–909 (2015)
20. Feng, J., Song, L., Huo, X., Yang, X., Zhang, W.: An optimized pixel-wise weighting approach for patch-based image denoising. IEEE Sig. Process. Lett. **22**(1), 115–119 (2015)
21. Bhujle, H., Chaudhuri, S.: Novel speed-up strategies for non-local means denoising with patch and edge patch based dictionaries. IEEE Trans. Image Process. **23**(1), 356–365 (2014)
22. Cao, M., Li, S., Wang, R., Li, N.: Interferometric phase denoising by median patch-based locally optimal wiener filter. IEEE Geosci. Remote Sens. Lett. **12**(8), 1730–1734 (2015)
23. Chatterjee, P., Milanfar, P.: Patch-based near-optimal image denoising. IEEE Trans. Image Process. **21**(4), 1635–1649 (2012)
24. Zhong, H., Han, P.P., Zhang, X.H., Yu, Y.Q.: Hybrid patch similarity for image denoising. Electron. Lett. **48**(4), 212–213 (2012)
25. Boulanger, J., Kervrann, C., Bouthemy, P., Elbau, P., Sibarita, J.B., Salamero, J.: Patch-based nonlocal functional for denoising fluorescence microscopy image sequences. IEEE Trans. Med. Imaging **29**(2), 442–454 (2010)
26. Deledalle, C.A., Denis, L., Tupin, F.: Iterative weighted maximum likelihood denoising with probabilistic patch-based weights. IEEE Trans. Image Process. **18**(12), 2661–2672 (2009)
27. Khalilabad, N.D., Hassanpour, H.: Employing image processing techniques for cancer detection using microarray images. Comput. Biol. Med. **81**, 139–147 (2017)
28. Wang, X.H., Istepanian, R.S.H., Song, Y.H.: Microarray image enhancement by denoising using stationary wavelet transform. IEEE Trans. Nanobiosci. **2**(4), 184–189 (2003)
29. Stefanou, H., Margaritis, T., Kafetzopoulos, D., Marias, K., Tsakalides, P.: Microarray image denoising using a two-stage multiresolution technique. In: 2007 IEEE International Conference on Bioinformatics and Biomedicine (BIBM 2007), Fremont, CA, pp. 383–389 (2007)
30. Mastriani, M., Giraldez, A.E.: Microarrays denoising via smoothing of coefficients in wavelet domain. World Acad. Sci. Eng. Technol. Int. J. Electron. Commun. Eng. **1**(2) (2007)
31. Howlader, T., Chaubey, Y.P.: Noise reduction of cDNA microarray images using complex wavelets. IEEE Trans. Image Process. **19**(8), 1953–1967 (2010)
32. Vishakha, P.S, Supriya, S.T.: Study and analysis of microarray denoising using systholic boolean orthonormalizer network in wavelet domain. In: Advancement in Electronics and Telecommunication Engineering (2012)
33. Fouad, I.A., Mabrouk, M.S., Sharawy, A.A.: A new method to grid noisy cDNA microarray images utilizing denoising techniques. Int. J. Comput. Appl. **63**(9), February 2013

34. Srinivasan, L., Rakvongthai, Y., Oraintara, S.: Microarray image denoising using complex gaussian scale mixtures of complex wavelets. IEEE J. Biomed. Health Inf. **18**(4), 1423–1430 (2014)
35. Nykter, M., Aho, T., Ahdesmäki, M., Ruusuvuori, P., Lehmussola, A., Yli-Harja, O.: Simulation of microarray data with realistic characteristics. BMC Bioinf. **7**(1), 349 (2006)
36. Stanford Microarray Database (SMD), http://smd.stanford.edu/
37. Luo, E.: Statistical and adaptive patch-based image denoising. A Doctorial Dissertation of University of California, San Diego (2016)
38. Buades, A., Coll, B., Morel, J.: A review of image denoising algorithms, with a new one. SIAM Multiscale Model Simul. **4**(2), 490–530 (2005)
39. Dabov, K., Foi, A., Katkovnik, V., Egiazarian, K.: BM3D image denoising with shape-adaptive principal component analysis. In: Signal Process with Adaptive Sparse Structured Representations (SPARS 2009), pp. 1–6, April 2009
40. Zhang, L., Dong, W., Zhang, D., Shi, G.: Two-stage image denoising by principal component analysis with local pixel grouping. Pattern Recogn. **43**, 1531–1549 (2010)
41. Dabov, K., Foi, A., Katkovnik, V., Egiazarian, K.: Image denoising by sparse 3D transform-domain collaborative filtering. IEEE Trans. Image Process. **16**(8), 2080–2095 (2007)

Hypervisors Comparison and Their Performance Testing

Maxim Polenov$^{(\boxtimes)}$, Vyacheslav Guzik, and Vladislav Lukyanov

Department of Computer Engineering, Southern Federal University, Taganrog, Russia
{mypolenov,vfguzik}@sfedu.ru, sith@pochta.ru

Abstract. This paper is devoted to comparing approaches to the implementation of virtualization based on the use of the most common hypervisors. The paper considers advantages of using virtual machines while creating modern network infrastructure; as well as this, it describes a number of experiments on starting different hypervisors in one selected host along with the installation of two virtual machines with identical configurations. The results of CPU utilization and statistics on the use of a random access memory by hypervisors without virtual machines and with virtual machines are presented; conclusions on using different hypervisors are drawn.

Keywords: Virtualization · Hypervisor · Virtual machine
Guest operating system · Hypervisor performance

1 Introduction

Nowadays, virtualization technology is actively developing and is being applied in virtually all industries where information infrastructures and technologies are involved. It can be a small, medium and large enterprise or company, or a regular user or a group of users. This topic is very relevant, many articles have been devoted to it, where the authors define the virtualization technology, describe its different types and compare popular software for organizing a virtual infrastructure [1, 2].

There is no single definition of the term "virtualization"; however, its purpose is to isolate computing processes and computing resources from each other. Practically, it looks like this. First, a special operating system, called a hypervisor, is installed on the physical server. Then one or more guest operating systems are installed "over" the hypervisor; an application can be deployed in each of them. From the standpoint of the guest operating system, the server with the hypervisor looks like a server that consists of "virtual" standardized server components (processors, memory, disk subsystem controllers, hard disks, and so on), although the real components of the physical server may be very different. A collection of such virtualized server components, the guest operating system, and the application is called a virtual machine (VM). Multiple virtual machines can be hosted by the same physical server [3].

Virtual machines have several advantages over conventional computers: hardware independence, isolation, compatibility and encapsulation.

© Springer International Publishing AG, part of Springer Nature 2019
R. Silhavy (Ed.): CSOC 2018, AISC 763, pp. 148–157, 2019.
https://doi.org/10.1007/978-3-319-91186-1_16

2 The Most Popular Hypervisors

A hypervisor is a program or hardware scheme that provides or allows simultaneous parallel execution of multiple operating systems on the same host computer. A hypervisor also provides isolation of operating systems from each other, protection and security, partitioning of resources between different running OSs, and resource management.

The hypervisor itself is a minimal operating system. It provides the operating systems running under its control with the virtual machine service while emulating the real (physical) hardware of the specific machine. The hypervisor manages these virtual machines as well as the allocation and release of resources for them. The authors give an example of the use of virtual machines in smart home systems and wireless user authorization [4, 5].

Now the most popular hypervisors are Hyper-V (Microsoft), XenServer (Citrix) and vSphere (VMware). Let us consider their main characteristics.

The developers of the above hypervisors position them as products for implementation on various platforms and are not always requiring special knowledge of users, and most of hypervisors can work steadily using standard (default) settings. The authors in this paper will try to find out whether there are features that have to be considered by users when implementing hypervisors.

The client-server applications working on installed hypervisors are considered also to identify their key features for users. In some products developers go for cardinal changes of the operation principles of client-server applications trying to make it the universal and cross-platform solution. The authors also try to find out how effective and convenient such changes are.

Hyper-V
Microsoft Hyper-V technology is a hypervisor-based hardware virtualization system that provides guests with direct access to the server devices without the participation of intermediate virtual drivers. This is a next-generation virtualization technology based on a 64-bit hypervisor that offers a platform with reliable and scalable capabilities. It is distributed in two ways: as a Windows Server role or as a separate product of Microsoft Hyper-V Server [6].

XenServer
Xen is a cross-platform hypervisor. It is the product of the University of Cambridge and is distributed under the terms of the GPL license. From the very beginning, the two branches were supported: the open source and the commercial version. In 2007, Citrix acquired Xen Source, and in 2009, it announced that commercial versions of XenServer would be completely free, which implied its fully free and open sources.

Xen belongs to platforms that use hardware virtualization in their solutions. It should be noted that, while not fundamentally different from program virtualization, the hardware virtualization provides performance that is quite comparable to the performance of a real machine, which has provided it with such a wide application in the industrial environment [7].

vSphere

Hypervisor VMware ESXi is a functional and unique in its segment software product for server virtualization. Its popularity resulted in mass use in the corporate sector allowed the developers of the 6th version to correct the errors typical for hypervisor-competitors. The main functionality that the hypervisor implements is creating virtual machines on a physical server. As guest operating systems, all major versions are supported. One of the key features of the hypervisor is that it is written almost from scratch; its distribution size is 348 MB, most of which are used for the drivers of multiple operating systems (VMware tools) [8].

3   Testing Hypervisors

Let us consider the work of three hypervisors on the same physical machine with deployed identical virtual machines. For comparison, we use obtained graphics of the CPU usage and the consumption of RAM by the hypervisor in the idle mode (without virtual machines) and in the active mode of virtual machines. The characteristics of the used hypervisors are shown in Table 1, the characteristics of the physical test computer are shown in Table 2, and the characteristics of virtual machines (VM1, VM2) are given in Table 3.

Table 1. Hypervisors characteristics

Hypervisor	Version	Note
VMware vSphere	Standard 6.5	A free version of the VMware hypervisor is used. The full operation of the hypervisor requires additional network equipment. Access to hypervisor management is via the web interface and does not require the installation of client software
Citrix XenServer	Standard 7.2	A free version of Citrix's hypervisor is used. This hypervisor does not require the installation of additional hardware. To administer the hypervisor, you need to install additional client software
Microsoft Hyper-V	Windows Server 2016	This hypervisor is used as a role in the Microsoft server operating system

Table 2. Test computer characteristics

CPU	Motherboard	Network card	RAM
AMD-FX 4300 3800 MHz	Gigabyte 78LMT-S2PV	Intel(R) PRO/1000 GT Desktop Adapter	Crucial CT51264BA160BJ.C 8F 8 GB

Table 3. Virtual machine characteristics

Characteristics	VM1	VM2
Allocated virtual processors	2	2
Allocated RAM	3 GB	3 GB
Installed operating system	Microsoft Windows 10 × 64	Linux mint 18.2

It should be noted that the authors have already conducted experiments to find out the possibility for hypervisors to function on an ordinary physical computer [9].

3.1 VMware VSphere

As noted in Table 1, the deployment of this hypervisor required the installation of an additional network card, since the hypervisor could not identify the card integrated into the motherboard.

In Fig. 1, you can see how many resources the hypervisor consumes without installed and running virtual machines.

Fig. 1. Using resources with the VMware vSphere hypervisor (without virtual machines)

As you can see from this figure, the hypervisor itself uses 1.39 GB of memory for its needs and practically does not use processor resources.

Next, we installed two virtual machines whose characteristics are given in Table 3, and the total consumption of hypervisor resources with running virtual machines is shown below (Figs. 2, 3 and 4).

Fig. 2. The total workload of the VMware vSphere hypervisor with deployed virtual machines

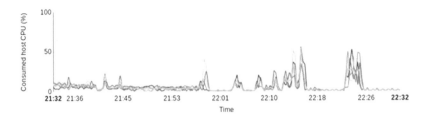

Fig. 3. The graph of the CPU usage by the VMware vSphere hypervisor

Fig. 4. The graph of the use of RAM by the VMware vSphere hypervisor

Let's consider the graph of the hypervisor processor and RAM usage separately.

As you can see in Fig. 2, the hypervisor with working virtual machines uses the processor by 14%, and in Fig. 3, it can be seen that in certain time intervals, the processor was loaded by almost 50%; it was caused by the installation and configuration of virtual machines. Figure 4 shows that the hypervisor with working virtual machines uses about 5 GB of memory, which is not bad. Three gigabytes of virtual memory was allocated to virtual machines, and the hypervisor itself, as seen in Fig. 1, consumes 1.39 GB of memory; the current memory consumption indicates that the hypervisor does not immediately take the memory allocated to the virtual machine but consumes it as needed.

Starting with version 6.5, the management of the hypervisor is done only via the web interface implemented on the HTML 5 platform. This eliminates the need to install client software on the computer and also makes it a multiplatform interface; however, a number of problems with the speed of management tools and its incorrect work may arise on computers with low performance.

3.2   Citrix XenServer

This hypervisor was installed on a computer which characteristics are presented in Table 2. During the installation, the hypervisor was able to recognize and operate both on-a-motherboard and the external network card, in contrast to VMware vSphere. The consumption of resources by the hypervisor itself is shown in Fig. 5.

Fig. 5. Using resources with the XenServer hypervisor (without virtual machines)

As you can see in the Fig. 5 the hypervisor uses for work 989 MB of RAM and loads the processor by 2%.

Next, two virtual machines whose characteristics correspond to the parameters specified in Table 3 were deployed on the hypervisor. The overall hypervisor load as well as the CPU utilization and memory usage can be seen in Figs. 6, 7, and 8.

Fig. 6. Total hypervisor load (with virtual machines)

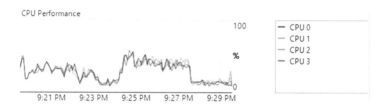

Fig. 7. XenServer CPU hypervisor load graph

Fig. 8. The graph of the RAM use by the XenServer hypervisor

As seen in Fig. 7, the processor was loaded by about 50% during the installation of operating systems on virtual machines. Figure 8 shows that almost all the RAM is used; the XenServer hypervisor, in contrast to VMware vSphere, completely takes the allocated RAM for a virtual machine: 6 GB for two virtual machines and 1 GB of memory for the needs of the hypervisor itself; in total, 7 GB are produced as shown in Fig. 6.

The hypervisor is managed through the XenCenter client application installed on the computer, from which the hypervisor will be controlled. The application can be installed on the operating systems of the Windows family; there is an analogue of OpenXen-Manager for other operating systems. This feature can cause a number of difficulties for the user; however, XenCenter works well on computers with low performance.

3.3   Microsoft Hyper-V

The latest version of the server operating system of Microsoft – Windows Server 2016 was installed on the computer whose characteristics were presented in Table 2. After installing the operating system, the Hyper-V role was installed on it. Nothing else was installed on the operating system; the system remained in its pure form with the established role of the hypervisor. The total workload of this system is shown in Figs. 9 and 10.

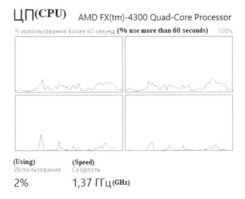

Fig. 9. The schedule of the Windows Server 2016 processor with the Hyper-V role (without virtual machines)

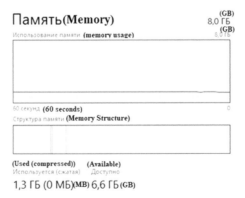

Fig. 10. The graph of the RAM use in Windows Server 2016 with the role of Hyper-V (without virtual machines)

As you can see from the figures, the system with the hypervisor role uses 1.3 GB of RAM and loads the processor by 2%. It should be noted that at the time of the tests, the operating system did not need updates, so they were not downloaded and installed.

As in the previous cases, virtual machines with the characteristics specified in Table 3 were deployed on the hypervisor. However, when they started, the system reported a lack of RAM to run one of the machines, perhaps the system used a certain memory reserve for itself.

That is why, despite the availability of 6.6 GB of RAM, it was not possible to run both virtual machines at once. To solve this problem, it was decided to reduce the amount of allocated memory to 2.5 GB for the virtual machine VM2 in Table 3; all the other characteristics remained unchanged. After that, both virtual machines were successfully started. The graphs of the use of hypervisor resources are shown below in Figs. 11 and 12.

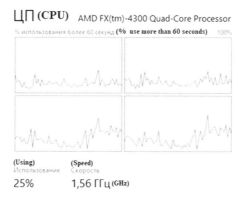

Fig. 11. The schedule for the Windows Server 2016 processor with the Hyper-V role (with virtual machines)

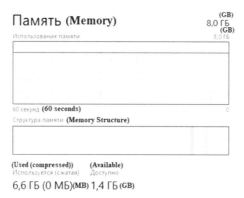

Fig. 12. The graph of the RAM use in Windows Server 2016 with the role of Hyper-V (with virtual machines)

As seen from the above figures, the entire system with virtual machines uses the processor by 25%, while the amount of memory consumed is 6.6 GB, and the available memory is 1.4 GB.

Hyper-V hypervisor management is carried out with the help of special software, which is installed on the server together with the hypervisor. It only works on the Windows operating system, but you can connect to the server with appropriate settings using the RDP protocol and thus manage the hypervisor. Therefore, any operating system with the ability to install software with support of RDP can be used to manage this hypervisor.

4   Conclusion

Let's review the results of the startup and operation of all the three hypervisors displayed in Table 4.

Table 4. Summary of the work of hypervisors

Hypervisor	Using resources without virtual machines	Using resources with virtual machines
VMware vSphere	CPU: 0% RAM: 1.39 GB	CPU: 14% RAM: 4,48 GB
Citrix XenServer	CPU: 2% RAM: 989 MB	CPU: 11% RAM: 7 GB
Windows Server 2016 with the role of Hyper-V	CPU: 2% RAM: 1.3 GB	CPU: 25% RAM: 6,6 GB

Hypervisors were tested with the default settings, and no additional settings were used; it was done to provide the ability to take into account the inexperience of the user who first encountered virtualization and, in particular, with a hypervisor. One of the virtual machines used the operating system of the Windows family, which is included in the list of operating systems supported by hypervisors, as well as the operating system of the Linux family, which was not included in the list of operating systems supported by hypervisors in order to see how it will work. All the three hypervisors coped with the launch of both operating systems; problems arose only with Hyper-V, which refused to run the virtual machine due to lack of RAM. However, here it is worth taking into consideration that the hypervisor is installed here as a role, i.e. on top of the operating system, so it has to share its resources with the OS. There is a separate version of Hyper-V that is installed directly on a physical computer.

Unlike the other two, the VMware vSphere hypervisor does not immediately take all of the RAM allocated to virtual machines, but uses it as needed, while the hypervisor itself uses a larger amount of RAM in comparison with others participating in the test. GI-PERSONSER supports a large number of guest operating systems. The free version has the necessary basic administration tools; to obtain advanced administration tools it is necessary to purchase a paid version.

To test the performance of client software supplied with hypervisors, we used a netbook with an Intel Atom N 455 processor, 2 GB of RAM with the Windows operating system installed. As noted above, starting from version 6.5, the VMware vSphere hypervisor is managed only through the web interface implemented on the platform of HTML 5. This web-interface did not work stable enough on the mentioned above netbook; there were long downloads and sometimes a drop in the interface itself. In general, its work can be assessed as satisfactory. This cannot be said about the application XenCenter, which is needed to manage the hypervisor XenServer; the application worked perfectly on the client device without falling and freezing. Microsoft Hyper-V is managed from the physical machine itself, but when using the RDP protocol to connect the client machine to the hypervisor, no special problems were also observed.

Summing up the results of the carried-out analysis it is possible to tell that the majority of hypervisors are really available to users, but there are some features discussed in this paper, which users need to consider at hypervisors' installation and configuration. Users who encountered virtualization for the first time can use the results obtained in this paper.

Acknowledgments. This work was carried out within the State Task of the Ministry of Education and Science of the Russian Federation (Project part No. 2.3928.2017/4.6) in Southern Federal University.

References

1. Graniszewski, W., Arciszewski, A.: Performance analysis of selected hypervisors (virtual machine monitors – VMMs). Int. J. Electron. Telecommun. **62**(3), 231–236 (2016)
2. Prakash, B., Bhatia, A., Bhattal, G.: A comparative study of various hypervisors performance. Int. J. Sci. Eng. Res. **7**(12), 65–71 (2016). https://www.ijser.org/researchpaper/A-comparative-study-of-Various-Hypervisors-Performance.pdf
3. Virtual machines. https://www.digitaltrends.com/computing/what-is-a-virtual-machine/
4. Polenov, M., Kostyuk, A., Muntyan, E., Guzik, V., Lukyanov, V.: Application of virtualization technology for implementing smart house control systems. In: Advances in Intelligent Systems and Computing, vol. 465, Software Engineering Perspectives and Application in Intelligent Systems, vol. 2, pp. 329–339. Springer (2016)
5. Polenov, M., Guzik, V., Lukyanov, V.: Using virtualization technology for the user authorization system. In: Advances in Intelligent Systems and Computing, vol. 575, Software Engineering Perspectives and Application in Intelligent Systems, vol. 3, pp. 192–200. Springer (2017)
6. Hyper-V. https://docs.microsoft.com/en-us/virtualization/index
7. Citrix XenServer. https://docs.citrix.com/content/dam/docs/en-us/xenserver/xenserver-7-0/downloads/xenserver-7-0-quick-start-guide.pdf
8. vSphere Hypervisor. https://www.vmware.com/products/vsphere-hypervisor.html
9. Kostyuk, A., Polenov, M., Lukyanov, V., et al.: Research of virtualization deployment possibility in smart house control systems. Inf. Commun. **3**, 72–77 (2015). (in Russian)

The Multicriteria Model Support to Decision in the Evaluation of Service Quality in Customer Service

Tatyana Belém de Oliveira Barreto, Plácido Rogério Pinheiro[✉],
and Carolina Ferreira Gomes Silva

Graduate Program in Applied Informatics, University of Fortaleza, Fortaleza, CE, Brazil
tatyanabelem@yahoo.com, placido@unifor.br, carolfgs@yahoo.com

Abstract. Understanding the needs of the customer is one of the key factors in achieving competitive advantage. For that reason, the expectation associated to the management of the relationship with the client are very challenging. In context, we observed the need to identify the factors that go in a successful customer service in order to monitor the quality of such service. The purpose of this paper is to identify and classify influencing elements of the customer's satisfaction with after sales service using a multi-criteria model of decision. We contextualize customer relationship marketing and then we use cognitive mapping to pinpoint the factors that are more relevant and that affect the client's overall satisfaction with the post sales service.

Keywords: Software release planning · Multi-objective optimization
Verbal decision analysis · ZAPROS III-i

1 Introduction

Companies must not focus solely on delivering a good final product but also on its customer service and the relationship with the client. In fact, it is very difficult to maintain positive levels of client satisfaction because of the client's notion of what is good customer service and his expectations of it. For that reason, it is fundamental for the companies to learn which aspects of post-sale service their clients consider. The companies in all its levels should prioritize the customer's needs and wishes. In order to do so they need to develop and manage a successful long-term relationship with their customers, aiming to build trust and loyalty. According to [20] one of the roles of marketing is to form bonds between the client and the company, by listening to the customers and learning their needs and fears.

Monitoring their clients' levels of satisfaction and understanding their experiences and interactions with the company are big challenges faced by the companies. Customer satisfaction surveys are meant to get feedback from the clients providing data that helps management and planning that brings improvement of both processes and final product. By increasing their customers' satisfaction, companies hope to obtain not only loyalty but also the increase of sales by word of mouth marketing.

Satisfaction results from expectation, desire and product\service performance and as so it can vary in time. Besides satisfaction, companies need to focus also on the

© Springer International Publishing AG, part of Springer Nature 2019
R. Silhavy (Ed.): CSOC 2018, AISC 763, pp. 158–167, 2019.
https://doi.org/10.1007/978-3-319-91186-1_17

customer's perception since perceived value stems from the equation of the benefits attained and sacrifices made in order to acquire the product or service. [4] define client satisfaction as the concept of product or service that pleases the consumer by meeting its quality expectations and emotional needs.

In 1990, Terry G Vavra named the post-sale interval as "aftermarketing". Post sales management along a good marketing strategy becomes a process capable of influencing the organizational development and bringing successful results and promoting changes.

Demands that arise after the finalizing of a sale can generate conflict if not well handled because of the customer's expectations. It is fundamental that the company monitors and learn from all the instances of client-company interaction, identifying the pivotal moments of perception that arise from these interactions. Each interaction has its own value and the quality of it can bring positive or negative results upon customer's satisfaction levels. According to [23], it is very important that the company handle all employee-client encounter, which he names contact points.

Any flaws in a contact point can have impact on client satisfaction and consequently on the company's image, for that reason it is important that all people connected to customer support be committed to it. Client satisfaction and product quality affect profits directly and are pivotal for its success and prosperity.

According to [10] service quality is harder to measure than product quality and thus expectation plays a deeply important role becoming a marker for performance evaluation. The author also stresses the importance of identifying factors that are considered by the client, listing them by order of importance and preference in a hierarchical scale.

Furthermore, by knowing their clients' needs and expectations, companies can identify which attributes are more important and put their efforts to the improvements of these. According to [12], consumers always consider the same basic criteria of service evaluation, no matter the nature of the service. The following are what he calls determinant factors in service quality: technical knowledge, reliability, readiness, accessibility, courtesy, communication, credibility, safety, tangible aspects and understanding of the customer's needs.

Moreover, [24] affirms that exploratory and quantitative surveys established that for bank service, insurance, maintenance, realtors, phone service, auto repair and retail service in general there are five most expressive factors: trustworthiness, promptness, safety, empathy and tangible aspects. Because of how crucial it is to identify and rank which attributes are valued by costumers becomes relevant to know the process by which the clients create their expectations in order to provide the means for coming up with new and more effective marketing strategies.

2 Multi-criteria Decision Analysis Method

Decision-making is intrinsic to humans, every day we make decisions both personal and work related. We live with decisions all the time, especially in the workplace, where these decisions can affect the future of the company and of thousands of people. [19] affirms that a to make a decision most of the time is to evaluate a set of alternatives in relation to a problem, taking into account some factors. Choosing the wrong alternative can lead to wasting time and money. An important tool to support decision-making are multi-criteria method because they are helpful when conflicts of preference involving many factors arise. Being frequently used in subjects related to strategically planning, service quality and industrial projects. According to [12] a multiple criteria analysis makes it easier to choose, classify and arrange actions, analyzing a complex or controversial decision problem by selecting alternatives of action and point of view, considering the existing criteria and the structure of the problem by grouping the agents' judgments.

Even though there are many different ways of classifying the multi-criteria method, one of the main is the one that divides it between an American and a French school.

From the American school we can note the AHP multi-criteria method (Saaty 1980), the UTA (Jaquet-Legreze e Siskos 1982) and MAUT (Keeney e Raiffa 1993). From the French school: ELECTRE (Roy 1968, 1978; e Roy e Shalka 1984) and PROMÉTHÉE (Brans et al. 1985), (Mello et al. 2003).

Additionally, some multi-criteria methods can't be placed into only one school because they contain mixed technical elements from both, such as: TODIM (Gomes, Araya e Cariggnano, 2004) and Método MACBETH (Bana e Costa, Corte e Vansnick, 2005).

According to [9] the MCDA methods differ in the way the idea of multiple criteria are employed. Each method shows its own characteristics regarding the way to evaluate the criteria, the application and computation of the weights, the mathematical algorithm used, the model to describe the system of preferences of the individual that faces the decision making, the level of uncertainty incorporated in the and the capacity of stakeholders to participate in the process.

Multi-criteria methods may be applied to many different decision-making contexts aiming to help the person choosing to better understand the situation through judgment and assessment. [1] affirms that the choice of method depends on the nature of the problem and the decision-making process.

[18] breaks the choosing process in three stages:

- Structuring – Contextualizing, identifying the problem, labeling, defining the primary evaluation elements (EPA), EPA's action orientation and argument by area, creation of relation-means\ends maps, clusters, fundamental point of view determination and describers;
- Evaluation – this stage contains the functions of value, substitution levels, potential actions impact profiles, additive aggregation formula and sensibility analysis;
- Prescription – This stage is based on the reading of the previous stages, allowing prescriptions and final considerations of the model.

The process of making decisions involves many people, called actors. There is the chooser, the analyst and the specialist. According to [15], the analyst provides methodological support and may act as facilitator. The specialist provides effective data about the problem being analyzed since the knowledge, context and behavioral artifice may influence the decision process. [2] stresses the importance of the chooser's involvement in the process of identification of the most significant aspects, so that the model may replicate its strategical objectives.

After establishing the actors involved on the decision process, it is necessary to define the potential alternative or action that should be analyzed during this process. According to [15] potential action is the object of the decision or that is directed to give support to the decision.

Once you know the actions that are going to be evaluated on the decision-making process it's important to study the reference issues, so you can later put together the multi-criteria model that is going to be used. [8] defines four kinds of issues:

- Description Issue ($P.\delta$) – the alternatives are described.
- Classification Issue ($P.\beta$) – The alternatives are ranked.
- Order Issue ($P.\gamma$) – The alternatives are put in order.
- Choice Issue ($P.\alpha$) – The best alternative is chosen.

There are many techniques that help structuring the problems such as: brainstorm, priority matrix, fish-spine diagram, decision tree, cognitive mapping.

Cognitive maps have a hierarchical structure in the form of means-ends and help to express the objectives that are deemed important by the actors making it easier to understand the problem. Cognitive maps may generate relevant interpretation of the problem through the identification of element connections related to the subject. [6] says that the use of cognitive maps allows managers to describe and explore the analyzes of the difficulties that the organization is facing, focusing on structuring the problem in order to find appropriate action options to solve these issues.

After established the scenery, one must define the objectives and actions to be reached, and identify the primary elements of evaluation, which represents the relation of the main attributes considered important by the chooser pertaining the objective. The primary elements of evaluation the characteristics of the actions deemed more important to the decision are defined, these are called fundamental points of view. [9] affirms that these fundamental point of view (PVF) specify the values that the chooser considers necessary and sufficient to evaluate the potential actions and its unfolding and they are measured by ordinal scales called describers. These scales measure the possible levels of impact that a certain action has. The definition of the fundamental point of view, its respective describers, reference levels and relative weight are the minimum data necessary for the finalizing of the structuring stage.

3   Methodology

The paper relies on conceptual and theoretical approaches, resorting on academic papers developed by several researchers, with the purpose of providing a set of knowledge that will serve as ground research for the proposed theme.

The methodology incorporated in this study seeks to solve the questions and confirm the elaborated targets. Considering this fact, the paper relies on qualitative research. Qualitative research is the understanding of the meaning that people attach to things in their lives. It is a way of approaching the empirical world [21].

Considering its purpose, this research characterizes itself as being an exploratory research. Exploratory research aims to familiarize oneself with a phenomenon or obtain new knowledge [7].

The proposed study is characterized by a research of an applied character. The applied researches aim to find a solution to some practical problem [7]. Aiming to find a solution to an immediate problem facing a society or an industrial/business organization.

It is noteworthy that, due to the exploratory character of the paper, the results and conclusions are not definitive, carrying only an initial perspective about the theme. Additional conclusive research can be conducted.

In order to structure the problem, interviews with experts and decision makers of the process of customer relationship management of medium-sized developers were conducted and to map the solutions to the problem in question. Initially, the topic was contextualized, and the main objectives of the research were made clear, and then the following question was asked: **what criteria are determinant in customer satisfaction with after-sales services of developers?**

4   Multicriteria Paradigm for Identification of Improvement Opportunities for Customer Service

Here we will detail the steps of constructing the paradigm for a real situation. The structuring phase begins with the identification of the actors involved, who in this work were: the author of the paradigm - in the role of facilitator; managers of incorporators from Fortaleza - as a decision maker; and clients as agents who, even if they do not get involved in the decision-making process, will suffer consequences from them.

The label defined to portray the interests of the decision maker in regard to the problematic was: to identify which criteria are determinant in the clients' satisfaction with the after-sales service of developers.

In an interview with the decision-maker, the Primary Elements of Evaluation (PEA), which he deems relevant in the customer service department, are identified: promptness, confidence, empathy, safety and physical structure. As a structuring tool for the construction of the value chain, the method of cognitive mapping was used. The Decision Explorer software was used to relate the information provided by the decision maker, according to Fig. 1.

After the construction of the cognitive map, it was possible to identify the Fundamental Points of View (FPV) that were organized in the tree structure shown in Fig. 2. Based on these points of view, the construction of the descriptors gets started. This is a very important step for structuring the problem, as it will list which actions lead to a better range of the objectives and serves as an evaluation of the impact of each action on each objective.

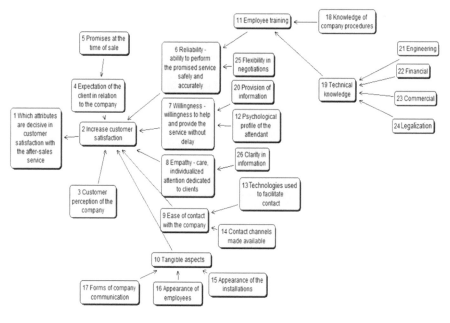

Fig. 1. Cognitive mapping of the problem.

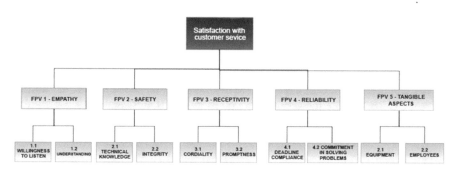

Fig. 2. Tree structure of the problem

Tangible Aspects

This fundamental point of view was operationalized through a qualitative, constructed and discreet descriptor, with the purpose of evaluating the physical evidences, such as facilities, appearance of the personnel and equipment used in the service. For this point of view, four levels with different impacts in the business were defined. These levels were defined according to the actors' experience in the process. Table 1 shows the key-worry descriptors.

Table 1. Descriptors for FPV1 – Tangible Aspects

LI	Description	Order
L4	Appearance of the employees	1°
L3	Convenience of service hours	2°
L2	Appearance of physical facilities	3°
L1	Modern equipment appearance	4°

Reliability

This fundamental point of view was operationalized through a qualitative, constructed and discreet descriptor, with the objective of evaluating the company's commitment to meeting the clients' interests, as well as meeting the agreed commitments. For this point of view, four levels with different impacts in the business were defined. Table 2 shows the key-worry descriptors.

Table 2. Descriptors for FPV2 – Reliability

LI	Description	Order
L4	Compliance with deadlines provided to customers	1°
L3	Keep customers informed of when services will run	2°
L2	Service efficiency	3°
L1	Concern about solving problems	4°

Receptivity

This fundamental point of view was operationalized through a qualitative, constructed and discreet descriptor, with the objective of evaluating manners, respect, promptness and speed in performing the service, as well as the skills and knowledge necessary to perform the services. For this point of view, four levels with different impacts in the business were defined. Table 3 shows the key-worry descriptors.

Table 3. Descriptors for FPV3 – Receptivity

LI	Description	Order
L3	Agility in resolving the customer's request	1°
L2	Availability of the attendant in serving the client	2°
L1	Convenience in making contact with the company	3°

Safety

This fundamental point of view was operationalized through a qualitative, constructed and discreet descriptor, with the objective of evaluating physical and financial safety and confidentiality. For this point of view, four levels with different impacts in the business were defined. Table 4 shows the key-worry descriptors.

Table 4. Descriptors for FPV 4 – Safety

LI	Description	Order
L4	Technical knowledge of the attendants	1°
L3	Politeness at the time of service	2°
L2	Justification of information provided to the customer	3°
L1	Transparency at the time of service	4°

Empathy

This fundamental point of view was operationalized through a qualitative, constructed and discreet descriptor, with the objective of evaluating the effort to understand the needs of the clients, providing individual attention, ease of contact, and concern to keep the clients informed in clear language. For this point of view, six levels with different impacts in the business were defined. Table 5 shows the key-worry descriptors.

Table 5. Descriptors for FPV 5 – Empathy

LI	Description	Order
L4	Personalized customer service	1°
L3	Interest in knowing and understanding the customers' needs	2°
L2	Honest interest in listening to the customer	3°
L1	Service communication	6°

5 Final Considerations

The expectations related to a good customer relationship management are very challenging. Given that the offered products have similar prices and characteristics, what can differentiate one company from another is the service provided to customers. Companies are increasingly in need of continuous monitoring of their customers. The differential of companies will be how they use the opportunities in this context to maximize their results.

This work focused on the importance of customer satisfaction and the main aspects related to the management of customer service. The objectives of this research came from the perception of the company managers about the importance of customer satisfaction for organizational growth. We sought to identify the main determinants of customer satisfaction with the care service through the use of cognitive mapping in the structuring phase of the problems that influence the decision process.

Although restricted to quantitative studies directed to customer satisfaction using the multicriteria methodology, academic papers and publications on the subject were used as reference. The limitations found derive from the emotional conditions of the actors, both clients and managers/employees, making it difficult to create a measurement scale using subjective variables and complicated standardization.

As a proposal for the continuity of this work, we suggest applying the fundamental points of view listed in this company as a way to validate the contribution of this work in management of crisis on decision making process.

Acknowledgements. The second author thanks the National Council for Technological and Scientific Development (CNPq) through grants # 304747/2014-9. The authors thank the Edson Queiroz Foundation/University of Fortaleza and FINEP (Funding Agency for Studies and Projects) for all the support provided.

References

1. de Almeida, A.T.: Multicriteria decision model for outsourcing contracts selection based on utility function and ELECTRE method. Comput. Oper. Res. **34**, 3569–3574 (2007)
2. De Azevedo, R.C., de Oliveira Lacerda, R.T., Ensslin, L., Jungles, A.E., Ensslin, S.R.: Performance measurement to aid decision making in the budgeting process for apartment building construction: a case study using MCDA-C. J. Constr. Eng. Manag. **139**, 225–235 (2013)
3. e Costa, C.A.B., Corrêa, É.C., De Corte, J.M., Vansnick, J.C.: Facilitating bid evaluation in public call for tenders: a social-technical approach. Omega **30**, 227–242 (2002)
4. Boone, L.E., Hurtz, D.L.: Contemporary Marketing, 16th edn. South-Western Cengage Learning, Mason (2012)
5. Brasil, A.T.F.: A Novel approach based on multiple criteria decision aiding methods to cope with classification problems. Masters dissertation in applied informatics, University of Fortaleza, Brasil (2009)
6. Eden, C.: Analyzing cognitive maps to help structure issues or problems. Eur. J. Oper. Res. **159**, 673–686 (2004)
7. Kothari, C.R.: Research Methodology – Methods and Tecniques, 2nd edn. New Age International Publishers, New Delhi (2004)
8. Lacerda, R.T., Ensslin, L., Ensslin, S.R.: A performance measurement framework in portifolio management. A constructivist case. Manag. Decis. **49**, 648–668 (2011)
9. De Montis, A., De Toro, P., Droste-Franke, B., Omann, I., Stagl, S.: Criteria for quality assessment of MCDA methods. In: 3rd Biennial Conference of the European Society for Ecologica Economics, Viena, 3–6 May 2000
10. Oliver, R.L.: Satisfaction: A Behavioral Perspective on the Consumer, 2nd edn. Routledge, Nova York (2010)
11. Parasuraman, A., Zeithaml, V.A., Berry, L.L.: A conceptual model of service quality and its implications for future research. J. Market. **49**, 41–50 (1985)
12. Pinheiro, P.R., de Souza, G.G.C.: A multicriteria model for production of a newspaper. In: The 17th International Conference on Multiple Criteria Decision Analysis, vol. 315, p. 325 (2004)
13. Roy, B.: Multicriteria Methodology for Decision Aiding. Springer-Science + Business Media, Dordrecht (1996)
14. Roy, B.: Paradigms and challenges. In: Figueira, J., Greco, S., Ehrgott, M. (eds.) Multiple Criteria Decision Analysis: State of the Art Surveys Series: International Series in Operations Research & Management Science, vol. 78, XXXVI, pp. 3–24 (2005)
15. Silva, C.F.G., Nery, A., Pinheiro, P.R.: Multi-criteria model in information technology infrastructure problems. Proc. Comput. Sci. **91**, 642–651 (2016)

16. Silva, C.F.G., Pinheiro, P.R., Barreira, O.: Multicriteria problem structuring for the prioritization of information technology infrastructure problems. In: Silhavy, R., Silhavy, P., Prokopova, Z. (eds.) Applied Computational Intelligence and Mathematical Methods, 1st edn., vol. 662, pp. 326–337. Springer (2017)
17. Silveira, C.F., Ensslin, L., Petri, S.M.: Performance appaisal and relationship marketing: a case study in na industry. Bus. Manag. Rev. **05**, 176–186 (2015)
18. Simao Filho, M., Pinheiro, P.R., Albuquerque, A.B.: Verbal decision analysis applied to the prioritization of influencing factors in distributed software development. In: Rocha, Á., Reis, L.P. (eds.) Developments and Advances in Intelligent Systems and Applications, 1st edn., vol. 718, pp. 49–66. Springer (2017)
19. Simão, M.F., Pinheiro, P.R., Albuquerque, A.B.: Applying verbal decision analysis to task allocation in distributed software development. In: 28th International Conference on Software Engineering & Knowledge Engineering, San Francisco, pp. 402–407 (2016)
20. Solomom, M., Bamossy, G., Askegaard, S., Hogg, M.K.: Consumer Behavior – A European Perspective, 3rd edn. Prentice-Hall Inc, New Jersey (2006)
21. Taylor, S.J., Bogdan, R., Devault, M.: Introduction to Qualitative Research Methods: A Guidebook and Resource, 4th edn. Wiley, New Jersey (2015)
22. Vasconcelos, M.F., Pinheiro, P.R., Simao Filho, M.: A multicriteria model applied to the choice of a competitive strategy for the printed newspaper. In: Silhavy, R., Silhavy, P., Prokopova, Z. (eds.) Cybernetics Approaches in Intelligent Systems, 1st edn., vol. 1, pp. 206–215. Springer (2017)
23. Vavra, T.G.: Improving your Measurement of Customer Satisfaction: A Guide to Creating, Conducting, Analyzing and Reporting Customer Satisfaction Measurement Programs. ASQ Quality Press, Milwaukee (1997)
24. Zeithaml, V.A., Bitner, M.J., Gremler, D.D.: Services Marketing Integrating Customer Focus Across the Firm, 7th edn. McGraw Hill Education, New York (2013)

Intelligent Method of Reconfiguring the Mechanical Transport System

Stanislav Belyakov and Marina Savelyeva[✉]

Southern Federal University, Taganrog, Russia
beliacov@yandex.ru, marina.n.savelyeva@gmail.com

Abstract. This paper presents the problem of controlling the transportation of cargo in the mechanical transport system. The overall efficiency of the system includes the costs of reconfiguration, involving both software and manual changes in the parameters and relationships of equipment components. The decision-making phase about choosing the reconfiguration method is preceded by the analysis of the utility of possible methods. The complexity of the solution of the considered problem consists in the ambiguous estimation of the network state due to the considerable number of parameters and incompleteness of information about their values. Except the status, the reconfiguration effect depends on the dynamics of the input flows and the effect of the external environment on the network after reconfiguration. The way of the solution of the problem, based on the image representation of the reconfiguration experience, is considered. The model of the image representation of knowledge and reasoning on their basis is described. The advantage of model of representation of precedents of reconfiguring by images is analyzed. The example of the image representation of situations reflecting the set of knowledge about the input flow, the degree of congestion of the subnet, and the forecast of the behavior of the reconfigured network is given. The boundaries of application of the proposed method are analyzed.

Keywords: Mechanical transport system · Reconfiguration · Intelligent system
Image analysis of precedents

1 Introduction

Transportation of certain types of cargo carried out by systems in which the role of the vehicle is played a set of conveyors. The transported object is placed on the conveyor and supplied with a label containing information about the destination point and transport quality parameters. The conveyors are interconnected by switches for the direction of movement of the cargo. The cargo movement from the point of departure to the destination consists in passing a sequence of conveyors and direction switches that provide the required quality of service. These devices form a mechanical transport system (MTS). Examples of such systems are MTS of baggage handling at airports.

MTS is represented by quite complex objects. The number of their components can reach several hundred. This generates a multivariate transport: each unit of cargo can reach the destination, redirecting to different conveyors. The choice of a particular option depends on the state of the system.

© Springer International Publishing AG, part of Springer Nature 2019
R. Silhavy (Ed.): CSOC 2018, AISC 763, pp. 168–177, 2019.
https://doi.org/10.1007/978-3-319-91186-1_18

The required quality of transportation in the MTS is provided by structural or parametric adaptation to external influences on the system. The means of parametric adaptation is the change in the behavior of MTS components. For instance, the variation the routing tables in the direction switches changes the flows in the network and, as a result, the cost and time of transportation. Parametric adaptation is implemented programmatically by controllers of the control network in the MTS. Structural adaptation is based on change of connections and equipment elements. Example is the connection of a new branch of conveyors or additional engines on separate conveyors, which changes the capacity of the MTS. Structural adaptation (reconfiguration) is implemented by switching equipment of MTS. Obviously, structural adaptation requires more time and resources, but in case of accidents and failures, its use has no alternative.

In the optimization of transportation costs through the MTS, an important role is played by the choice of the reconfiguration option in the pre-emergency states of any subnet. Deterministic strategies of software or hardware reconfiguration, as practice shows, as the complexity of the MTS is less effective. The reason is the ambiguity and uncertainty of the factors influencing the external environment on the state and dynamics of the MTS. The possible way to solve this problem can be to use the reconfiguration experience that experts have. In this paper, the method is presented that allows choosing the most effective variant of reconfiguration by using knowledge. Knowledge is represented by specific conceptual model of the image of the reconfiguration situation.

2   Known Reconfiguration Control Methods

Reconfiguration of the MTS is part of the general task of minimizing the costs of transfer a consignment, which is set as follows:

$$\begin{cases} E_T(t) + E_R(f) \rightarrow min \\ \quad\quad t < t^* \\ \quad\quad f \subseteq F \end{cases} \tag{1}$$

where $E_T(t)$ is the average cost of transportation of the consignment, which is determined by the transport time t, t^* is restriction on the transportation time. Function $E_R(f)$ is average damage from emergency situations that require reconfiguration, F is the set of known defects leading to accidents, f is a subset of the defects that occurred during the transportation of the consignment. The task (1) reflects the feature of transportation with the help of the MTS: the time limit defines a certain mode of operation of the system, in which it is translated by structural or parametric reconfiguration. Operation in any selected mode is accompanied by the appearance of defects, due to which the reconfiguration of the equipment should be performed. The resulting damage can be reduced by identifying a pre-emergency condition and choosing the reconfiguration option that will allow to continue effectively work under the new conditions.

The analysis of network capacity [1] allows to predict the possibility of the accident in case that the intensity of the input flow to the maximum network capacity. It is difficult to use this in practice because the real capacity of separate subnets changes in time. It arises because of power decrease of the supply voltage network, exhaustion of a resource of the executive mechanisms, changes in the geometry of the conveyors, the dimensions and mass of the transported cargo, etc. Therefore, a formal estimate of the balance of the intensity of the input flow and the capacity is obviously not enough for eliminate the threat of MTS overload.

Theoretical researches of reconfiguration methods are carried out for a long time [2]. The subject of research is reconfiguration mechanisms and decision-making strategies for choosing the best reconfiguration option. The important role in the latter case is acquired by researches of new high-level conceptual models of knowledge representation about reconfiguration [2–5].

The use of intelligent reconfiguration control methods is of particular interest due to the difficulty of detecting a pre-emergency situation [6]. It is possible to identify a pre-emergency situation only by the critical values of the parameters determined in practice [7]. In the presence of representative training samples, the application for classifying situations of neural networks is effective [4]. However, in the task of reconfiguring MTS this approach does not bring the expected result because of the great role of the logical conclusions of the expert.

The need to represent complex logically organized knowledge leads to the study of models of image thinking of the expert [8]. In [9, 10], the method of image description of knowledge and the principle of logical inference was proposed. As the analysis shows, the application of the proposed method requires the analysis of the features of the MTS.

Traditionally, the main method of parametric adaptation of MTS is routing. The change in the generalized cost of transportation on separate segments of the networks makes it possible to change their capacity. However, this mechanism does not allow to identify pre-emergency states that are possible at the observed values of the intensity of the input flows.

The use of a protective reduction of intensity of the flow [11] can be less effective than reconfiguring MTS equipment.

3   Intelligent Reconfiguration Control Method

The basis of the proposed method is the following principles.

The practical precedent of reconfiguration is analyzed by the expert, which leads to the appearance in his mind of the image. The image integrally connects the state of the MTS and the external environment with the reconfiguration procedures used in practice and estimates of the effectiveness of the system after reconfiguration. The existence of such mental images in the mind of the expert is beyond doubt [12]. The distinctive feature of the image is that it reflects not a single situation, but many similar situations in meaning to the observed situation.

The knowledge base of an intelligent reconfiguration system (IRS) is filled with image descriptions. The logic of the further use of images is based on two operations:

comparisons of couple of images and display the image to a given area. The first operation allows to evaluate the semantic closeness of the two images. The second allows to simulate the known precedent in randomly given conditions for the operation of the MTS. The result of the application of any operation is the conclusion about the semantic equivalence of the images.

In the process of functioning of the MTS, the monitoring means fix critical values of the system state parameters. If there is an abnormal situation, the IRS allocates the appropriate subnet and evaluates the perspective of re-applying previously implemented reconfiguration options. It is possible in two ways:

– dialog method, assuming the construction of the user's image of the observed situation and comparing it with known IRS. The decision to apply this or that variant of reconfiguration is made in the case of the semantic closeness of the images;
– automatic method based on the transfer of known reconfiguration options to the dedicated subnet and evaluation of the semantic integrity of the result. If the semantic integrity is saved, then the decision is made to use the appropriate reconfiguration option.

If the reconfiguration option is not found, the IRS fixes the lack of knowledge for the observed situation and the need for machine learning by adding new images.

4   Model of the Image

The image of the situation observed and analyzed by the expert is described as

$$J = <c, H(c)>, H(c) \subseteq c^2$$

where c is the center of the image corresponding to the precedent with the real values of the observed parameters of the situation. For MTS such parameters are:

– MTS subnet $n \subset N$, in which there was the precedent for reconfiguration;
– the number of cargo units m in the subnet n;
– the intensity of incoming (v_{in}) and outgoing (v_{out}) flows for subnet n;
– linguistic assessment of the speed (V) of the arrival of cargo at the input of the network N;
– the used variant of reconfiguration $r \in R$ from the set of admissible variants R;
– linguistic evaluation of the work g network N after reconfiguration.

The second component of the image $H(c)$ is a set of admissible transformations of the center of the situation, which does not change its meaning. The introduction of admissible transformations increases the quality of expert knowledge. Preservation of the meaning of situations when comparing them is the only way to provide the reliability of decisions. $H(c)$ contains the analytical experience of the expert, reflecting his in-depth knowledge of the reconfiguration of the MTS. Let us emphasize that the transformations are not equivalent to the accuracy of the description of the situation, since they represent a lot of the same precedents in the sense, to which the solution proved by practice is applicable.

The accuracy of describing the boundaries of admissible transformations is an independent element of the image, therefore, a more general one should be considered a fuzzy model of the image

$$\tilde{J} = \; <\mu_H <c, H(c)\gg,$$
$$c \in H(c), H(c) \subseteq c^2, \mu_H: H(c) \rightarrow [0, 1].$$

Here μ_H is the degree of the expert's confidence that the admissible transformation $H(c)$ reflects the meaning of the situation.

As an example, consider the clearly described image of the reconfiguration of MTS after detecting the pre-emergency situation in one of the subnets. Figure 1 shows the subnet schema, the parameters of the precedent that occurred have the values $m = 100, v_{in} = 4, v_{out} = 4, V =$ "Will be the same". The reconfiguration option r is shown in Fig. 2 by dashed lines of the connected conveyors $g =$ "improved". The reconfigured switch is indicated by the symbol R. The example of the admissible subnet conversion is shown in Fig. 3. Dashed lines denote conveyors and switches, the presence of which does not fundamentally change the situation.

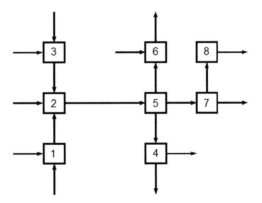

Fig. 1. The subnet schema that requires reconfiguration

Note that the operation of comparison of images aims to evaluate their proximity not as it is accepted in the case-based reasoning (CBR) [13]. Instead of the proximity metric, it is proposed to use the topology of the mutual arrangement of centers and areas of transformations. Variants of the mutual arrangement are divided into classes that reflect the preferences in decision making. For example, in Fig. 4 shows the class of "closest images", where pairs of images can be assigned that have common fragments of domains of admissible transformations and at least one precedent in this area. The decision function for the class has the form

$$F = c_1 \in H(c_1) \; \cap H(c_2) \lor c_2 \in H(c_1) \; \cap H(c_2)$$

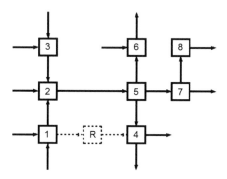

Fig. 2. The schema of the reconfigured subnet

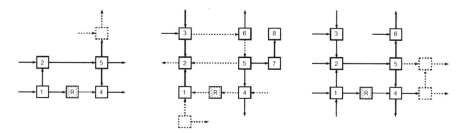

Fig. 3. Admissible subnet transformations

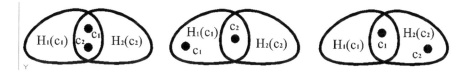

Fig. 4. The class of "closest" images

Substantially the indicated mutual arrangement means that the images have a common sense, and the meaning is confirmed by practice. The number of classes can be increased if more complex decision-making procedures are applied.

5   Transformation of the Images

Transferring the reconfiguration experience to the situation for which the similar image is not found is to reproduce the known image in a given area of image space.

We call the operation of displaying the image into a given area by transformation. Formally, the transformation is described as

$$<\bar{c}, H(\bar{c})> \ = F_{TR}(<c, H(c)>)$$

where F_{TR} is the transformation operator, $\bar{c}$ is the center of the transformed image, $H(\bar{c})$ is its admissible transformations. The condition for the existence of the transformed image is

$$h_i(\bar{c}) \neq \emptyset, i = \overline{1, |H(\bar{c})|}$$

The possibility of performing the transformation operation is determined by the topology of the image space, i.e. ratios and rules for preserving the continuity of the displaying in image space. The rules specify the transformation invariant. The topology of the image space (A) is formally represented as

$$A = \bigcup_i A_i, A_i \cap A_j \neq \emptyset,$$
$$\forall c \in A: H(c) \subseteq A_i$$

From the informative point of view, the new (transformed) image J_{TR} must preserve the intuitively understood by the expert the meaning put in the initial image in the form of transformations. The center of the transformed image $\bar{c}$ represents the real situation, the previously constructed solution of which can be reused.

The example of the application of the transformation may be the following. In some subnet of the MTS, the task of reconfiguration was previously solved in the conditions of the loaded schedule of arrival of consignments. The arisen overload of one of the conveyors in accordance with the recommendations for operation required its complete stop for the time of reconfiguration. In the situation that arose, this could lead to serious losses due to the blocking of the adjacent conveyor. To reduce losses, it was decided to use a mobile conveyor. Part of the cargo flow was redirected, which allowed to perform repairs without stopping the conveyor completely. Connecting the mobile conveyor to the MTS was made possible by the access road to the failed conveyor. It is not always possible to apply the similar solution, because it makes sense in conditions of intensive input cargo traffic with the high cost of potential damage and mandatory availability of the access road for the mobile conveyor. Trying to apply this solution to another subnet, it is necessary to be convinced of saving an essence of the situation described above. This possibility is given by the transformation operation.

6 The Operator of Transformation for MTS

It is impossible to define the operator F_{TR} universally because of the difference in the topologies of image spaces. For each component of the center of the image of the reconfiguration situation of the MTS $c = \ <n, m, v_{in}, v_{out}, T, r, g>$ parameters defining the expert's representation of the component and the invariants of the transformation of the parameter values must be set. As the example, consider some of them.

The concept of the subnet component $n \subset N$, as analysis shows, is characterized by its spatial location. The configuration is heavily influenced by the relative location of such objects as adjustable blocks and elements of MTS, ports for receiving and delivering cargo to external systems, power supply sources, technical driveways and passages, emergency protection systems. The spatial relations of proximity, contiguity and inclusion should be considered as an invariant of the transformation [14]. For example, the experience of reconfiguring the subnet located close to the reception of passenger luggage cannot be transferred to any other subnets in the depths of the airport buildings.

The idea of the component of the number of cargo units is formed by its boundary values, the schedule of arrival and sending of transports with cargo, type of cargo and destination points. The transformation invariant should be located in the range of the boundary values characteristic for the current schedule, as well as the observance of the list of cargo types and destinations. For example, the experience of reconfiguration under conditions of the significantly busy network and the rigid time schedule for the arrival and departure of transport does not make sense to transfer to the weakly loaded network with the same schedule and the same nomenclature of cargo.

The description of the invariants for the transfer of each component can be considered as a set of transformation rules. The rule takes a true value if the transformed component preserves the invariant values of the parameters.

7 Discussion

The advantage of the proposed method is manifested in increasing the reliability of the decisions made about configuring the MTS. Comparing the transformation operation with the estimation of the distances between precedents in case based reasoning [13], we should note the following:

- CBR is based on the assumption that close (by the accepted metric) situation allows close solutions. This is plausible, but it assumes that the metric considers the semantic equivalence of situations. The analysis shows [13] that metrics do not consider the meaning invested by experts in the description of precedents. Accordingly, there is a high probability that the metrically close precedent will be meaningless in practice;
- the image includes the precedent as the element of the area that contains not a single set of reliable situations. The proximity of the images is assessed not by the proximity of the relevant precedents, but by the presence of the subset of the same reliably possible situations. For this reason, as a result of the comparison of images, the probability of choosing the plausible and reliable solution increases.

8   Conclusion

Intelligent control of reconfiguration of MTS gives the greatest effect in complex networks. Necessary condition is the existence of experience in operating such networks. In this case, the decision to configure is based on the transfer of experience.

The reliability of the made decisions is determined by the topology of the image space. Substantially and indisputably given topology guarantees preservation of the sense of the transferred experience and its experimental confirmation. Accordingly, the reuse of solutions has the reliability that does not exceed the reliability of the description of the image space topology.

The effectiveness of any intelligent system depends on the quality of the knowledge put in it, and knowledge accumulates as the experience of reconfiguration increases. Therefore, further research is advisable to continue in the direction of studying the mechanisms of image representation of knowledge about the reconfiguration of the MTS.

Acknowledgment.   This work has been supported by the Council for Grants (under RF President) and State Aid of Leading Scientific Schools (grant MK-521.2017.8).

References

1. Cormen, T.H., Leiserson, C.E., Rivest, R.L., Stein, C.: Introduction to Algorithms, 3rd edn. 1312 p. MIT Press, Cambridge (2009)
2. Ma, Y., Liu, F., Zhou, X., Gao, Z.: Overview on algorithms of distribution network reconfiguration. In: 2017 36th Chinese Control Conference (CCC) (2017)
3. Shariatzadeh, F., Kumar, N., Srivastava, A.K.: Optimal control algorithms for reconfiguration of shipboard microgrid distribution system using intelligent techniques. IEEE Trans. Ind. Appl. **53**(1), 474–482 (2017)
4. Srivastava, I., Bhat, S.S.: Soft computing techniques applied to distribution network reconfiguration: a survey of the state-of-the-art. In: 2016 8th International Conference on Computational Intelligence and Communication Networks (CICN) (2016)
5. Brennan, R.W., Vrba, P., Tichy, P., Zoitl, A., Sünder, C., Strasser, T., Marik, V.: Developments in dynamic and intelligent reconfiguration of industrial automation. Comput. Ind. **59**(6), 533–547 (2008)
6. Özdamar, L., Ekinci, E., Küçükyazici, B.: Emergency logistics planning in natural disasters. Ann. Oper. Res. **129**(1–4), 217–245 (2004)
7. Azab, A., ElMaraghy, H., Nyhuis, P., Pachow-Frauenhofer, J., Schmidt, M.: Mechanics of change: a framework to reconfigure manufacturing systems. CIRP J. Manuf. Sci. Technol. **6**, 110–119 (2013)
8. Kuznetsov, O.P.: Kognitivnaya semantika i iskusstvennyy intellekt. Iskusstvennyy intellekt i prinyatie resheniy **4**, 32–42 (2012)
9. Belyakov, S., Bozhenyuk, A., Rozenberg, I.: The intuitive cartographic representation in decision-making. In: World Scientific Proceeding Series on Computer Engineering and Information Science, vol. 10, pp. 13–18 (2016)

10. Belyakov, S., Belyakova, M., Savelyeva, M., Rozenberg, I.: The synthesis of reliable solutions of the logistics problems using geographic information systems. In: 10th International Conference on Application of Information and Communication Technologies (AICT), pp. 371–375. IEEE Press, New York (2016)
11. Protective Correction of the Flow in Mechanical Transport System наша публикация
12. Kaplan, R., Schuck, N.W., Doeller, C.F.: The role of mental maps in decision-making trends. Neuroscience **40**(5), 256–259 (2017)
13. Lenz, M., Bartsch-Spörl, B., Burkhard, H.-D.: Case-Based Reasoning Technology: From Foundations to Applications. Springer, Heidelberg (2003)
14. Longley, P.A., Goodchild, M., Maguire, D.J., Rhind, D.W.: Geographic Information Systems and Sciences, 3rd edn. Wiley, New York (2011)

Efficient Load Balancing and Multicasting for Uncertain-Source SDN: Real-Time Link-Cost Monitoring

Thabo Semong[✉] and Kun Xie

College of Computer Science and Electronics Engineering, Hunan University,
Changsha, Hunan 410082, China
semongt@biust.ac.bw, xiekun@hnu.edu.cn

Abstract. Software-Defined Networking (SDN) is the next generation network architecture with a new approach to design, build and manage computer networks. Multicasting is a way of transmitting the same content to different destinations. In this paper, we propose an efficient load balancing and multicasting scheme for uncertain-source SDN, using current link cost monitoring. We consider the balance of bandwidth utilization, tree building, and the process of uncertain-source selection. We name this scheme RTLMUS (Real-Time Link Monitoring Uncertain-Source). We then ensure that each destination node connect to just one potential source node. Simulation results demonstrate that our algorithm can improve the network bandwidth utilization and successfully select the multicast source node from the set of potential uncertain-sources.

Keywords: Multicast routing · Software-Defined Networking
Uncertain-source · Load-balancing

1 Introduction

Software-Defined Network (SDN) is an innovative network architecture that was proposed by the researchers of Stanford University, and was highly welcomed in recent years. The main idea of SDN is to separate the control and forwarding planes of the network system, so as to ensure that the network administrators can program packet forwarding behavior to improve the innovation capabilities of network applications [1–5]. In a multicast transfer it is not always necessary that the source of the multicast transfer has to be in a specific location as long as certain constraints are satisfied. A major reason is the widely used content replica design for improving the robustness and efficiency in various networks, such as the content distribution network [6,7], IPTV networks and datacenter networks. When delivering a content file to multiple destinations, the source of such a multicast transfer can be any of the node with the replica.

Authors in [8], developed a Maximum Likelihood Estimator (MLE) for loss rates on internal links based on losses observed by multicast nodes. MLE

© Springer International Publishing AG, part of Springer Nature 2019
R. Silhavy (Ed.): CSOC 2018, AISC 763, pp. 178–187, 2019.
https://doi.org/10.1007/978-3-319-91186-1_19

exploited the inherent correlation between such observations to infer the performance of paths between branch points in the tree spanning a multicast source and its destinations. They derived its rate of convergence as the number of measurements increases, and establish robustness with respect to certain generalizations of the underlying model. Xie et al. [9], proposed a reliable multicast routing with uncertain sources, named ReMUS. Their goal was to minimize the sum of the transfer cost and the recovery cost, while at the same time finding such a ReMUS. They designed a source based multicast method to solve the problem by exploiting the flexibility of uncertain sources, when no recovery node exists in the network. Researchers in [10], focused on the uncertain multicast and constructed a forest with the minimum cost (MCF), so as to enable each destination to reach to one and only one source. Compared with the above mentioned works in the literature, our proposal tries to utilize the real-time link cost monitoring to balance the bandwidth utilization in an SDN. Researchers in [11] proposed a load balance for multicast traffic based in SDN. The paper used the feature of SDN to monitor real link state on-time and assign the weight of the link based on current utilization of the link. They showed that modification of link-cost based on current link utilization is an efficient way of approaching load balance for multicast traffic. Our approach in this paper is similar to theirs, we have used the features of SDN for monitoring network statistics in order to propose load balance approach for uncertain source multicast traffic. In our approach, we have introduced the concept of *"uncertain sources"*.

The equivalence multipath routing technology (ECMP) based on a hash algorithm is an effective load balancing solution. Multiple forwarding paths may exist within the targeted network in ECMP. The switching or routing devises in the network gets the header fields of the packets as they ingress to make hash calculations, and then use that hash value to determine the path to forward through. This simply mean that IP packets having the same header field are being forwarded along the same path. This results in what is regarded as elephant flows, meaning that bandwidth waste and load imbalance are being created by these large flows being forwarded to the same path [12]. As illustrated by Fig. 1, this ultimately result on some paths being congested while others remains unused. This disadvantage of ECMP results on an unutilized network resources.

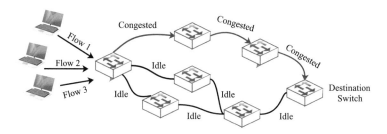

Fig. 1. Load imbalance for ECMP

Combining the processes of uncertain potential source node selection and that of the tree building based on real-time link cost adjustment is novel, and difficult to solve as these processes are closely related and impact on each other. The logically centralized advantage of the SDN control plane enable the real-time adjustment of link costs used for routing calculations to be more effective for implementing traffic engineering (TE). The properties of the SDN paradigm are ideally positioning it to overcome many limitations of real-time link cost adjustment in traditional network deployments. Protocols like Multicast Open Shotest Path First (MOSPF) [13] and PIM-SM [14], are used to implement multicast routing in traditional network deployment. The current Internet multicast standard like PIM-SM, employs a shortest path tree to connect the source-destination pairs, and TE is difficult for PIM-SM since the path from the source to each destination is the shortest one.

The convergence delays caused by the forwarding of link state updates means that routers will most of the time have differing views of the real-time network traffic state, and hence the routers are not well positioned to independently implement TE goals by modifying link costs. Most of networks which implement TE through link weight adjustment use an approach similar to the one used by Fortz et al. [15]. In this approach an external network manager with a global view of the network state is used to calculate changes to link costs. This external network manager can use different techniques to measure the network traffic state, including matrixes based on packet or flow level measurements at the network edge, polling of management information bases in routers, network tomography, or packet sampling. Adjustments to link costs by this network manager are considered as a significant change to the network which is performed over a coarse timescale. This is because each router must be individually updated with new link costs, and this process incurs convergence delays before all routers agree on a new, global set of link costs. Our proposed RTLMUS scheme demonstrate the effectiveness of utilizing more than one source node and real-time link cost adjustment for traffic load balancing in an SDN. RTLMUS ensures that each destination node connects to just one potential source node, and that the traffic from that particular source to its multicast nodes is evenly balanced according to the real-time available bandwidth. In a special setting where there is only one element in the source node set, our proposed RTLMUS scheme will be equivalent to a single-source multicast. Constructing the RTLMUS tree is more challenging when compared to the STM because of the flexibility of the elements of the source nodes set and the real-time link cost adjustment. Any node with the replica content can become the element of the source nodes set.

2   Methods

The total cost of a multicast tree, which spans the single source and all destinations is widely used as the performance metric of a multicast. In most cases, the cost of multicast tree depends on the amount of links occupied without considering the diversity that exists among the network links. To optimize and balance the traffic flow, given a multicast group, many approximation methods have been proposed to construct the SMT in the literature. Given the same set of destinations, the load balancing of a multicast group is sensitive to the real-time cost of each link and selection of the source. The content replica design [7] is widely used for improving the robustness and efficiency in various networks. Thus, any replica can participate in the multicast transmission by acting as a potential source. This can result in different multicast transfers that exhibit the diversity of load balances.

2.1 Problem Definition

RTLMUS will still be a deterministic if all destinations access any $s \in S$ as their common source even if $|S| > 1$. The traffic load balancing effects of RTLMUS is basically dominated by the strategies of using the potential sources, hence, in the following paragraphs we explain the process of constructing the RTLMUS trees. The RTLMUS scheme create trees spanning each destination $d \in D$ with only one source, such that the traffic load of the whole network is optimally balanced based on the real-time link cost adjustment. Any pair of source nodes s_1 and s_2 appearing in a multicast tree must be isolated into different trees. Isolating the source nodes into different trees ensure that no destination node can be reached by more than one source node. This will reduce the congestion at each potential source node. This, therefore means that if we have k potential sources in $|S|$ set, then the RTLMUS scheme will consists of k isolated trees, each of which roots at one of those k potential sources. The process of constructing the RTLMUS scheme is more similar to constructing a SMT, which is an NP-hard problem in a general graph. However, the construction of RTLMUS is more challenging than the SMT problem in general setting, because it involves considering multiple potential source nodes as well as the real-time cost of each link in other to balance the traffic flow.

Theorem 1: Given a RTLMUS problem with potential source node set $|S|$ and a destination node set $|D|$ in a network $G(V, E)$, the challenge of constructing its RTLMUS scheme is an NP-Hard.

Proof: We prove the NP-hardness of constructing our RTLMUS scheme by giving a polynomial time reduction from SMT, which is an NP-hard problem. The SMT problem for RTLMUS must be considered first, in other to solve this, meaning that we have to find an SMT for spanning all the destination node set $|D|$ and source node set $|S|$ of that RTLMUS. Clearly, the optimal solution of this SMT problem cannot be found within polynomial time. Note that $|S|^2$ source pairs exist in the optimal SMT and each source pair are connected by only one path. The path between any two sources contains at least one redundant link. A link is considered to be redundant only if each destination node $d \in D$ can still reach at least one source node $s \in S$ even after the removal of that link. Therefore, the SMT problem of a RTLMUS would be reduced to the RTLMUS scheme by removing all the redundant links in the optimal SMT. The process of removing potential redundant links can be completed within polynomial time. Then, the remaining links in the original optimal SMT form a RTLMUS scheme where each destination node $d \in D$ connects to only one potential source node $s \in S$. Therefore, the process of constructing the RTLMUS scheme is an NP-Hard, hence Theorem 1 is proven.

2.2   Mixed Integer Linear Programming

We design a Mixed Integer Linear Programming (MILP) formulation for the RTLMUS problem. Let N_v denote the set of all neighbor nodes of node v in G, and u is in N_v if (u,v) is a link from u to v. Let $|S|$ denote the potential source set of the RTLMUS scheme, and $|D|$ denote the destination set. Let P be a path from the multicast source node $s \in S$ to the destination node $d \in D$. Given an SDN network $G(V,E)$, where V and E denote the set of nodes (switches) and links respectively. Each link $e \in E$ is associated with a cost function $C_{(u,v)}$, bandwidth utilization $U_{(u,v)}$ (in Mbps) and maximum bandwidth capacity $M_{(u,v)}$ (in Mbps) i.e. for link (u,v). Given a destination set $D \in V$ and a potential source set $S \in V$, RTLMUS tries to deliver the content to the whole set of destinations in $|D|$ from partial or even all potential sources in set $|S|$, utilizing the real-time link cost adjustment to balance the traffic flows. There is a constraint that each destination node in set $|D|$ must just reach to one and only one potential source node in set $|S|$. This does not impose any constraint on the selection of used sources for each destination $d \in D$. Let σ denote the floating point maximum value of the controller platform and let δ denote the floating point minimum value of the controller platform. The RTLMUS scheme's task is to construct RTLMUS tree $T(V`,E`)$, such that $V` \subseteq V$, $E` \subseteq E$, and (i) $C_{(u,v)} = \delta$ when $U_{(u,v)} = 0$, (ii) $C_{(u,v)} = \sigma$ when $U_{(u,v)} \geq M_{(u,v)}$, (iii) $\forall d \in D \exists P$ such that P is a path from s to d in T, and $C_{(u,v)} = (\frac{U_{(u,v)}}{M_{(u,v)}})$ when $U_{(u,v)} > 0$.

To achieve this goal, our RTLMUS scheme includes the following binary decision variables. Let binary variable $\varpi_{(d,s)}$ denote whether a destination node d selects a source node s as the root node. Meaning, there is a path from d to s if $\varpi_{(d,s)} = 1$. In this setting, let binary variable $\pi_{(d,u,v)}$ denote whether link (u,v) is in the path P from the destination node d to the source node s. Let binary variable $\varepsilon_{(u,v)}$ denote whether link (u,v) is in the output tree T. We should then be able to find easily the path from each destination d to just one source node s with $\varpi_{(d,s)} = 1$. Thus, every link (u,v) in the path has $\pi_{(d,u,v)} = 1$. The routing of the resultant trees with $\varepsilon_{(u,v)} = 1$ for every link (u,v) in the RTLMUS scheme can be achieved by the union of the paths from all the destination nodes in $|D|$ to at least one source node in $|S|$. In the linear link cost scheme, the link cost function is defined as:

$$C_{(u,v)} = \begin{cases} \delta & : U_{(u,v)} = 0 \\ \frac{U_{(u,v)}}{M_{(u,v)}} & : U_{(u,v)} > 0 \end{cases} \qquad (1)$$

In the inverse proportional link cost scheme, the link cost function is defined as:

$$C_{(u,v)} = \begin{cases} \delta & : U_{(u,v)} = 0 \\ \frac{1}{1-(U_{(u,v)}/M_{(u,v)})} - 1 & : 0 < U_{(u,v)} < M_{(u,v)} \\ \sigma & : U_{(u,v)} \geq M_{(u,v)} \end{cases} \tag{2}$$

Shortest-path routing is implemented by uniformly setting $C_{(u,v)}$ for all links in the network to a constant value of 1. While these cost function definitions allow the possibility of $U_{(u,v)} = 0$, in practice $U_{(u,v)}$ is always greater than 0, as the Link Layer Discovery Protocol (LLDP) queries periodically generated by the controller ensure that all network links will carry some small amount of traffic. The object function of our RTLMUS scheme is as follows:

$$\min \sum_{(u,v)\in E} C_{(u,v)} \times \varepsilon_{(u,v)} \tag{3}$$

where $C_{(u,v)}$ denote a cost function of link (u,v). To find $\varepsilon_{(u,v)}$, our MILP formulation include the following constraints, which explicitly describe the routing principles for RTLMUS tree creations;

$$\sum_{s\in S} \varpi_{(d,s)} = 1, \forall d \in D \tag{4}$$

$$\sum_{v\in N_s} \pi_{(d,s,v)} = 1, \varpi_{(d,s)} = 1, \forall d \in D, \exists s \in S \tag{5}$$

$$\sum_{u\in N_d} \pi_{(d,u,d)} = 1, \varpi_{(d,s)} = 1, \forall d \in D, \exists s \in S \tag{6}$$

$$\sum_{v\in N_u} \pi_{(d,u,v)} = \sum_{v\in N_u} \pi_{(d,v,u)} = 1, \varpi_{(d,s)} = 1 \tag{7}$$

3   Architecture

Under this section we present the system architecture for our RTLMUS scheme as illustrated on Fig. 2. We then give detailed description of some of the components of our proposed architecture due to limited space.

Topology Discovery: This module implements topology discovery through Link Layer Discovery Protocol (LLDP) polling. By using the information obtained from the topology discovery module we can construct the network topology graph $G(V, E)$, where the node set V corresponds to the switches and the link set E corresponds to the links. Then, the data that are relative to the topology graph G would be sent to the tree construction module to build up the tree.

Multicast Routing Model: This module implements detection of multicast potential sources, multicast tree calculation, and installation/removal of Open-Flow rules to direct multicast traffic. Shortest path tree calculation is performed

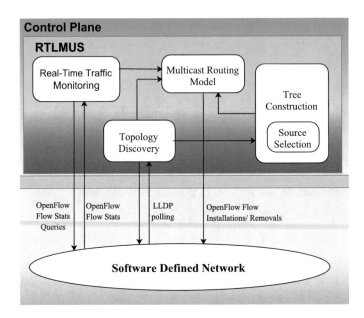

Fig. 2. Proposed architecture

for each combination of multicast potential source IP address and destination IP address using Algorithm 1. Whenever a multicast receiver joins or leaves a multicast group for which a potential source has been identified, the multicast tree calculation and flow installation is performed. Routing is considered separately for each multicast group, and no flow aggregation between separate multicast groups is implemented.

Real-Time Traffic Monitoring: This module is tasked with implementing real-time bandwidth usage tracking, and reports link bandwidth utilizations and capacities (i.e. $U_{(u,v)}$ and $M_{(u,v)}$, as defined in Sect. 2) every time a routing calculation is performed. Traffic estimation is implemented through periodic polling of all switches in the network (as defined in the OpenFlow v1.0 specification [16]. This query interval determines the frequency at which link cost functions are modified. Query times for each switch are randomly staggered, based on the time at which the switch first connected to the controller.

3.1   The Load-Balanced Algorithm

The Pseudocode of our proposed RTLMUS algorithm is presented on Algorithm 1. This RTLMUS algorithm utilizes the well known Dijkstra's algorithm to calculate the routing path based on the real-time link cost functions presented in Sect. 2.

Algorithm 1. *Pseudocode for RTLMUS algorithm*

Input: An undirected weighted graph $G(V, E)$, the set of source node S, the set of destination nodes D.

Output: A Load Balanced RTLMUS-Forest.

1: Create Graph G with a virtual source node.
2: Create set Q which include the virtual source node.
3: Connect all nodes in $|S|$ to the virtual source node, with cost function $C_{(u,v)} = \infty$.
4: **for** $\forall E \in G$ **do**
5:      $egress\_port \leftarrow adjacency\_map[link.egress\_node][link.ingress\_node]$
6:      $max\_link\_util\_mbps \leftarrow flow\_tracker.get\_max\_util(link.egress\_node, egress\_port)$
7:      $link\_util\_mbps \leftarrow flow\_tracker.get\_curr\_util(link.egress\_node, egree\_port)$
        $CalculateLinkCost$: According to the link cost function presented in Sect. 2.
8:      $link\_cost \leftarrow CalculateLinkCost(link\_util\_mbps, max\_util\_mbps)$
9:      $weighted\_graph \leftarrow weighted\_graph \cup (link\_cost, edge.egress\_node, edge.ingress\_node)$
10: **end for**
11: Add nodes to set Q.
    Calculate $RTLMUS\_Path\_Tree$ on set Q using Dijkstra's algorithm with $|S|$. Returns a map of lists where $path\_tree\_map[dst\_node]$ = Set of links from all $s \in S$ to its destinations.
12: $RTLMUS\_path\_tree\_map \leftarrow CalculateRTLMUS\_Path\_Tree(weighted\_graph, tree\_|S|)$
13: Return the RTLMUS-Forest

4   Results and Discussions

4.1   Simulation Setup

In this section, we employed a synthetic network to evaluate the performance of our proposed RTLMUS scheme by simulations. The implemented prototype was evaluated through network emulation of representative topologies and workloads using Mininet [17], which is popularly used for emulating OpenFlow network environments. In this paper, we used Inet-3.0 [18] to generate large synthetic networks with a mean degree of 4 by default. We then applied our algorithm to evaluate its load balancing effect. Network $|V|$ spans from 4000 to 10000, potential source nodes set $|S|$ from 2 to 6, destination nodes set $|D|$ from 20 to 100. All links are considered to be bi-directional with the capacity of 1Gbps for each direction. First, we randomly generate a number of matrices of variable size and select the link nodes (source and destination nodes) randomly. The performance metrics include (1) traffic concentration and (2) average link untilization across all the links. We compared the performance of our RTL-MUS scheme with the slightly modified recover aware edge reduction algorithm (RAERA) [19], which we renamed mRAERA and the shortest-path routing algorithm. Figure 3(a) shows that in general, as the number of source nodes increases from 2 to 6, the average link usage (Mbps) increase as well. The lower matric value suggest that the network traffic load is well balanced across all the links. As the number of source nodes increases, the assumption is that the destination nodes will be increasing as well and hence more links will be added. This will ensure that the average link utilization among all links in the network go down, which is very good for network performance. The reliability of the multicast

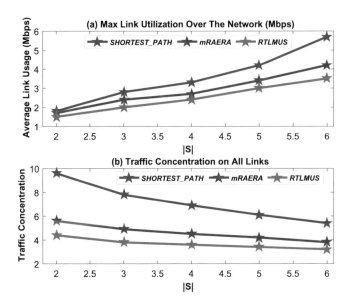

Fig. 3. Impact of the number of source nodes |S| on (a) Average link utilization across all links and (b) Traffic concentration across all links, respectively.

delivery will be high and hence avoiding packet loss. Our RTLMUS scheme performs much better than the mRAERA and the shortest path in this regard. We can deduce from Fig. 3(b) that increasing the number of source nodes reduces the traffic concentration across all the links. Low traffic concentration is desirable as it is a sign that flows are evenly distributed across the network links. RTLMUS scheme still manage to perform better when compared to mRAERA and the shortest path algorithms. This results in general indicates that RTLMUS scheme is very effective in terms of load balancing. Utilization of the multiple source nodes effectively reduces the congestion at the source, and the real-time link monitoring ensures that no link on the network would be over utilized while other links remain idling.

References

1. Shu, Z., Wan, J., Lin, J., Wang, S., Li, D., Rho, S., Yang, C.: Traffic engineering in software-defined networking: measurement and management. IEEE Access **4**, 3246–3256 (2016)
2. Li, Y., Chen, M.: Software-defined network function virtualization: a survey. IEEE Access **3**, 2542–2553 (2015)
3. Liu, J., Li, Y., Chen, M., Dong, W., Jin, D.: Software-defined internet of things for smart urban sensing. IEEE Commun. Mag. **53**(9), 55–63 (2015)
4. Semong, T., Xie, K., Zhou, X., Singh, H.K., Li, Z.: Delay bounded multi-source multicast in software-defined networking. Electronics **7**(1), 10 (2018)

5. Kitsuwan, N., McGettrick, S., Slyne, F., Payne, D.B., Ruffini, M.: Independent transient plane design for protection in OpenFlow-based networks. IEEE/OSA J. Opt. Commun. Netw. **7**(4), 264–275 (2015)
6. Chun, B.G., Wu, P., Weatherspoon, H., Kubiatowicz, J.: An anycast service for large content distribution. In: International Workshop on Peer-to-Peer Systems, February 2006
7. Presti, F.L., Petrioli, C., Vicari, C.: Dynamic replica placement in content delivery networks. In: 13th IEEE International Symposium on Modeling, Analysis, and Simulation of Computer and Telecommunication Systems, pp. 351–360. IEEE, September 2005
8. Cáceres, R., Duffield, N.G., Horowitz, J., Towsley, D.F.: Multicast-based inference of network-internal loss characteristics. IEEE Trans. Inf. Theor. **45**(7), 2462–2480 (1999)
9. Xie, J., Guo, D., Hu, Z., Wu, J., Chen, T., Chen, H.: Reliable multicast routing with uncertain sources. In: 2017 IEEE/ACM 25th International Symposium on Quality of Service (IWQoS), pp. 1–6. IEEE, June 2017
10. Hu, Z., Guo, D., Xie, J., Ren, B.: Multicast routing with uncertain sources in software-defined network. In: 2016 IEEE/ACM 24th International Symposium on Quality of Service (IWQoS), pp. 1–6. IEEE, June 2016
11. Craig, A., Nandy, B., Lambadaris, I., Ashwood-Smith, P.: Load balancing for multicast traffic in SDN using real-time link cost modification. In: 2015 IEEE International Conference on Communications (ICC), pp. 5789–5795. IEEE, June 2015
12. Dixit, A., Prakash, P., Hu, Y.C., Kompella, R.R.: On the impact of packet spraying in data center networks. In: INFOCOM, 2013 Proceedings IEEE, pp. 2130–2138. IEEE, April 2013
13. Moy, J.: RFC 1584: Multicast Open Shortest Path First (MOSPF). Network Working Group (1994)
14. Farinacci, D., Liu, C., Deering, S., Estrin, D., Handley, M., Jacobson, V., Thaler, D., Sharma, P., Wei, L., Helmy, A.: Protocol independent multicast-sparse mode (PIM-SM): protocol specification (1998)
15. Fortz, B., Rexford, J., Thorup, M.: Traffic engineering with traditional IP routing protocols. IEEE Commun. Mag. **40**(10), 118–124 (2002)
16. OpenFlow Consortium: OpenFlow Switch specification v1. 0., 11 August 2009. http://www.openflowswitch.org. Accessed 12 June 2015
17. Team, M.: Mininet (2014). http://mininet.org. Accessed 01 Jan 2016
18. Winick, J., Jamin, S.: Inet-3.0: Internet topology generator. Technical report CSE-TR-456-02, University of Michigan (2002)
19. Shen, S.H., Huang, L.H., Yang, D.N., Chen, W.T.: Reliable multicast routing for software-defined networks. In: 2015 IEEE Conference on Computer Communications (INFOCOM), pp. 181–189. IEEE, April 2015

Decision Trees Accuracy Improvement for Production Errors Classification

Michal Kebisek[1(✉)], Lukas Spendla[1], Pavol Tanuska[1], and Lukas Hrcka[2]

[1] Faculty of Materials Science and Technology, Slovak University of Technology,
Trnava, Slovakia
{michal.kebisek,lukas.spendla,pavol.tanuska}@stuba.sk
[2] PredictiveDataScience, s. r. o., Bratislava, Slovakia
lukas.hrcka@predictivedatascience.sk

Abstract. The paper is focused on improvement of classification accuracy of decision trees used in the data mining process. Real production data from the paint shop process serve as its basis. The proposal utilizes various approaches for selection of target attribute intervals and classes and key attributes for classification. The decision tree parameters are optimized to obtain the best possible combination. The results are evaluated across multiple decision tree algorithms.

Keywords: Accuracy improvement · Classification · Data mining
Decision tree

1 Introduction

The current trend, in various areas of business, is the utilization of large amounts of data across the various hierarchical levels, in accordance with the Industry 4.0 concept. Production and manufacturing area is therefore gaining more and more focus, to solve the challenges involved in quality, efficiency and sustainability of the production processes.

Knowledge discovery that is highly used in the business parts of companies is therefore more and more implemented for various challenges also in the production and manufacturing areas. Crucial part of this process is accuracy of predictions and classification. Since the accuracy of utilized knowledge discovery algorithms is influenced by various parameters, finding the most suitable ones presents a challenge [1].

2 Data Preparations and Data Mining Model Proposal

In our article we have focused on the improvement of accuracy of classification using decision trees. Before the analysis of suitable approaches and parameters, that could improve the accuracy of the classification it was required to obtain the data that will serve as a basis for our data mining model.

We have focused on the car body painting process, therefore we have analyzed the paint errors on car body. Data from this process are stored in various systems. For our proposal of classification accuracy improvement, we have selected 3 overlapping sets of data that are required to fully analyze the paint errors in the paint shop process, namely:

- Paint errors from the paint shop, that are recorded by the employees;
- Car position informations from the shop floor, to identify exact paint process times and positions;
- Weather data from weather stations situated in the vicinity of the production hall.

These data sets served as a start point for our proposed data mining model for decision tree accuracy evaluation, presented on Fig. 1. All data sources used in this paper are manually exported from existing systems. Since these data sets originate from different systems, various data transformation had to be performed, before they could be joined together. This includes removal of duplicate values, transformation of attributes, aggregation of multiple event occurrences, etc. Subsequently the attributes have to be further transformed to be used for classification using decision trees.

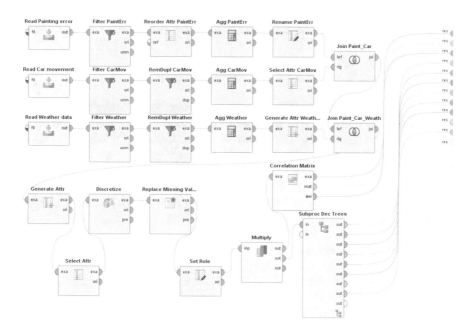

Fig. 1. Proposed data mining model

These data served as a basis for our decision trees evaluation in the final step. It should be noted that the subprocess "Subproc Dec Trees", containing the decision trees, contains also various types of transformations, statistical performance evaluations etc.

3 Definition of Decision Tree Parameters for Accuracy Improvement

3.1 Selection of the Target Attribute and Determination of the Classification Classes

The most important parameters for decision trees is the determination of the target classification attribute and its possible representations, i.e. classification categories determination. The target classification attribute may contain binary values (e.g. "yes" and "no") or multiple options. Result however have to be in a nominal form, i.e. discreet text, not continuous numeric values [2, 3]. In our paper we have selected attribute that will classify paint errors based on other parameters into five classification classes, based on the average number of paint errors per one bodywork.

The intervals for classification classes are the first issues that we had to deal with. Therefore, we have utilized three different methods to determine the classification classes:

- Manual user defined intervals based on the calculated average number of failures per one bodywork;
- Intervals defined using "Discretize by Binning" operator;
- Intervals defined using "Discretize by Frequency" operator.

Manual User Defined Intervals Based on the Average Number of Paint Errors
As a first method we have divided the target attributes interval, by generating new polynomial value "AverageErrorPerCar" for every record. Value 2.825 that represents the average number of paint errors per bodywork, which served as a basis for our classification, was calculated in the previous analysis.

Based on this value we have defined five non-overlapping intervals, according to defined error rate distribution, based on the following RapidMiner [4] function expression code:

```
IF (([AverageErrorPerCar] < 2.075), "Min error",
  IF (([AverageErrorPerCar] >= 2.075 &&
[AverageErrorPerCar] < 2.575, "Less average error",
    IF(([AverageErrorPerCar] >= 2.575
&&[AverageErrorPerCar] < 3.075, "Average error",
      IF (([AverageErrorPerCar] >= 3.075 &&
[AverageErrorPerCar] < 3.575, "More average error", "Max
error"))))
```

The distribution of individual intervals uses fixed range with value of 0.5. After the intervals definition we have utilized the data processing model to obtain the record distribution across the different classification classes, i.e. how many records were classified into each manually defined target parameter interval.

In the results, that can be seen in Table 1, the highest number of records have been categorized into the "Less than average error" classification class, with contains average number of paint errors for one bodywork in the range from 2.075 to 2.575. This classification class contained 54 records that represent nearly half of all recorded paint errors. The smallest number of paint errors are in the "Minimum error" and "More than average error" classification classes, each containing only three recorded paint errors.

Table 1. Manually user defined classification classes for paint errors.

Nominal value	Absolute count	Fraction
Less than average error	54	0.491
Maximum error	33	0.300
Average error	17	0.155
Minimum error	3	0.027
More than average error	3	0.027

Intervals Defined Using Discretize by Binning Operator

Another approach for defining the interval range of the target variable for classification is to use one of available algorithms in the RapidMiner software platform, namely "Discretize by Binning" in the first case.

This operator automatically suggests intervals that are used for classification [5]. Only required parameter is number of intervals (bins) that should be created. In our case we have required five intervals. The operator subsequently generated five intervals, using the split value range of 0.887. Number of records in each classification class have differ from the manually user defined intervals. The results with distribution of records across the individual classification classes are shown in Table 2.

Table 2. Classification classes defined using Discretize by Binning operator.

Nominal value	Absolute count	Fraction
Range1 [−∞–2.658]	69	0.627
Range2 [2.658–3.545]	8	0.073
Range3 [3.545–4.432]	4	0.036
Range4 [4.432–5.319]	13	0.118
Range5 [5.319–∞]	16	0.145

Intervals Defined Using Discretize by Frequency Operator

The third approach for defining the interval range of the target variable for classification is to use "Discretize by Frequency" operator, which uses different algorithm than the previously used operators. The "Discretize by Frequency" operator divides the target attribute into desired number of intervals, with interval ranges defined in a manner, that each interval contains similar number of records.

The results, presented in Table 3, clearly shows the same number of records in each interval. Different are however interval ranges for each classification class of the target attribute. This gives us different overview of possible interval ranges suitable for the classification.

Table 3. Classification classes defined using discretize by frequency operator.

Nominal value	Absolute count	Fraction
Range1 [−∞–2.295]	22	0.200
Range2 [2.295–2.474]	22	0.200
Range3 [2.474–2.643]	22	0.200
Range4 [2.643–5.072]	22	0.200
Range5 [5.072–∞]	22	0.200

3.2   Selection of Key Attributes for Classification Improvement

Another important step is the selection of right attributes that will be used for classification utilizing decision trees [6]. The data set, used in this paper, contains 19 different attributes with various data types. However, some of the attributes are not relevant or contain duplicate values. After removing them we have obtain 14 attributes that could be used for the classification decision making.

Subsequently we have selected key decision making attributes by determining their weights and information gain. To evaluate these parameters, we have utilized the "Weight by Information Gain" operator, which calculates the relevance of each attribute based on the information gain. The higher is the information weight of the attribute, the higher is its relevance and importance to the decision making, e.g. classification using the decision trees [7].

The weighs were calculated for every key decision making attribute, for all three methods of the target attribute interval selection, described in the previous section of this paper, namely:

- Manual user defined intervals based on the calculated average number of failures per one bodywork;
- Intervals defined using "Discretize by Binning" operator;
- Intervals defined using "Discretize by Frequency" operator.

For the better clearness and usability of the decision tree, we have considered only five most important attributes, i.e. attributes with the highest information gain.

The results, obtained for each method for selection of target attribute range, were very similar. For example, results for "Discretize by Frequency", are shown on Fig. 2. In all three cases however, the attribute "Day of the week" (Week day) was selected as the attribute with the highest weight, i.e. highest information gain. Weight of this attribute was also much higher than weight of any following attributes, with "Average air humidity" (Avg hum) attribute being the nearest one. This attribute also appeared as second highest in all three considered methods.

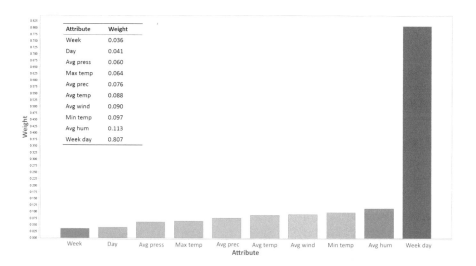

Attribute	Weight
Week	0.036
Day	0.041
Avg press	0.060
Max temp	0.064
Avg prec	0.076
Avg temp	0.088
Avg wind	0.090
Min temp	0.097
Avg hum	0.113
Week day	0.807

Fig. 2. Calculated attribute gain weights for discretize by frequency method.

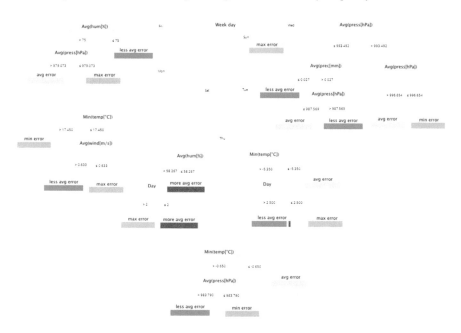

Fig. 3. Selected decision tree.

Setting the Decision Tree Parameters

The correct setting of decision trees parameters is necessary to achieve the best results. Due to the various capabilities of each parameter in the decision tree operator (mostly numerical), we have utilized the "Optimize Parameters" operator, to find optimal values for selected parameters, specifically: "Minimal leaf size", "Minimal size for split",

"Numbered of prepruning alternatives", "Maximal depth", "Criterion", "Apply pruning", "Confidence", "Apply prepruning" and "Minimal gain".

Using these parameters in the operator, we have created 28,344,976 possible combinations that took the model nearly five days to complete. The optimal decision tree parameters values obtained in this step, served as a basis for the classification model utilizing decision tree which is presented in next chapter of this paper (see Fig. 3).

4   Results

There are several procedures and criteria for evaluating experiment results [8]. We have chosen the following four:

- Prediction accuracy – the ability of the decision tree to correctly assign new sample data to the correct classification class;
- Velocity – the ability to apply a model on large volumes of data;
- Robustness – the ability of the decision tree to create correct classifications and predictions;
- Usability – one of the most important criteria, but being the only one of the above-mentioned criteria subjectively evaluable and provides a view of the tree and the interpretation of its results, hidden relationships and ways of finding solutions.

To evaluate the final experiments with different target attribute settings, decision attributes, and decision tree parameters, we have created the following table (see Table 4) that shows the precision of the model prediction (accuracy).

Table 4.   Obtained accuracy values for selected methods.

Model accuracy	Manual attributes selection					Attributes by weight				
	Non-optimized decision tree param.				Optimized param.	Non-optimized decision tree param.				Optimized param.
	Gain ration	Inform. gain	Gini index	Accuracy		Gain ration	Inform. gain	Gini index	Accuracy	
Manual selection	70.91	70.91	72.73	76.36	79.09	72.73	70.91	71.82	75.45	79.09
Discretize by Binning	69.09	70.00	67.27	70.00	80.00	68.18	68.18	70.00	73.64	80.91
Discretize by Frequency	46.36	44.54	47.27	46.36	56.36	48.18	46.36	44.55	51.82	59.09

The highest precision of the model with the used data was achieved by dividing the intervals of the target attribute by the operator "Discretize by Binning", the key attributes selected according to calculated weights by the operator "Weight by Information Gain", and triggered parameter optimization using the parameter "Optimize Parameters". The accuracy of the model was 80.91%. The problem, however, was the interpretation of the tree, since, apart from the accuracy, we also took into account other criteria and parameters as mentioned above.

The decision tree interpretation was inadequate and very simple after identifying the rules and graphical representation of the tree, due to the generation of only one level, by day of the week.

Because of the better interpretability of the results, we decided to state as a result the decision tree (shown on Fig. 3) with an accuracy of 79.09%.

We achieved this result with the following settings for each decision tree parameter:

```
Decision_Tree.Criterion        = accuracy
Decision_Tree.Confidence       = 0.15000006999999999
Decision_Tree.Minimal_gain     = NaN
Decision_Tree.Apply_pruning    = false
Decision_Tree.Apply_prepruning = false
```

To verify the decision tree results and the widest possible use of all types of decision tree algorithms, we have decided to choose the best decision tree settings and verify tree quality on other algorithms [9].

In this way, we have tested sequentially algorithms: "Random Forrest", "Gradient Boosted Trees", "ID3", "C4.5", and "Random Tree Algorithm". Optimization of all parameters was selected by the "Optimize Parameters (Grid)" operator, with some computations running for several days.

All combinations of the "Gradient Boosted Trees" algorithm could not be tested because the number of combinations was 1,188,625,088 which were not enough for our computing resources.

The accuracy of the "Random Forest" algorithm was 80.91% after optimizing all parameters, "Gradient Boosted Trees" achieved 77.27%. The accuracy of the "C4.5" and "Random Tree" algorithms was 82.73%, and the accuracy value of the algorithm "ID3" was 81.82%.

Table 5. Obtained results for selected decision tree algorithms.

Algorithm	Accuracy	Classification error	Kappa	Absolute error	Correlation
Decision Tree	80.91%	19.09	0.639	0.279	0.859
Random Forest	80.91%	19.09	0.622	0.409	0.845
Gradient Boosted Tree	77.27%	22.73	0.638	0.744	0.842
C4.5	82.73%	17.27	0.670	0.428	0.855
Random Tree	82.73%	17.27	0.639	0.300	0.770
ID3	81.82%	18.18	0.306	0.306	0.842

The comparison and the results of the decision tree algorithms can be seen in Table 5, below is just a brief explanation of the parameters selected for quality assessment:

- Classification error – displays the number of incorrectly classified cases, defines the percentage of incorrect predictions;

- Kappa – a parameter that is similar to the precision value but with the difference that it takes into account randomly assigned correct predictions, which ultimately makes this parameter more robust than the accuracy value;
- Absolute error – the average absolute deviation of the prediction from the actual value. The lower values means smaller deviation and more accurate classification;
- Correlation – the value of the correlation coefficient between the predicted and the actual values, expresses the value of the dependence between the two parameters.

Despite the somewhat higher accuracy achieved in the above-mentioned algorithms, and the highest accuracy of "C4.5" and "Random Tree" algorithms, we also took into account other evaluation criteria. The accuracy of the model was the first important criterion. However, the tree's interpretability was also an important criterion, but it was inadequate for the target attribute that was divided by the "Discretize by Binning" operator. Due to the classification of values in one level and only five leaf nodes, we retained the most appropriate and best-represented tree, created in the previous step with the result of 79.09%, which was taken as the final result.

5   Conclusion

In our paper we have described approach to increase accuracy of decision trees used for the classification of paint errors. Data sets that served as a basis for our proposal, are utilizing available data from the paint shop process. To increase the accuracy of classification we have evaluated options based on selection of target attribute intervals, key attributes used for classification, and selected decision tree algorithms.

From the results of our evaluation, presented in this paper, we can conclude, that in addition to the classification model accuracy, also other evaluation criteria have to be considered. This is mainly due to the fact that highest accuracy classification models do not provide the best classification results. Therefore it is also necessary to verify the results of evaluated decision model also using the calculated data, empirical knowledge and process knowledge itself, to decide on the form and representation of the results.

Therefore, the best selection of target attribute criteria and key attribute for classification approach differ, based on the production data used. Therefore it is not possible to make any general assumption, and different data sets needs to be evaluated on their own.

Acknowledgments. This publication is the result of implementation of the project VEGA 1/0272/18: "Holistic approach of knowledge discovery from production data in compliance with Industry 4.0 concept" supported by the VEGA.

This publication is the result of implementation of the project: "Increase of Power Safety of the Slovak Republic" (ITMS: 26220220077) supported by the Research & Development Operational Programme funded by the ERDF.

References

1. Trnka, A.: Big data analysis. Eur. J. Sci. Theol. **10**, 143–148 (2014)
2. Witten, I.H., Frank, E., Hall, M.A., Pal, Ch.J.: Data Mining: Practical Machine Learning Tools and Techniques, 4th edn. Morgan Kaufmann (2016)
3. Bonaccorso, G.: Machine Learning Algorithms: A Reference Guide to Popular Algorithms for Data Science and Machine Learning. Packt Publishing, Birmingham (2017)
4. RapidMiner: Data Science Platform. https://rapidminer.com. Accessed 19 Dec 2017
5. Zaki, M.J.: Data Mining and Analysis: Fundamental Concepts and Algorithms. Cambridge University Press, Cambridge (2014)
6. Rokach, L., Maimon, O.Z.: Data Mining With Decision Trees: Theory and Applications. World Scientific Publishing Company, River Edge (2014)
7. Larose, D.T., Larose, Ch.D.: Data Mining and Predictive Analytics. Wiley, New York (2015)
8. Provost, F., Fawcett, T.: Data Science for Business: What You Need to Know About Data Mining and Data-Analytic Thinking. O'Reilly Media, Sebastopol (2013)
9. Kelleher, J.D., Mac Namee, B., D'Arcy, A.: Fundamentals of Machine Learning for Predictive Data Analytics: Algorithms, Worked Examples, and Case Studies. The MIT Press, Cambridge (2015)

The Computational Structure of the Quantum Computer Simulator and Its Performance Evaluation

Viktor Potapov[✉], Sergei Gushanskiy, Vyacheslav Guzik,
and Maxim Polenov

Department of Computer Engineering, Southern Federal University,
Taganrog, Russia
vitya-potapov@rambler.ru, {smgushanskiy,vfguzik,
mypolenov}@sfedu.ru

Abstract. This paper describes the basics of performance, as well as the structural and functional component of the development and implementation of the quantum computer simulator. In accordance with this, the computational structure of the quantum computer simulator has been derived, taking into account all the available features of constructing a simulator of a quantum computing device. Also, a software implementation of the derived universal computational structure of such simulator that satisfies and operates according to the principles of this scheme is implemented.

Keywords: Quantum register · Quantum computer simulator · Complex plane
Qubit

1 Introduction

When creating a quantum computer simulator (QCS), developers pursue completely different goals (simulating quantum systems and qubit, the effects of individual gates or simulating quantum algorithms) and various approaches in implementing the interface (graphical, console), types of modeling.

At the moment, there are a lot of different modeling environments (both console and graphic), API libraries, as well as models of individual algorithms. Some models are built on the basis of existing mathematical modeling environments, for example, the Grover's algorithm model developed at the St. Petersburg State University. There are different models and approaches of modeling: QuIDDPro [1] has a mathematical core built on a graph approach, which, unlike the usual matrix approach, can save a lot of memory in the simulation. A separate mention is the APIlibquantum [2], Cove [3] libraries for constructing quantum computing simulation programs that provide ready-made functionality for building your own model.

© Springer International Publishing AG, part of Springer Nature 2019
R. Silhavy (Ed.): CSOC 2018, AISC 763, pp. 198–207, 2019.
https://doi.org/10.1007/978-3-319-91186-1_21

2 The Computational Structure of the Quantum Computer Simulator

The suggested computational structure of the quantum computing device simulator [4], is shown in Fig. 1. It is possible to develop a simulator of quantum computer, observing the requirements displayed in the scheme. This structural and functional illustration of the computational model of the simulator of a quantum calculator is universal for its construction; $k_{1,1}$ denotes the first qubit of the quantum register 1 (QR1); $k_{1,2}$ – second qubit of QR1; $k_{1,m}$ – the current qubit of QR1; $k_{1,n}$ – the last qubit.

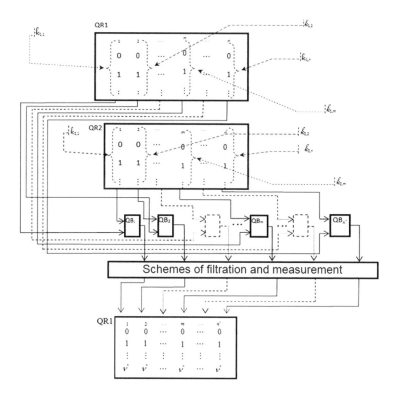

Fig. 1. The scheme of the computational structure of QCS

The computer structure of the QCS conditionally contains two quantum registers – QR1 and QR2. The only real register can be conditionally divided into a larger number of auxiliary parts. Such a separation can be done for the purpose, for example, of removing the quantum debris in preparation for the measurement process. The dimension of each qubit is the same: $\dim H_m = \hat{v}$.

That means in a single qubit different numbers of $\hat{v}$ can be written and stored. The number of qubits k_m in the QR1 and QR2 register is the same and equal $\hat{n}$. Thus, the registers QR1 and QR2 can be parts of one real QR. For example, all odd quantum bits belong to the QR1 register, and all even qubits to QR2. Each qubit k_m of the QR1 register is connected to a quantum block of intersections – a block QB_m (that is, an input of a block QB_m). The same applies to the qubits of the QR2 register. The output of each block QB_m is transmitted to the filtration and measurement circuits. The block QB_m itself realizes an integral operation of intersection in the form of a block-diagonal Roth matrix [5].

3   Software Simulation of Quantum Algorithms and Processes

Software simulation was implemented in accordance with described above the developed scheme of the computer model of QCS. This model can take a unitary function f and effectively determine its trace (direction vector) in accordance with the use of vector and matrix algebra. This trace of function f by n qubits can be found along one of the axes X or Y (Fig. 2).

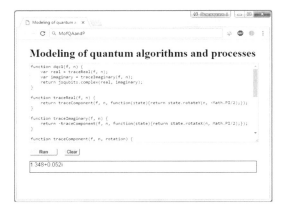

Fig. 2. The main form/output form of the result of the model simulation

Determination of the QCS Performance

What affects the performance of simulators of quantum computing devices?

Quantum circuit, its size and depth. Arguing about this important component of the simulator, it is possible to distinguish: the number of elements N_{gates} (quantum logic gates) and branches $N_{branches}$ of the quantum circuit, the number of interconnected branches N_{deep} (depth, some gates operate with binding of several branches, for example, CCNOT). A quantum scheme over a basis B is a sequence of unitary operators $U_1[S_1], U_2[S_2], \ldots, U_l[S_l]$, where $U_l \in B$, $S_l \subseteq \{1, \ldots, n\}$, n is the number of qubits. More formally, the matrix elements of the operator U[S] have the following form. Let

$$U = \sum_{x,y \in \{0,1\}^d} u_{x,y} |x><y|, \quad S = \{j_1, j_2, \ldots, j_d\}, \quad 1 \le j_1 < j_2 < \ldots < j_d \le n \quad (1)$$

We denote x[S] as a sub-sequence of bits that stand in places from the set S. Then the operator U[S] is written as

$$U[S] = \sum_{x,y \in \{0,1\}^n : x[\overline{S}] = y[\overline{S}]} u_{x[S],y[S]} |x><y|, \quad (2)$$

$$U_1[S_1], U_2[S_2], \ldots, U_l[S_l] = \frac{N_{gates}}{N_{branches} + N_{deep}}. \quad (3)$$

Quantum registers and their number. From the point of increasing the productivity, two QRs for parallelizing the tasks lying on them, for example, in the scheme of the computational structure of QCS described in Sect. 1.

Initial basic state of QCS (IBS QCS). A quantum state is any possible state in which a quantum system can exist. Let the state of a quantum system be given by a vector $|\psi\rangle$. When decomposed in accordance with the complete sets of vectors of the system $|\phi_i\rangle$ and its environment $|\theta_j\rangle$: $|\psi\rangle = \sum_{i,j} c_{ij} * |\phi_i\rangle * |\theta_j\rangle$, where c_{ij} is the amplitude of the probability of finding the system in the j-th state. It is obvious that the performance of SLE in terms of NBS depends on the number of branches of the quantum circuit $N_{branches}$ and the type of the basic state (BS). It is worth noting that in itself IBS QCS does not affect its performance. However, when considering the IBS within the life cycle of a quantum circuit, one cannot ignore the amount of manipulation of the user when working with QCS in the framework of its initial preparation and tuning (QCS refers to the type of automated systems).

We obtain an equation that reflects the amount of productivity in the initial baseline SLE state:

$$|\psi\rangle = N_{branches} * k, \quad (4)$$

where k is the coefficient of the form of the basic state. Since the value $N_{branches}$ can be infinitely large, it is not possible to evaluate this parameter in any way except for a percentage.

4 Running Various Algorithms Using the Framework of the Developed QCS

The developed QCS also assumes the implementation of a set of quantum algorithms, which in turn affects the performance of all QCS. The performance costs in this case are directly proportional to the complexity class of a particular quantum algorithm, which is considered in [6]:

$$\sum_n S = \frac{O(n)}{N_o + N_q} * (P \mid NP \mid ZPP \mid BPP \mid BQP), \tag{5}$$

where $O(n)$ is the time complexity of the quantum algorithm, or, simply, the time of operation of the particular algorithm; N_o – the number of operations performed in the algorithm, N_q – the number of queries to the quantum oracle during the operation of the quantum algorithm, $P \mid NP \mid ZPP \mid BPP \mid BQP$ – the complexity classes of the quantum algorithms (| – the logical OR operation).

In the system below, mathematical formalizations of some complexity classes are presented. The algorithm for polynomial time (P) is equivalent to a deterministic Turing machine (TM), which calculates the response from a given word to the input tape from the input alphabet. The complexity of the function f is the function C, depending on the length of the input word and equal to the maximum of the machine's operating time for all input words of fixed length (1). If for a function f there exists an TM such that (6a) for some number C, then we say that it belongs to the class P.

$$\begin{cases} P : C_M(n) = \max_{x:|x|=n} T_M(x)(1), \, C_M(n) < n^c & \text{(a)} \\ NP : \cup_{i=0}^{\infty} NTIME(i * n^i) = \cup_{i=0}^{\infty} \cup_{k=0}^{\infty} NTIME(i * n^k) & \text{(b)} \\ NTIME(f(n)) = \{L \mid \exists m : L(m) = L, T(m,x) \leq f(|x|)\} & \text{(c)} \\ \quad ZPP : 0 \mid 1 \\ \quad BPP : P(\text{polynomial runtime}), \; error = (2\sqrt{p(1-p)})^n \\ \quad BQP : P(\text{polynomial runtime}), \; 0 \leq error \leq \frac{1}{3} \end{cases} \tag{6}$$

NP is the class of problems that can be answered in a time P, and the class NTIME (f) is the class of problems for which there exists a nondeterministic TM whose work stops when the length of the input is exceeded (6c). The formal definition of class NP through the class NTIME looks like this (6b).

Summarizing all of the above, we get:

$$P = \left[\frac{\left[\sum_l U_l[S_l] * |\psi\rangle \right]}{2} \right] * \sum_n S \quad P = \left[\frac{N_{gates} * k}{2N_{deep} + 2} \right] * \frac{O(n)}{N_o + N_q} * (P \mid NP \mid ZPP \mid BPP \mid BQP) \tag{7}$$

5 Development of a Quantum Algorithm for Solving Systems of Linear Algebraic Equations

The solution of large systems of linear algebraic equations (SLAE) is a very common problem in scientific calculations with many fields of application. Until recently, it was believed that quantum algorithms could not reach the necessary speed for solving these problems, because the matrix of coefficients A is of size N^2, and all or most of the coefficients in A can be accessed to compute x, which requires a time $\Omega(N^2)$.

In addition to N, the running time of algorithms for systems of linear equations (both classical and quantum algorithms) depends on another parameter κ, the number of conditions of the matrix A.

5.1 Preparing the Parameters for the Operation of the Algorithm

Let's imagine the best quantum algorithm with the time of work is $O(k * \log^3 k * \log N)$. Speaking of the above "O", we mean the development of the most successful, in terms of time, quantum algorithm for the purpose described in this paper, namely, the solution of systems of linear algebraic equations. Since it is assumed that there is a huge number of quantum algorithms not only for this particular (private) practical problem, but also for general ones. To build the algorithm, we introduce a new tool – quantum-amplitude amplification with variable time, which allows us to strengthen the probability of success of quantum algorithms in which some branches of computation stop before other branches. The usual amplification of the amplitude will wait for the stoppage of all branches, which can lead to significant inefficiency. The new algorithm enhances the probability of success in several stages and takes advantage of the parts of the calculations that stop earlier. We expect that the new method will be useful for constructing other quantum algorithms.

We can construct a quantum Algorithm A' that calls algorithm A several times within the total time

$$O(T_{\max} \sqrt{\log T_{\max}} + \frac{T_{av}}{\sqrt{p_{succ}}} * \log^{1.5} * T_{\max}), \tag{8}$$

which gives the state $\alpha|1\rangle \otimes |\psi_{good}\rangle + \beta|0\rangle \otimes |\psi'\rangle$ with probability $|\alpha^2| \geq 1/2$.

Conversely, the usual amplitude gain will work in time $O(\frac{T_{\max}}{\sqrt{p_{succ}}})$. The developed algorithm A' provides an improvement, while T_{av} significantly less than $T_{\max}$. Repeating A' $O(\log \frac{1}{\varepsilon})$ times, you can get $|\psi_{good}\rangle$ with probability not less than $1 - \varepsilon$.

The algorithm A' is optimal, up to a coefficient $\log T_{\max}$. If algorithm A has only one stop time $T = T_{av} = T_{\max}$, amplitude amplification [7] cannot be performed with less than $O(\frac{T_{\max}}{\sqrt{p_{succ}}})$ steps. Thus, the expression $\frac{T_{av}}{\sqrt{p_{succ}}}$ is necessary. The expression $T_{\max}$ is also necessary, since in some branch of computation A can work for $T_{\max}$ steps.

Consider the system of linear equations $Ax = b$, where $A = (a_{i,j})_{i,j \in [N]}, x = (x_i)_{i \in [N]}, b = (b_i)_{i \in [N]}$. Suppose that A is a Hermitian. Let $|\vartheta_i\rangle$ is the eigenvectors of A and λ_i is their eigenvalues. Suppose that all λ_i satisfy $1 \leq \lambda_i \leq 1$ for some known κ. Then we can convert the state $|b\rangle = \sum_{i=1}^{n} b_i|i\rangle$ in $|x\rangle = \sum_{i=1}^{n} x_i|i\rangle$ the following way:

1. If in terms of eigenvectors $|\vartheta_i\rangle$ of A we have $|b\rangle = \sum_i c_i * |\vartheta_i\rangle$, then $|x\rangle = \sum_i \frac{c_i}{\lambda_i} * |\vartheta_i\rangle$.

2. By estimating the eigenvalues, we can create a state $\left|b'\right\rangle = \sum c_i * \left|\vartheta_i\right\rangle * \left|\lambda_i^-\right\rangle$ where are the estimates of the true eigenvalues.

3. Then create the state $\left|b''\right\rangle = \sum c_i * \left|\vartheta_i\right\rangle * \left|\lambda_i^-\right\rangle * (\frac{1}{k*\lambda_i^-}\left|1\right\rangle + \sqrt{1 - \frac{1}{k^2*(\lambda^-)^2}}\left|0\right\rangle)$

   Provided that the last bit is 1, the rest of the state is $\sum_i \frac{c_i}{\lambda_i} * \left|\vartheta_i\right\rangle\left|\lambda_i^-\right\rangle$ which can be converted into an approximation of $\left|x\right\rangle$ by performing an estimation of the eigenvalues in the opposite direction and computing of λ_i^-.

4. Then amplify a part of the state in which the last qubit is equal to 1 (using amplitude amplification) and with good probability we get a good approximation of $\left|x\right\rangle$.

   Algorithm A produces a quantum state $\left|\psi\right\rangle$ with probability p if:

- The algorithm has two output registers R and S;
- The measurement of R gives 1 with probability p and, depending on this measurement result, S register is in the state $\left|\psi\right\rangle$.

There is an procedure Estimate (A, c, p, k), which for a given constant c, $0 < c \leq 1$ and quantum Algorithm A, yields probability estimate ε^- such that, with probability at least $1 - \frac{1}{2^k}$ we have

1. $\left|\varepsilon - \varepsilon^-\right| < c\varepsilon^-$, if $\varepsilon \geq p$;
2. $\varepsilon^- = 0$, if $\varepsilon = 0$.

When we use the «Unit evaluation» as a subroutine in the state generation algorithm, we need a unique result, not one of the results with a high probability. We replace H to $H + \frac{\delta*\pi}{2^n} * I$ by a random choice of $\delta \in [0, 1]$.

5.2 Quantum Algorithm for SLAE Solving

Let us begin with a description of the algorithm for the execution time of a variable. The following registers are used in this algorithm:

- Input register I, which contains the input state $\left|x\right\rangle$ (and is also used for the output state);
- A register of results O with base states |0>, |1> and |2>;
- Register of steps S with basic states |1>, |2>, ..., |2m> (to prevent interference between different branches of computation).
- Evaluation register E, which is used to estimate the eigenvalues (which is a subprogram for the algorithm).

We will call ε without an index an error parameter for the algorithm subroutines.

5.3 Algorithm for Generating States

Input: parameters $x_1, \ldots, x_m \in [0, 1]$, Hamiltonian H.

1. Initialize O to |2>, S to |1> and E to |0>. Set j = 1.
2. Suppose that $m = \left\lceil \log_2 \frac{k}{\varepsilon} \right\rceil$

3. Repeat to j > m:

Step j:

(a) Suppose that $H' = H + \frac{x_j * \pi}{2^j} * I$. Using registers I and S, run the «Unit evaluation (H', 2^j, ε)». Let λ' be the output of the Unit evaluation and let $\lambda = \lambda' - \frac{x_j * \pi}{2^j}$.

(b) If $\lambda > \frac{1}{2^{j+1}}$, then conversion is performed

$$|2\rangle_O \otimes |1\rangle_S \rightarrow \frac{1}{k * \lambda} |1\rangle_O \otimes |2 * j\rangle_S + \sqrt{1 - \frac{1}{(k * \lambda)^2}} |0\rangle_O \otimes |2 * j\rangle_S$$

(c) Run a «Unit evaluation» in reverse order to erase the intermediate information.
(d) Check if the register E is in the correct initial state. If not, apply $|2\rangle_O \otimes |1\rangle_S \rightarrow |0\rangle_O \otimes |2 * j + 1\rangle_S$ in the resulting register O.
(e) If the result register O is in the state $|2\rangle$, increase j by 1 and go to step 2.

5.4   The Basic Algorithm

In Fig. 3 is shown the quantum scheme of a quantum algorithm for solving systems of linear algebraic equations for a particular problem, namely, the solution of 2×2 linear equations with four qubits and 15 gates.

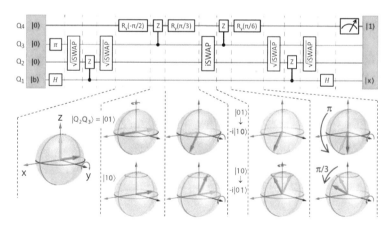

Fig. 3. Quantum scheme of the algorithm for solving 2×2 linear equations with four qubits and 15 gates

Input: Hamiltonian H.

1. Create a uniformly random $x_1, \ldots, x_m \in [0, 1]$.
2. Apply amplitude amplification of the time variable to the algorithm for generating states with H and $x_1, \ldots, x_m$ as an input.

3. Apply the transformation $|2*j\rangle_S \rightarrow |j\rangle_S$ to the S register. After this, apply the Fourier transform F_m to the S register and measure. If the result is 0, print the state to register I. Otherwise, stop the measurement without outputting the quantum state.

The number of conditions is defined as the ratio between the largest and the smallest singular value of A:

$$k = \max_{i,j} \frac{|\mu_i|}{|\mu_j|}, \tag{9}$$

where μ_i are the singular values of A.

In the case of sparse classical matrices, the best classical algorithm works within the time frame of $O(\sqrt{k}*N)$, whereas the eponymous quantum algorithm [8] implemented within the framework of this article operates in time $O(k^2 * \log N)$ with an exponentially better dependence on N.

The described temporal advantage of the quantum algorithm over the classical algorithm is not unique, since the performance of the quantum algorithm for solving systems of linear algebraic equations is constant from the point of view of the scale (number of equations) of the SLAE itself. Indeed, quantum parallelism [9], characterized primarily by parallelizing the entire computational set of a particular problem, produces, from the point of view of productivity, the computation of one equation (of all SLAE equations simultaneously).

6 Conclusion

Built on the basis of the proposed computer structure, the quantum computer simulator allows to optimize the computational process of the simulator in accordance with the set of components: a dual quantum register, a filtering and measuring scheme, quantum blocks of intersections QB_m and the number of qubits k_m. To find new ways of applying this universal structural scheme for modeling any parameters in the task being performed.

Such research should be applied before realizing quantum algorithms and computing devices. For example, in order to assess the required level of entanglement for the effective operation of the algorithm [10] or the optimal number of QCS elements and their interaction.

Acknowledgments. This work was carried out within the State Task of the Ministry of Education and Science of the Russian Federation (Project part No. 2.3928.2017/4.6) in Southern Federal University.

References

1. QuIDDPro: High-Performance Quantum Circuit Simulation. http://vlsicad.eecs.umich.edu/Quantum/qp/
2. Libquantum 1.1.1. http://www.libquantum.de/api/1.1/index.html

3. Cove: A Practical Quantum Computer Programming Framework. http://cove.purkeypile.com/
4. Guzik, V., Gushanskiy, S., Polenov, M., Potapov, V.: Models of a quantum computer, their characteristics and analysis. In: Proceedings of the 9th International Conference on Application of Information and Communication Technologies (AICT 2015), pp. 583–587. IEEE Press (2015)
5. Pravilshchikov, P.: Quantum parallelism and a new model of computation. In: Proceedings of the XII All-Russian Meeting on the Control Problems, pp. 7319–7334 (2014). (in Russian)
6. Potapov, V., Guzik, V., Gushansky, S.: About the performance and computational complexity of quantum algorithms. Inf. Commun. **3**, 24–29 (2017). (in Russian)
7. Brassard, G., Høyer, P., Mosca, M., Tapp, A.: Quantum amplitude amplification and estimation. In: Contemporary Mathematics 305, Quantum Computation and Information, pp. 53–74. American Mathematical Society (2000)
8. Smith, J., Mosca, M.: Algorithms for quantum computers. In: Handbook of Natural Computing, pp. 1451–1492. Springer (2010)
9. Potapov, V., Gushansky, S., Guzik, V., Polenov, M.: Architecture and software implementation of a quantum computer model. In: Advances in Intelligent Systems and Computing. Software Engineering Perspectives and Application in Intelligent Systems, vol. 465, pp. 59–68. Springer (2016)
10. Potapov, V., Gushansky, S., Guzik, V., Polenov, M.: Development of methodology for entangled quantum calculations modeling in the area of quantum algorithms. In: Advances in Intelligent Systems and Computing. Software Engineering Perspectives and Application in Intelligent Systems, vol. 575, pp. 106–115. Springer (2017)

Integration of Production Line with the Wonderware Platform

Andrea Vaclavova and Michal Kebisek[✉]

Faculty of Materials Science and Technology, Slovak University of Technology, Trnava, Slovakia
{andrea.vaclavova,michal.kebisek}@stuba.sk

Abstract. The aim of this paper is to create communication between Wonderware and model of the production line using DAServer. The paper consists of three parts. The first part deals with the introduction of Wonderware and the options it provides. The second part provides a brief overview of the Wonderware modules that we have used. And the third part shows the results we have achieved in creating communication.

Keywords: ArchestrA · DAServer · System management console
Wonderware

1 Introduction

This paper is focused on the process of setting up a communications link between model of the production line and the Wonderware software using SMC (System Management Console) and ArchestrA. Currently, the production line is functional and capable of production, but the data is not stored anywhere, that means it is not possible to analyze this data, since we only know the current condition of values contained in the PLCs. Once the production line is switched off, these data are lost and after re-energizing, the PLC will re-load the current status [1].

The production line consists of several stations. The production process begins at stations in Zone 1, either by continuous or discrete material. Zone 1 A contains three stations, namely the Shaker conveyor station, the Quality sorting station and the Corn dosing station, where the discrete material is prepared for filling the bottles. Zone 1 B contains four stations. Filtering station, Mixing station, Reactor station and Quality-probe station are used to filter and mix the correct liquid ratios. In Zone 2, there is the Bottling (Filling) station where are the individual bottles filled with liquid or discrete material, and the bottles are also closed on the rotary table. Subsequently, in the Packaging station, the filled bottles are placed in containers with a capacity of 2×3 to transfer to the conveyor, which is provided by a two-axis automated industrial manipulator. The bottles can then proceed to Zone 4, where an automated rack stacker, i.e. a Storage station, is located. The station is used for automated storage of containers. It is able to store up to 16 crates on 4 floors and storage of crates is conducted by a tree-axis Cartesian manipulator [2].

© Springer International Publishing AG, part of Springer Nature 2019
R. Silhavy (Ed.): CSOC 2018, AISC 763, pp. 208–215, 2019.
https://doi.org/10.1007/978-3-319-91186-1_22

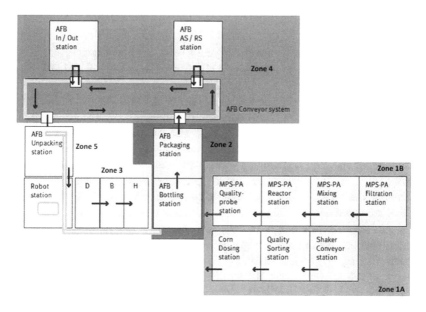

Fig. 1. Model of production line [2]

In Zone 4 there is also an In/Out Station which is used for storage filled crates at two loading ramps and to return empty crates through an input conveyor. Manipulation is provided by a three axis pneumatic manipulator with pneumatic linear clamp. Zone 5 provides bottles from crates to Unpacking Station. Bottle unloading is secured by a two-axis industrial manipulator equipped with a clamp capable of grabbing three bottles at once. The bottles are transported with the Recycling Station via a conveyor. Here the bottles are opened and their content is emptied. The station contains an industrial robot that removes the bottle cap and put it in the Distribution station. The station contains a pump to suck the liquid content of the bottles and the vacuum pump with suction of discrete material. Empty bottles are conveyed via a conveyor to Zone 3. The first is the Distributing Station, which provides the dosing of semi-finished products for the manufacture of shutters. The station is able to pick individual pieces of the product from the gravity tank and, through a simple manipulator, handles them to an associated station for further processing. The process continues on the Buffer Station, which provides building up the stock and to separate the semi-finished products in the production process. The insertion of the semi-finished products into the storage is detected by the optical sensor. The activity of the storage is controlled by light barriers placed in front of and behind the separator. When the output position is released, the separator can separate one piece of the semi-finished products. The last station in Zone 3 is the Handling Station, which is capable of recognizing and sorting two types of semi-finished products according to color. The sorting is done with a dual-axis pneumatic manipulator with a pneumatic clamp. This clamp includes an optical sensor capable of recognizing the black color. After recognizing the color, it can place the blanks in the appropriate tray. The following figure shows the model of the production line described (see Fig. 1) [2].

2   Methods

Wonderware is the extensive software for real-time operations management, including Supervisory Human Machine Interface (HMI), Supervisory Control and Data Acquisition (GeoSCADA), Mobile Operations, Production Management, Manufacturing Execution System (MES), Performance Management, and Enterprise Manufacturing Intelligence (EMI) work flow. It is also the leader in integration with asset management, supply and demand chain, and Enterprise Resource Planning (ERP) applications.

Wondereare software deliver significant cost reductions associated with designing, building, deploying, and maintaining secure and standardizes applications for manufacturing and infrastructure operations. Wonderware solutions enable companies to synchronize their production and industrial operations with business objectives, obtaining the speed and flexibility to attain sustained profitability [3].

The ArchestrA technology is an open and extensible system of components based on a distributed, object-oriented design. By leveraging Microsoft.NET and other Microsoft technologies, it provides a productive toolset for building critical operations management software solutions. Rather than having to program applications, they can be assembled using software objects, reducing development time.

ArchestrA is essentially a group of services that run in the background on any computer to witch the software has been installed. Of these services, the most important is the aaBootstrap.exe service. The bootstrap manages all of the basic layers of communication between all computers that have the bootstrap running [4] (Fig. 2).

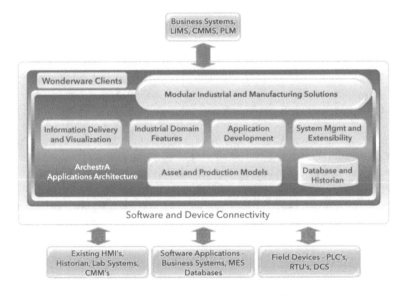

Fig. 2. Model of ArchestraA application architecture [5]

The ArchestrA Framework supports core services that are required by the most of the different types of manufacturing systems. These core services include the following:

- Integrated Development Environment (IDE);
- Version management;
- License management and centralized deployment;
- System diagnostics and system administration;
- Internationalization;
- Data visualization and monitoring;
- Event-based processing, scripting, and calculation capabilities;
- Alarm and event management, historization, and security;
- Data acquisition and field device integration;
- Interobject communications and name service;
- Reporting and ad-hoc query capability;
- Support for industry standards, such as OPC and SQL [4].

The System Management Console (SMC) provides ArchestrA Application Server application diagnostic by allowing to view the runtime status of some system objects and to perform actions upon those objects. The SMC is Microsoft Management Console (MMC) container snap-in for all of the diagnostic and management utilities for Galaxy application. Actions include setting platforms and engines in an executable or idle mode and starting and stopping platforms and engines.

The console tree shows the items that are available in the console. Other snap-ins that may appear below the ArchestrA SMC node includes the Galaxy Database Manager, the Log Viewer, and the DAServer Manager.

DAServer Manager allows local or remote configuration of the DAServer and its device groups and items, and can monitor and perform diagnostic on DAServer communication with PLCs and other devices. DAServer Manager is also a Microsoft Management Console snap-in. Many high-level functions and user-interface elements of the DAServer Manager are universal to all DAServers [4].

3   Results

The following figures show the BPMN diagrams of set up process of communication between the production line and Wonderware via SMC and ArchestrA.

After creating a connection by adding PLC IP addresses to the SMC module (see Fig. 3), it is also necessary to create communication in the ArchestrA by creating individual topics and importing signals from CSV files (see Fig. 4).

The communication was created by the Wonderware DAServer, where first it was necessary to connect each PLC by adding its own specific IP address to the SMC module, more precisely to tree item ArchestrA.DASSIDirect followed by their configuration (see Fig. 5).

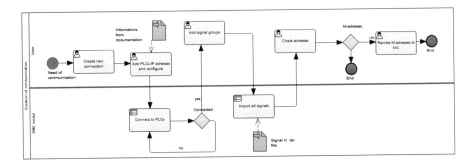

Fig. 3. Creating of communication

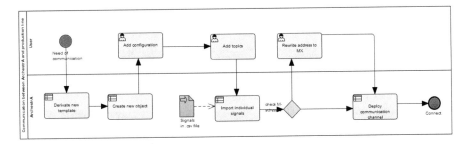

Fig. 4. Completing of communication in ArchestrA

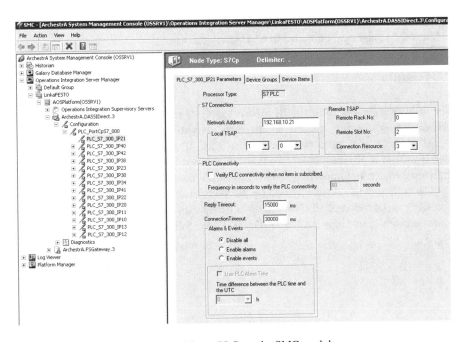

Fig. 5. Adding a PLCs to the SMC module

From the documentation for the each stations, it was necessary to add every group of signals that the production line contains. Alternatively, to set the update interval according to requirement and importance (see Fig. 6).

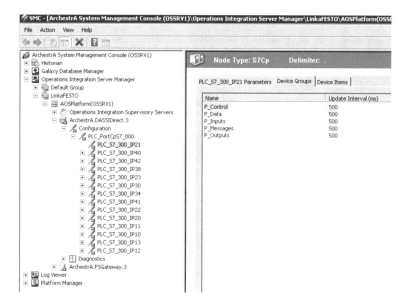

Fig. 6. Adding groups of signals of each station

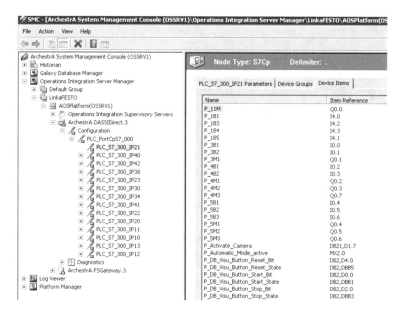

Fig. 7. Import of signals

The next step was the import of specific signals that each station contains. The import is conducted by using a CSV file from the original documentation to the production line. The only change that had to be made was to overwrite M-address of some signals to MX-address considering Wonderware did not support signals with such a reference (see Fig. 7).

The last step was to implement a communications link in ArchesteA using the template from the toolbox called DDESuiteLinkClient. First, we need to derivate a new template and then derivate a new object that will be used for the production line model. After the addition of the individual signal groups, the import of specific signals takes place, but not for the whole station, as in the SMC module but directly for the signals belonging to specific groups (see Fig. 8).

Fig. 8. Implementation of communication in ArchestrA

4   Conclusion

Thanks to the communication between the production line and the Wonderware, we can use the other options that Wonderware provides. The next step will be to create a complete production line model, including individual objects and their attributes. While setting the properties of these objects, we will also use Historian to store history of data. Wonderware provides tools for processing of huge data collection and data storage in Historian, as well as trending, which also contributes to better evaluation for further analysis. Afterwards, we create visualization of the entire line and individual objects using InTouch [6, 7]. Stored data we will be able to analyze [8]. We will see the state of the production line status thanks to visualization and then the next step is to create a production line control using Wonderware MES.

Acknowledgments. This publication is the result of implementation of the project VEGA 1/0272/18: "Holistic approach of knowledge discovery from production data in compliance with Industry 4.0 concept" supported by the VEGA.

This publication is the result of implementation of the project: "Increase of Power Safety of the Slovak Republic" (ITMS: 26220220077) supported by the Research & Development Operational Programme funded by the ERDF.

References

1. Stenerson, J., Deeg, D.: Programming Siemens Step 7 (TIA Portal), a Practical and Understandable Approach. CreateSpace Independent Publishing Platform (2015)
2. Tempest: Production line description. User manual. Tempest (2014)
3. Wonderware: Wonderware software. https://www.wonderware.com. Accessed 01 Oct 2018
4. Wonderware: Application Server. Training manual. Schneider Electric Software (2015)
5. Direct industry: Wonderware. http://www.directindustry.fr/prod/wonderware/product-33737-1657215.html. Accessed 01 Oct 2018
6. Zhang, X., Jie, Z., Ke, L.: Design and implementation of control system for beer fermentation process based on SIMATIC PLC. In: 27th Chinese Control and Decision Conference, CCDC, Qingdao, pp. 5653–5656 (2015)
7. Wonderware: Wonderware InTouch. https://www.wonderware.com/hmi-scada/intouch/. Accessed 01 Oct 2018
8. Simoncicova, V., Tanuska, P.: Creating a management view on key indicators using business intelligence in small and medium enterprises. In: 28th International Conference on Cybernetics and Informatics (K and I), IEEE Czechoslovakia Sect CS Chapter, Levoca (2016)

The Semantic Models of Arctic Zone Legal Acts Visualization for Express Content Analysis

A. V. Vicentiy[1,2(✉)], V. V. Dikovitsky[1], and M. G. Shishaev[1,3]

[1] Institute for Informatics and Mathematical Modeling – Subdivision of the Federal Research Centre "Kola Science Centre of the Russian Academy of Science", 24A, Fersman st., Apatity, Murmansk region 184209, Russia
alx_2003@mail.ru, dikovitsky@gmail.com
[2] Apatity Branch of Murmansk Arctic State University, Lesnaya st. 29, Apatity, Murmansk region 184209, Russia
[3] Murmansk Arctic State University, Egorova st. 15, Murmansk 183038, Russia
shishaev@arcticsu.ru

Abstract. Currently, large amounts of data are available in text form. However, due to the characteristic features of the text in natural languages, the development of fully automatic methods for analyzing the semantics of texts is a difficult task. This paper describes the composition, structure and some areas of application of the developed technologies of semantic analysis and visualization of semantic models of text documents. Also, methods for visual express content analysis of documents are described. These methods are part of the technology for visualizing semantic models of text documents and implemented as independent software tools. To demonstrate the main features of the technology, the experience of using the visualization of semantic document models for visual express content analysis of legal acts regulating the development of spatially-distributed systems of various levels and analysis of the results is described in detail. The final part of the paper identifies some promising areas of application of the developed technologies, as well as determines the main directions for further work and the possibilities to expand the functionality of the methods of visual express content analysis of text documents.

Keywords: Documents visual analysis · Content analysis
Human-computer interface · Management of spatially-distributed systems
Tensorflow · TF-IDF

1 Introduction

With the increasing number and volume of documents used to solve various kinds of applied problems, it becomes increasingly difficult to allocate sufficient time for their study. This leads to spreading of situations, when one has to give up the detailed study of the document, confining himself to a quick review in order to "grasp" the basic meaning. Thus, we sacrifice the accuracy of understanding the meaning of the document in favor of the speed of its interpretation. With respect to text documents, this process is often called "diagonal reading". With diagonal reading we get some set of main

© Springer International Publishing AG, part of Springer Nature 2019
R. Silhavy (Ed.): CSOC 2018, AISC 763, pp. 216–228, 2019.
https://doi.org/10.1007/978-3-319-91186-1_23

thoughts generalizing meaning of the document. Having received a general idea of the meaning of the document, we put it as an integral object into our mental system of concepts and, if necessary, we can "extract" it for more detailed study. In this work we make next step in this direction: obtaining a most general view of the document's meaning which can be expressed in one image interpreted by one look. We shall call this process a "quick sense catching". By this we can apply visual analysis approach to analyze semantic of text documents.

Working with electronic documents incarnates the task of quick sense catching in different situations. In this paper three cases are considered:

1. Presenting a search results. Even the most complicated search engine guarantees only the relevance of the result, that is, its correspondence to the formulated query, but not to the actual expectations of the user. So, the last is an ultimate property of the search result, it is called pertinency. For its a priori assurance, it is possible to use different approaches to the organization of search engines, including those proposed by the authors [1]. However, for this it is necessary to have some model of mental stereotypes of each user, which is extremely difficult to provide in mass-use information systems. Therefore, a potentially effective approach may be to accompany the document in a search result list by an image, characterizing its generalized meaning. This will provide a posteriori rapid assessment of the matching of search results to user expectations. In this case, the main task of visualization is to make sure that the user briefly understands the meaning of the document.

2. Identification of conflicting documents. This case is based on the assumption that documents that are identical in terms of a set of concepts but have different semantic structure will probably contradict each other. The technique of quick sense catching will help to identify such potential conflicts in the case of a large number or large amount of source documents. According to initial assumption the main task of visualization in this case is to reflect similarity of terms bases used within documents under consideration, as well as a strength of relations between terms.

3. Division a set of documents into groups of similar in meaning. Although the usual practice is the a priori distribution of similar documents to different folders, in many cases a folder may collect a huge count of documents which forces a user to cluster the set of documents. This case is similar to the previous one, except that it is not necessary to understand the meaning as such for splitting a lot of documents into similar clusters - it is enough to identify just similarity of meaning of two or more documents. Potentially, this makes other methods of visualization of the generalized sense of the document more effective.

In this work we consider two stage way to obtain general semantic view of the document. At the first stage in the automated mode, the semantic analysis of the document is carried out. This is done by applying content analysis and lexicographic analysis methods to text documents. The first is implemented by counting TF and other measures while the second - by neural network trained for semantic resolving of text sentences. The first stage is resulted in a weighted semantic network that characterizes the frequency of use of concepts in a document, as well as the presence and strength of semantic links between concepts. This semantic image of the document is visualized as a single whole view on the second stage. Then it is being presented to user for quick catching the sense of the document. So, proposed approach is a combination of semantic and visual analysis.

Needs to quickly process constantly increasing volumes of information within text documents appears in very different spheres of activity. In this paper we look at the problem from regional administrative management perspective. There are a lot of objects of management in this field with quite complicated relations caused by administrative hierarchy, territorial references and other factors. This creates a fertile ground for the emergence of a large number of documents with overlapping, as well as contradictory meanings. As an example for the study, we took two documents that reflect Russia's strategic priorities in the Arctic region:

1. Strategy of social and economic development of the Murmansk region until 2020 and for the period until 2025 (hereinafter referred to as the «MR Strategy») [2].
2. Strategy for the development of the Arctic zone of the Russian Federation and ensuring national security for the period until 2020 (hereinafter referred to as the «AZRF Strategy») [3].

Having been issued at different administrative levels and, obviously, having intersecting subjects of consideration (the Murmansk region is entirely part of the Arctic zone of the Russian Federation), these documents are a good example for consideration.

The rest of the article is divided into two main parts. In the first part approaches, methods and software tools for forming the semantic model of a document are considered. The second part presents the results of the formation of the semantic images of the above documents, as well as preliminary conclusions about the perspectives and problems of using different visualization methods for quick sense catching in the context of three problems listed before.

2   The Technology of Document Semantic Model Construction

There are various methods of extracting formalized semantics are used for text analysis. They are based on apparatus of linguistics, statistical methods, mathematical logic, cluster analysis, methods of artificial intelligence and other methods and techniques. Approaches to the text processing and analysis can be divided to statistical and linguistic. The statistical approach is based on the assumption that the content and, partly, semantic of the text is reflected by the most frequently encountered words. The essence of statistical analysis is the accounting of occurrences of words in a document or sentence. A common approach is to assign each term t in the document of a non-negative weight. Weights of terms can be calculated in many different ways. The simplest of them is to put "weight" equal to the number of occurrences of term t in document d, denoted by tf (term frequency) [4]. This weighing method does not take into account the discriminatory power of the term. Therefore, when the statistics of the use of terms in collection of documents are available, works best scheme tf-idf weight computation, defined as follows:

$$tf - idf_{i,d} = tf_{i,d} \times idf_i, idf_i = \log \frac{N}{df_i} \qquad (1)$$

df - document frequency, defined as the number of documents in the collection, property term t, idf - inverse document frequency of term t, N - the total number of documents in the collection. The main drawback of statistical methods of text analysis is the impossibility of taking into account a deep semantics. Using of linguistic methods of text analysis along with statistical ones allows overcoming this drawback.

The proposed technology for text analysis includes several stages: graphematic, morphological, syntactic and semantic analysis. The results of each level are used by the next level of analysis as input. The purpose of the morphological analysis is to determine the morphological characteristics of the word and its initial form. The purpose of syntactic analysis is to determine the syntactic dependence of words in a sentence. In connection with the presence in the Russian language of a large number of syntactically homonymous constructions the procedure for automated text parsing based on syntax is insufficient to determine the dependencies between the concepts of a sentence. The complexity increases exponentially with an increase in the number of words in the sentence and the number of used rules. The semantic stage determines the formal representation of the meaning of words and constructions composing the input text. The formation of the semantic domain model (SDM) as a structure of weighted semantic relations on the basis of a collection of documents allows to account for and store the context of concepts, and take into account the various forms of syntax transmission, and solve the problem of the equivalence of words. SDM makes it possible to implement procedures for extracting and storing the contexts of words, partially solving the problem of compatibility of new information with knowledge already accumulated, and also to reveal contradictions in the semantic images of documents, in case new information contradicts the accumulated information. The procedure for formation of the SDM is presented below.

First the text is divided into sentences and subjected to a graphematical analysis and lemmatization. For the syntactic analysis and definition of the morphological characteristics of words, we use the grammatical dictionary of the Russian language [5], thesaurus WordNet [6], and also based on the tools for neural networks TensorFlow [7] SyntaxNet library. A feature of the application of neural networks is the ability to analyze the morphology and syntax for words that are not presented in the thesaurus. TensorFlow is an open-source software library. The library functions at the level of specifying the architecture of the neural network and its parameters. The data in TensorFlow are represented as multidimensional data sets with variable size - tensors. The set of morphological features defined by SyntaxNet, grammatical categories, and types of dependencies are specified in the Universal Dependency notation [8]. Universal Dependencies (UD) is a framework for cross-linguistically consistent grammatical annotation and an open community effort with over 200 contributors producing more than 100 treebanks in over 60 languages. Sentences are sent at the input of the neural network. Words of sentences are converted into a vector by the Word2Vec library. In the preliminary training phase Word2Vec [9] get the text corpus and provides vocabulary vectors as output. Vector word representations allow us to calculate the semantic closeness between words. Since Word2Vec algorithms are based on training a neural network, in order to achieve effective work, it is necessary to use large text corpus for training. Pre-prepared vectors obtained on the Google News [11] dataset part are available. The model contains vectors for 3 million words and phrases. Phrases were obtained using the skip-gram approach described in [10]. Further sentences in vector form are submitted to the input layer of the neural network which implemented in TensorFlow and trained on the Universal Dependences corpus. The Russian text corpuses in the Universal Dependen-cies are represented by the converted SinTagRus [12] and Google Russian Treebank [13]. The result of SyntaxNet is a dependencies tree of sentence and morphological characteristics of words. The result of the analysis is a weighted semantic image of document. The semantic image of the document is a semantic net, the set of concepts and the set of edges - a set of relationships over concepts:

$$D = \{C^D, L^D\}, C^D \subset C, L^D \subset L \tag{2}$$

where C^D – set of concepts in document, L^D – set of relation.

SDM is formed as a result of the integration of the semantic images of documents. SDM formally is represented as an n-ary weighted semantic network:

$$KB = \{C, L, Tp\}, L = \{l\}, l = <c_i, c_j, tp, w>, c_i, c_j \in C, tp \in Tp \tag{3}$$

where C – a concepts set, L – a relations set, w – an importance parameter of relation, Tp – a relations types set. The weight of relationship characterizes the significance of the semantic relation between concepts, determined on the basis of statistics of the occurrence of concepts. The structure of the semantic analysis service is shown on Fig. 1.

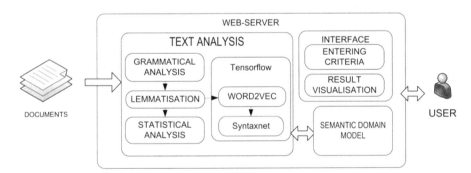

Fig. 1. The structure of the semantic analysis service

With this service, we analyzed two documents - "AZRF Strategy" and "MR Strategy". In this work we received two weighted semantic networks. In total it amounted to 1993 words and 2957 relations for the first document and 7847 words and 18044 relations for the second. The next step was to visualize them in a way providing an express content analysis and allowing a quick sense catching.

3   Experience in Visualization of Semantic Models of Documents in the Context of Various Tasks of Visual Express Content Analysis

Some examples of using methods of visual express content analysis of documents are given below to demonstrate the main features of the visualization technology of the text documents semantic models. These methods are part of the visualization technology of semantic models of text documents. The methods can be used both in a combination, and separately. Using the methods and interpretation of the results depends on the problem solved by the user.

For visual content analysis of the above-mentioned "MR Strategy" and "Strategy of the AFRF" we used three methods that were implemented by us as an independent software tool:

1. "Top 100 Words" - this method visualizes one hundred most significant concepts that characterize the document. Visualization is carried out in the form of a list of concepts of the document, ordered by the weight of concepts;
2. "Semantic Network" - this method visualizes the semantic model of the document in the form of a semantic network. The semantic network of the document consists of the concepts of the document and the various types of relationships between them;
3. "Semantic Networks Imposition" - this method visualizes the semantic models of documents in the form of a circle. The concepts of the document are placed on the circumference, and the relationships between them are displayed inside the circle.

Experience in the practical application of these methods for visual express content analysis of "MR Strategy" and "Strategy of the AFRF" and a brief interpretation of the results are described below.

3.1   Experience of Using the "Top 100 Words" Method

The first method for which we describe our experience of using for visual express content analysis of documents is the "Top 100 Words" method. This method, on the one hand, ensures the rapid creation of a document visual image, even for relatively large documents, and, on the other hand, is well suited to assess the subject matter of the analyzed document in general. In fact, this method visualizes the most important words of the document, which can be considered key to describing its content.

A statistical measure TF-IDF is used to calculate the importance or weight of a word. In the current implementation of the method, the absolute value of the importance of words is used. That is, when calculating the weight of a word, the volume of the document is not taken into account.

We applied the "Top 100 Words" method to "MR Strategy" and "AZRF Strategy" independently of each other. A fragment of the results of the method is shown in Fig. 2. The figure shows only eleven concepts (words) with the maximum weight from each document.

"MR Strategy"				"AZRF Strategy"		
	WORD	TF-IDF			WORD	TF-IDF
1	development	161		1	Arctic	190
2	region	141		2	Russian	173
3	area	126		3	Federation	161
4	Murmansk	120		4	development	90
5	state	93		5	provision	79
6	population	79		6	zone	68
7	rise	65		7	state	61
8	provision	63		8	system	52
9	Russian	62		9	security	33
10	creation	48		10	activities	30
11	system	42		11	population	26

Fig. 2. Results of document analysis using the "Top 100 Words" method (fragment)

Depending on the analysis task, the user can interpret the results in different ways. But even at first glance, it is clear that from a set of words with a maximum weight of more than 50% of the words are the same in both documents. This may indicate that these documents have a common theme and similar terminology. On the basis of this conclusion, we can say that when we split a collection of documents into several clusters, these documents can get into the same cluster with high probability, since they have some commonality.

The results of document analysis using the "Top 100 Words" method can be used to improve the efficiency of document search tasks at the stage of search results analysis. To improve the effectiveness of the search results analysis, each document should be supplemented with a visual image of the document. In this case, the visual image of the document is a list of the main concepts of the document. The list of concepts is ordered by the value of the concepts weights.

However, to solve the problems of identifying potential contradictions in documents, the "Top 100 Words" method is not suitable, as it does not provide any additional information about the document.

This method has some limitations. In fact, the "Top 100 Words" method allows to configure the number of concepts that are displayed on the screen. This raises the question of how many concepts are optimal. On the one hand, the number of concepts should be sufficient to solve the user's problem, but on the other hand, the number of concepts should not be too large. If display too many concepts, then the complexity of the problem of visual analysis of these concepts becomes comparable to the complexity of the problem of analyzing the source document. In this case, the application of the "Top 100 Words" method for visual express content analysis of the document becomes meaningless.

If to take into account the peculiarities of perception of visual information, in particular the "magic number Miller" [14], and display from 5 to 9 concepts with the maximum weight in the case of large documents, there is a high probability that the results of the visual analysis of a small number of concepts will be useless to the user.

At the moment, we do not have an unambiguous answer to the question about how many concepts are optimal for visual express content analysis of a document. We will try to find a solution to this question in the future work.

3.2 Experience of Using the "Semantic Network" Method

The second method for which we describe our experience of using for visual express content analysis of documents is the "Semantic Network" method. The result of the visualization is the semantic network of the document.

This semantic network is not a static visual image of the document. The user can interact with it interactively: deploy network nodes, investigate the links between the document's main concepts, visualize the semantic network of the document, starting with the user-defined concept, limit the number of displayed concepts, change the display parameters of the semantic networks, etc.

To demonstrate the capabilities of the "Semantic Network" method, we visualized the semantic models of the documents "MR Strategy" and "AZRF Strategy". The "development" concept was chosen as the "entry point" for visualizing the semantic network of both the first and second document. We chose this concept as the "entry point" of visualization for two reasons. First, based on the results of using the "Top 100 Words" method, this concept was included in the list of the most important words in both "MR Strategy" (first place in the list) and in "AZRF Strategy" (fourth place in the list). Secondly, the sum of weights of "development" concept for the two documents is the highest in comparison with the sum of the weights of other pairs of matching concepts.

For comparison, the visual images of the semantic networks "MR Strategy" and "AZRF Strategy" obtained using the "Semantic Network" method are shown in Fig. 3 (view with zooming).

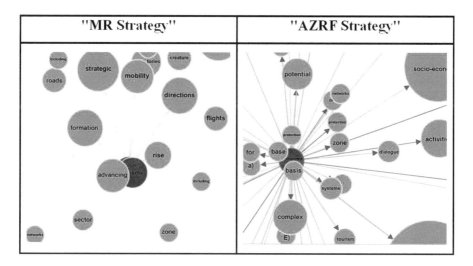

Fig. 3. Visual representation of the analyzed documents semantic networks (with zooming).

Visual analysis and work with semantic networks of documents is carried out by the user in accordance with predetermined analysis tasks. For more detailed analysis, the user can use a variety of options for interacting with semantic networks (zooming, opening nodes, mapping the types of links, rebuilding the semantic network, etc.).

Even without the calculations, relying only on the visual analysis of semantic networks of documents can be done many different conclusions. In particular, it can be found that the number of concepts associated with the concept of "development" in "MR Strategy" significantly exceeds the number of concepts associated with the concept of "development" in "AZRF Strategy". This may indicate that more attention is paid to the development of various activities in "MR Strategy" than in "AZRF Strategy". By zooming in on the visual display of semantic networks, it is possible to examine in more detail the links between the concept of "development" and other concepts. Also, it is possible to determine the development of which activities are given special attention in the document.

This example of visual express content analysis of documents is not the only option for using the "Semantic Network" method. In particular, this method can be used to solve problems of increasing the pertinence of the search results for a collection of documents and other tasks for which the weight and connections of concepts in the analyzed document are important. As for the task of identifying potential contradictions in documents, the «Semantic Network» method can only provide a preliminary answer about the potential for contradictions in documents. To obtain a more accurate answer, additional efforts are required from the user, connected with a deeper analysis of the meaning of concepts and the types of connections between them.

The "Semantic Network" method can also be used for visual clustering of documents. To effectively solve this problem, it is important to ensure the same arrangement for the same concepts. In general, the task of visualizing concepts in space relative to each other is a non-trivial task. In particular, the Word2Vec library, which is used in semantic document analysis technology, uses 300-dimensional vectors for words and phrases. Displaying words in a 300-dimensional space does not make sense for a visual express content analysis of a document. Therefore, it is necessary to reduce the visualization space of the semantic model of the document depending on the user's task. The method of reducing the visualization space of the semantic model of the document, depending on the user's task, will be developed in the future work.

As for the "Top 100 Words" method, the question of the optimal number of concepts and links that should be displayed to the user for visual express content analysis of the document remains also relevant for the "Semantic Network" method. It is worth noting that, due to the user's interaction with the visual image of the semantic network of the document when using the "Semantic Network" method, this question is not extremely important. First of all, this is due to the fact that the user can independently control the visualization settings, such as the number of concepts to be displayed, the minimum weight of the concept to be displayed, the folding and unfolding of network nodes, etc. Thus, the user can independently configure the optimal display of the semantic network of the document, taking into account the task to be solved and the individual features of the perception of visual information.

The decision on the optimal way to visualize the semantic network for the user requires additional research. This question will be considered in more detail in the future work.

3.3   Experience of Using the «Semantic Image Imposition»

For a visualization of the differences in the structure of semantic relations between the concepts of the two documents, we used the library D3.js Hierarchical boundary binding. D3.js is a JavaScript library for manipulating documents based on data. The concepts that are present in both the compared documents are arranged identically in both diagrams. A line between concepts represents semantic relations between these concepts in the semantic image of each document. Visualization of the semantic images of the documents "MR strategy" and "AZRF Strategy" are presented in Fig. 4.

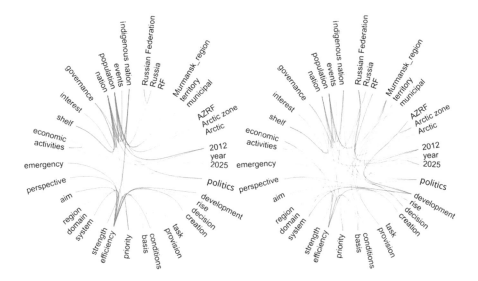

Fig. 4. Visualized semantic images of the documents "MR strategy" and "AZRF Strategy"

Analysis of visualized semantic images allows us to identify the most frequently encountered relations between the concepts. At the same time, it is possible to quickly determine the difference between the structures of relations in different documents. Based on this analysis it is possible to form hypotheses about the importance of the relations between the most important concepts of the analyzed documents. For example, the obvious conclusion arises about the greater detail and elaboration of the second document, while the former operates with more general concepts. Let's consider the structure of the relations of the concept "development" in the visualization of semantic images of both documents. The concept of development is considered in the first document in the context of the performance indicator, while in the second document development is considered in the context of specific places and time. It might be an indirect indication of the presence of inconsistencies and contradictions in those documents.

However, in order to confirm this assumption, it is necessary to conduct a deeper analysis and comparison of the semantics of the documents.

4   Conclusions

In this paper, we described our experience in using the visualization of document semantic models for visual express content analysis. As methods for visual express content analysis of documents the "Top 100 Words", "Semantic Network" and "Semantic Networks Imposition" methods were considered. Using the methods of visualization described in the article, the following conclusions about the analyzed documents were obtained:

1. In the analyzed documents, the list of the most significant concepts is similar. The experience of applying the "Top 100 Words" method, described above, showed that the coincidence of concepts is more than 50%.
2. The number of connections between the concepts in these documents varies significantly. The visualization of the semantic model of documents, carried out using the "Semantic Network" method allows us to estimate the difference in the number of connections visually, without performing calculations.
3. In the document "MR Strategy", the connections between the concepts are described in more detail than "AZRF Strategy". This conclusion can be made on the basis of analysis of visualization, carried out by the "Semantic Networks Imposition" method.

Also we can make some conclusions about the methods:

1. Based on our experience of document analysis, it can be concluded that the methods of visual express content analysis considered in this paper are not universal.
2. The effectiveness of a particular method depends on the type of problem being solved (clustering, increasing the pertinence, revealing contradictions).
3. The examples of visual analysis of documents described in the paper allow us to say that all considered methods of visual express content analysis can be successfully applied in solving problems for which it is sufficient to understand the general meaning of a document based on a visual display of its most significant semantic aspects.
4. Based on our experience, we can make a preliminary conclusion that the "Top 100 Words" method is better used in document clustering tasks, "Semantic Network" - for quick understanding of the document essence and increasing the pertinence of search results, and "Semantic Networks Imposition" - for comparison and identification of potential contradictions in the document structure.

But the final conclusions can be made only after verification of these methods on a large collection of documents.

5   Future Work

In general, our experience in using the visualization methods for express content analysis of the legal acts regulating the development of spatially distributed systems can be assessed as positive. The visualizations produced by these methods make it possible to obtain new knowledge about the analyzed documents without a detailed study of the contents of the documents.

In the future work, we plan to conduct a number of more detailed experiments for large collections of documents to find answers to the following questions:

1. For what types of tasks the visualization of document semantic models for express content analysis might be the most effective?
2. How does the effectiveness of visual analysis depend on the characteristics of the analyzed document?
3. How to present the results of visualization to the user in the best way?
4. How can visualization be reduced taking into account the characteristics of the analysis task?

Acknowledgements. The reported study was funded by RFBR and Ministry of Education and Science of Murmansk region (projects № 17-47-510298 p_a, 17-45-510097 p_a) and by RFBR according to the research project № 18-07-00132 A.

References

1. Shishaev, M.G.: Architecture and technologies of knowledge-based multi-domain information systems for industrial purposes. In: Dikovitsky, V.V., Shishaev, M.G., Nikulina, N.V. (eds.) Automation Control Theory Perspectives in Intelligent Systems. Proceedings of the 5th Computer Science On-line Conference 2016 (CSOC2016), vol. 3, pp. 359–369 (2016)
2. Strategy of social and economic development of the Murmansk region until 2020 and for the period until 2025. http://docs.cntd.ru/document/465602093
3. Strategy for the development of the Arctic zone of the Russian Federation and ensuring national security for the period until 2020. http://docs.cntd.ru/document/499002465
4. Salton, G., McGill, M.J.: Introduction to Modern Information Retrieval. McGraw-Hill, New York (1983). ISBN 0-07-054484-0
5. Zaliznyak, A.A.: Grammatical dictionary of the Russian language. http://odict.ru/
6. Thesaurus of the Russian language WordNet. http://wordnet.ru/
7. TensorFlow. https://www.tensorflow.org/
8. Universal Dependencies. http://universaldependencies.org
9. Word2Vec. https://code.google.com/archive/p/word2vec/
10. Mikolov, T., Sutskever, I., Chen, K., Corrado, G., Dean, J.: Distributed representations of words and phrases and their compositionality. In: Proceedings of NIPS (2013)
11. Google News. https://drive.google.com/file/d/0B7XkCwpI5KDYNlNUTTlSS21pQmM/edit?usp=sharing

12. SinTagRus. http://www.ruscorpora.ru/search-syntax.html
13. Google Russian Treebank. https://old.datahub.io/dataset/universal-dependencies-treebank-russian
14. Miller, G.: The Magical Number Seven, Plus or Minus Two. The Psychol. Rev. **63**, 81–97 (1956)

Proposal of a DTN Routing Scheme for Educational Social Networks in Developing Countries

Takahiro Koita[✉] and Shin Harada

Doshisha University, Kyoto, Japan
tkoita@gmail.com

Abstract. In developing countries where education is lacking, there are many regions without telecommunications infrastructure. The DTN (Delay and Disruption Tolerant Network) is a method of communication that does not rely on communications infrastructure, and one that can be used for reliable data transfer even in adverse environments. In this study, we aim to promote educational support in developing countries without telecommunications infrastructure using DTN and the educational SNS, Edmodo. We propose a system of data communication without waste, which, by considering the time in which children are free, limits distribution of content to those children who do not have free time, and distributes content as a priority to children who do have free time. Additionally, by comparing the proposed system with the conventional system in which data communication was conducted indiscriminately for all children met, we demonstrate the effectiveness of the proposed system in promoting educational support in developing countries that lack telecommunications infrastructure.

Keywords: DTN · Educational social networks · Routing scheme

1 Introduction

In developing countries, there are a number of reasons why children are unable to attend school [1]. In this study, we envisage providing support for the education of children who cannot go to school by using a SNS for educational purposes, known as Edmodo [2], to publish classroom videos and allow for the submission of homework. Edmodo, which can be used as a SNS, is the world's largest educational platform, and is used by more than 70 million users in 190 countries worldwide. A wide variety of content is shared around the world and can be used freely. Edmodo has two features. First, it can be used anytime and anywhere; since Edmodo does not depend on a specific device or OS and can be used wherever there is a communications environment, as long as the communications environment is set up, learning can take place anytime and anywhere. Second, it can be used free of charge. Since there is no financial burden involved in its deployment, it can be used on a casual basis. The fact that Edmodo can be used free of charge, and anytime and anywhere, makes it effective when providing educational support to children who are unable to go to school in developing countries. However, in developing countries, the telecommunications networks required for running Edmodo are not set up [3], and it is therefore difficult to share distributed content information.

© Springer International Publishing AG, part of Springer Nature 2019
R. Silhavy (Ed.): CSOC 2018, AISC 763, pp. 229–238, 2019.
https://doi.org/10.1007/978-3-319-91186-1_24

For this reason, DTN (Delay Tolerant Network) has been proposed as a communications system that can be used even when continuous telecommunications routes are not established [4, 5].

The purpose of this study is to provide educational support, through the use of Edmodo and DTN, in developing countries that lack telecommunications infrastructures. In developing countries, a communications environment like DTN is required to use Edmodo. However, DTN is not a dedicated communications environment for Edmodo, and since communications environments that use DTN have many interruptions and are unable to grasp the status of other nodes, when sending data to the destination node, a method in which the node with data sends indiscriminately to all nodes it meets is the fastest way to ensure that the data arrives at the destination [6]. The main form of data used by Edmodo are high-volume classroom videos, and where a wide variety of data is distributed daily, if reproduced data is transmitted indiscriminately to all nodes it meets, the nodes with a lot of free time will delete the data after using it; therefore, this will not have much impact on storage capacity. However, if too much data is sent to nodes without much free time, unused data will accumulate in the terminals, and this will stretch the storage capacity. When the storage capacity is stretched, there is a possibility that necessary data will not be communicated. To prevent such a lack of storage capacity, it is necessary to restrict indiscriminate communication and delete the reproduced data. By taking free time into consideration, communication to nodes with less free time is restricted, and this is considered to be effective in preventing a shortage of storage capacity. In this study, we propose a method of restricting the communication of nodes by considering the free time of children.

2   DTN

DTN is referred to as a delay-tolerant network, and is a communications system that has been proposed for communications environments in which continuous telecommunications routes cannot be established. A precondition for the legacy Internet communications protocol TCP/IP (Transmission Control Protocol/Internet Protocol) was that the intermediate networks can all be used constantly, making it vulnerable to interruptions due to movement of terminals, etc., and meaning that networks could only be constructed within a limited range. DTN is used as a communications system that can be utilized even in cases where there are frequent interruptions to communication. With DTN, a node with a fixed range of communication moves around while holding data, and transmits this data to the node it meets. Here, node refers to a user with a communication terminal, such as a mobile telephone, or a drone. The node receiving the data moves around while holding the data, communicates with other nodes when they enter its own range of communication, and transmits the data. The communication environment is constructed by repeatedly moving and transmitting. Since the data is held by each node and is only transmitted when it is possible to communicate, the communication environment becomes resilient toward interruptions. An overview of DTN is shown in Fig. 1. With DTN, a DTN architecture known as the bundle layer is proposed for data transmission. Through the introduction of the bundle layer, it becomes possible to transmit data even under environments wherein there are many interruptions and delays. If there is bundle layer support, transmission occurs from one node to a separate node, and by the receiving node continuing to transmit to a separate node, the data can be sent to remote areas, even in communication environments with many interruptions and delays.

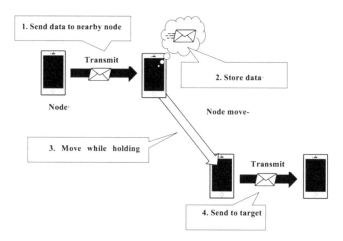

Fig. 1. Overview of DTN

2.1   Routing Scheme

With DTN, when the source node transmits data to the destination node, the data is transmitted via several nodes. Since, basically, all the nodes move, routing, which determines with what node you communicate, is very important. DTN routing schemes [7] include the following.

- Direct Routing
- Two-Hop Routing
- Epidemic Routing

Direct Routing is a system in which communication does not take place until there can be direct communication between the source node and the destination node without going via a relay node. Since there is no relay node, data transmission only occurs when the two nodes meet directly, and where there is a large distance between the two nodes in respect to the node communication range and movement range, communication often does not take place. With Two-Hop Routing, the transmission source node first creates n copies and sends those to n relay nodes it meets. The n data node receiving the data stores this in the bundle layer and moves, transmitting the data if it meets the destination node. The relay node, even if it meets other relay nodes, will not transmit the data, and will transmit only to the destination node. Epidemic Routing is a system by which nodes holding data transmit the data to all nodes they meet. This is a system in which nodes holding data will always transmit data to nodes without the data.

2.2   Epidemic Routing

Epidemic Routing is a typical DTN routing system that is also called infectious routing. Since Epidemic Routing is the fastest method with the highest arrival rate, in DTN routing studies many systems restrict communications based on Epidemic Routing. Epidemic Routing will now be explained. Epidemic Routing is a system for sending all kinds of communicable data, and is a simple routing system that does not place restrictions on communication. The routing procedure for Epidemic Routing is as follows.

1: The source node sends data to all nodes within its own communication range.
2: The node receiving the data stores the data in the bundle layer.
3: The node receiving the data, while moving, confirms whether the nodes existing in its own range of communication have the data or not.
4: It transmits data to those nodes within its range of communication that do not have the data.
5: By repeating 2 to 4, data is transmitted to the destination node.

Since Epidemic Routing does not restrict communication, it is able to distribute data on to the network more swiftly than other routing systems, and is thus suitable for a system in which data is transmitted to all nodes. Since dispersion of the data is fast, even when focusing on a specific node, it has the highest data arrival rate to specific nodes and the shortest time to arrival. On the other hand, when transmitting data only to specific nodes, even data that has no likelihood of reaching the transmission destination node is sent. Therefore, the network is often bursting with nodes storing large quantities of reproduced data. Whereas, for DTN, it is important for the data arrival rate to be high, and the increase in reproduced data is a large problem, so an improved form of Epidemic Routing is often used to resolve this issue.

2.3   Current Issues

Routing improvements are being made to resolve the problem of reproduced data inherent in Epidemic Routing. There are two main improvements to routing. The first is a system [8, 9] in which a lifetime is given to the data stored by the node. The lifetime refers to the survival time of the data, and after the data is received by the relay node, if the time set for the lifetime is exceeded, the data will be automatically deleted. It is possible to control the increase in reproduced data through the automatic deletion of data. The second is a method in which, based on the coordinates and movement history of each node, the data is only sent to nodes that have a high possibility of being able to reach the transmission destination node; however, the issue when using DTN Epidemic Routing as a communications system for Edmodo is that it does not take the free time of children into consideration, and so causes a shortage of storage capacity in nodes without much free time.

A communications environment such as DTN is necessary when using Edmodo in developing countries; however, DTN is not a dedicated communications environment for Edmodo, and sends reproduced data indiscriminately to all the nodes it encounters. Because the data primarily consists of classroom videos, the data size tends to increase.

As a wide variety of content is distributed on a daily basis, if this is communicated indiscriminately without considering free time, even though nodes with a lot of free time can delete data after using and will hardly be affected, if too much data is communicated to nodes without much free time, the unused data will accumulate in the node, which will stretch storage capacity and may prevent necessary data from being communicated. To prevent the stretching of storage capacity, it is necessary to control indiscriminate communication and delete reproduced data. Therefore, it is necessary to take the free time of nodes into consideration, by employing tactics such as restricting the communication of data to nodes with less free time.

Although improving the routing of Epidemic Routing may cause a reduction in reproduced data, it is possible the arrival rate will decrease. So, with current DTN, there is a trade-off relationship in which, as the reproduction data decreases, the arrival rate also decreases.

3   Proposed Scheme

It is necessary to consider the free time of the node in order to reduce reproduced data, which can be the cause of stretched storage space if sent indiscriminately. In this study, we propose a DTN routing system that takes the free time of children into consideration, in order to reduce the number of reproduced data items. The transmission point of the data is determined based on free time. Data communication is carried out on the upper-level nodes, who are the nodes among the communicable nodes that have a large amount of free time. Additionally, data communication is restricted by setting a threshold for capacity based on free time. Data communication takes place as long as the capacity of the communication destination node is not exceeded. If the threshold is exceeded, only data for the communication destination node is transmitted. The threshold differs according to the level of free time on the node; the threshold is smaller the lower the amount of free time is on the node, and increases as the node has a greater amount of free time. Figures 2 and 3 demonstrate actual examples of the proposed system. For example, in Fig. 2, the free time is set in terms of levels ranging from 1 to 6. When the five nodes with free time levels of 6, 5, 5, 3, and 1, respectively, are within the communication range of node A, through node A communicating with the upper-level nodes with free time of 6, 5, and 5, the amount of reproduced data is reduced. Correspondingly, in Fig. 3, when node B is within the range of communication of node A, if the threshold restricted according to the free time of node B is exceeded, node A does not transmit data to node B. However, where the recipient of the data held by node A is node B, node A transmits to node B regardless of the threshold. In the proposed system, by restricting the communication of nodes with less free time, it is possible to reduce the amount of reproduced data.

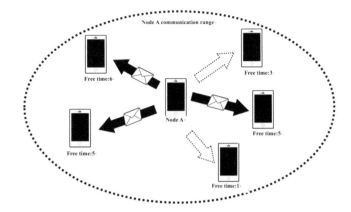

Fig. 2. Proposed Scheme: Communication with nodes with a large amount of free time

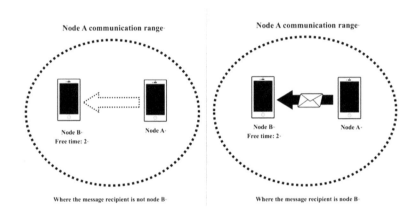

Fig. 3. Restrictions on transmitting to nodes with less free time

4   Simulation Experiments

In this chapter, there is a discussion of the simulation environment used in this experiment, the system of movement for the nodes in this simulation, and, lastly, the experimental results. In this experiment, we compare the proposed system with the traditional system of Epidemic Routing, by using a simulation, in terms of node behavior in consideration of the number of reproduced data items, the number of arrived data items, and free time, to demonstrate the effectiveness of the proposed system. The evaluation standard was set to the average number after simulating the number of reproduced data items and arrival data when using the legacy system and the proposed system.

For the simulation, we used one created in the Java language by ourselves. The basic information for the simulation mode is as shown in Table 1 below. The area size was set to 5 km × 5 km. It is assumed that drones would be used for the moving relay nodes, and the movement speed was set to 30 km/h (constant value). The movement speed of the other nodes was set to 5 km/h (constant value). The simulation time was set to 6 h. Data was generated by moving the relay nodes, and the transmission destination was set at random. The number of moving relay nodes was set to one. The initial position of the moving relay node was set to the vicinity of the cluster at the top left. The clusters were positioned in the four corners of the respective areas, and the distance between adjacent clusters was set to 3 km. A cluster is defined as a location wherein people gather. The number of nodes in each cluster was set from 12 to 13, respectively, totaling 51.

Table 1. Simulation model

Area size	5×5 (km^2)
Total no of nodes	51 (nodes)
Moving speed of nodes	5 (m/s)
Moving speed of moving relay nodes	30 (m/s)
Node communication range	10 (m)
Communication range of moving relay nodes	100 (m)
No. of data items	30 (items)
Simulation time	6 (hours)

The number of reproduced data items and arrived data items for the legacy system and proposed system are shown in Table 2. The number of reproduced data items was 1500 items for the legacy system and 903 items for the proposed system. Additionally, the numbers of arrived data items were 30 with the legacy system and 29.5 with the proposed system.

Table 2. Number of reproduced data items and arrived data items

	No. of reproduced items (items)	Number of arrived items (items)
Legacy system	1500	30.0
Proposed system (average value over 10 times)	903	29.5

The number of reproduced data items for ten experiments using the proposed system are shown in Fig. 4 and Table 3. The results were that the maximum value was the 1015 items in the second experiment, and the minimum value was 738 items on the sixth experiment, with the median value at 907, the mean value at 903, and the standard deviation at 66.93. Although the number of duplicated data items could be reduced when compared to the legacy system, per Table 3, the maximum value was 1015, the minimum value was 738, and the standard deviation was 66.93, indicating a variation result.

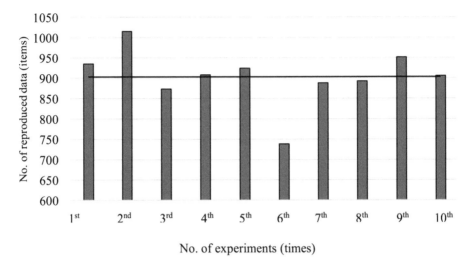

Fig. 4. Number of reproduced data items

Table 3. Number of reproduced data items and arrived data items

	Max. value	Min. value	Median value	Mean value	SD
Proposed system	1015	738	907	903	66.93

Figure 5 shows the ratio of nodes in the proposed system. There is no major bias in the ratio of nodes with randomly-set free time. The reason that there is a disparity in the reproduced data is considered to be due to the ratio of nodes for which free time is set. In the second experiment showing the maximum value, it can be seen that the ratio of nodes with a large amount of free time increases, and the ratio of nodes with less free time decreases. On the other hand, in the sixth experiment showing the minimum value, it can be seen that the ratio of nodes with a large amount of free time decreases, and the ratio of nodes with less free time increases. These results show that, since the proposed system preferentially communicates with nodes with a large amount of free time, there is variation in the number of duplicated data items depending on the node ratio.

In the experiment, we compared the behavior of the nodes by considering the number of reproduced data items, the number of arrived data items, and free time, between the legacy system epidemic routing, and routing systems such as the proposed system that considers the free time of the node.

In terms of the behavior of the node considering free time, in the legacy system, since data was transmitted unconditionally, even the nodes that did not have much free time performed a significant amount of communication. In the proposed system however, since the number of node communications decreases for nodes with less free time, the result achieved is that the number of reproduced data items is reduced.

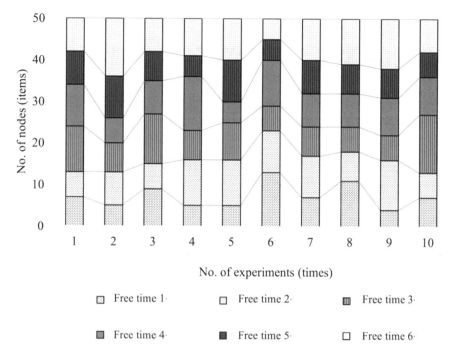

Fig. 5. Ratio of nodes

5   Summary

In this study, we envisage educational support in developing countries without tele-communications infrastructure using DTN and Edmodo for in areas where children cannot go to school. However, the increase in reproduced data and stretching of the capacity of nodes due to this increase, are problems. As a method of maintaining the arrival rate while decreasing the quantity of reproduced data, we proposed a method of DTN routing that considers free time, and in which data communications are restricted to nodes with less free time, and transmitted in a prioritized manner to nodes with more free time. With the proposed system, it is possible to reduce the reproduced data items and maintain the data arrival rate, and, by restricting the data communications for nodes with less free time, prevent stretching of their storage capacity.

References

1. J-Stage Practical Material Collection: Know the world! Let's think! JICA global square, p. 19 (2013)
2. Edmodo. https://www.edmodo.com/
3. Alliance for affordable internet: The Affordability Report 2013, pp. 16–17 (2013)
4. Farrell, S., Cahill, V., Geraghty, D., Humphreys, I., McDonald, P.: When TCP breaks: delay and disruption tolerant networking. IEEE Internet Comput. **10**(4), 72–78 (2006)

5. Farrell, S., Cahill, V.: Delay and Disruption Tolerant Networking. Artech House, Norwood (2006). IEEE Internet Comput. **13**(6), 82–87
6. Vahdat, A.: Epidemic routing for partially connected Ad Hoc networks, Duke Tech Report CS-2000-06 (2000)
7. Jones, E.P.C., Ward, P.A.S.: Routing Strategies for Delay-Tolerant Networks, pp. 12–17 (2006)
8. Zhanga, X., Negliab, G., Kurose, J., Towsley, D.: Performance modeling of epidemic routing. Comput. Netw. **51**(10), 2867–2891 (2007)
9. Haas, Z., Small, T.: A new networking model for biological applications of ad hoc sensor networks. IEEE/ACM Trans. Netw. **14**(1), 27–40 (2006)
10. Fujiwara, A., Minami, H.: Bluetooth & Wi-Fi mobile wireless communication experiments and passing frequency distribution power law. Commun. Soc. Tech. Study Rep. Inf. Netw. **110**(449), 139–144 (2011)

Fetal Hypoxia Detection Based on Deep Convolutional Neural Network with Transfer Learning Approach

Zafer Cömert[1(✉)] and Adnan Fatih Kocamaz[2]

[1] Bitlis Eren University, Bitlis, Turkey
comertzafer@gmail.com
[2] İnönü University, Malatya, Turkey

Abstract. Electronic fetal monitoring (EFM) device which is used to record Fetal Heart Rate (FHR) and Uterine Contraction (UC) signals simultaneously is one of the significant tools in terms of the present obstetric clinical applications. In clinical practice, EFM traces are routinely evaluated with visual inspection by observers. For this reason, such a subjective interpretation has been caused various conflicts among observers to arise. Although the existing of international guidelines for ensuring more consistent assessment, the automated FHR analysis has been adopted as the most promising solution. In this study, an innovative approach based on deep convolutional neural network (DCNN) is proposed to classify FHR signals as normal and abnormal. The proposed method composes of three stages. FHR signals are passed through a set of preprocessing procedures in order to ensure more meaningful input images, firstly. Then, a visual representation of time-frequency information, spectrograms are obtained with the help of the Short Time Fourier Transform (STFT). Finally, DCNN method is utilized to classify FHR signals. To this end, the colored spectrograms images are used to train the network. In order to evaluate the proposed model, we conducted extensive experiments on the open CTU-UHB database considering the area under the receiver operating characteristic curve and other several performance metrics derived from the confusion matrix. Consequently, we achieved encouraging results.

Keywords: Biomedical signal processing · Fetal monitoring
Deep convolutional neural network · Classification

1 Introduction

Electronic fetal monitoring (EFM) device is a useful tool that is used for recording Fetal Heart Rate (FHR) and uterine contraction (UC) signals simultaneously. Cardiotocography (CTG) also often called EFM helps the obstetricians to determine the fetal status and early detecting of several potential adverse outcomes such as hypoxia or acidaemia [1]. In order to decrease the rate of these adverse outcomes, EFM has been adopted as a quite useful tool, and it is used routinely for monitoring fetal welfare during antepartum and more importantly intrapartum periods [2]. The primary clinical goal is to detect as

© Springer International Publishing AG, part of Springer Nature 2019
R. Silhavy (Ed.): CSOC 2018, AISC 763, pp. 239–248, 2019.
https://doi.org/10.1007/978-3-319-91186-1_25

early as possible the symptoms of fetal distress so as to undertake timely actions during pregnancy [3].

In clinical practice, the maternal lies in a supine position, and two different probes, Doppler ultrasound and external pressure transducer are fitted to the maternal abdomen in order to record FHR and UC activities. During the test, EFM device products a paper strip (sometimes here called CTG trace), and this strip is checked only visually by observers. Observers try to recognize specific FHR patterns in accordance with several guidelines such as International Federation of Gynecology and Obstetrics (FIGO) [4]. The basal level of FHR signal, FHR variability, and temporal changes of FHR such as acceleration and deceleration are the basic parameters according to the guidelines. The baseline expresses the mean FHR value when accelerations and deceleration patterns are isolated from the signal [4]. FHR variability indicates the beat-to-beat effect of fetal sympathetic and parasympathetic nervous systems and existing of this variability points the fetal welfare while the absence or weakness of the variability is associated with adverse conditions such as umbilical cord compression, fetal hypoxia or fetal acidaemia [5]. As for transient changes of FHR, accelerations are adopted as the desired patterns and point the fetal movements whereas decelerations indicate the stressful conditions [6]. Although existing of the guidelines, a high-level disagreement between inter- and even intra-observers has been reported in numerous studies due to such a visual inspection [7, 8]. Furthermore, it is well-known that CTG has a poor reproducibility value. As a result of these reasons, the automated CTG analysis has been adopted as the most encouraging solution to tackle against these drawbacks [9].

The first studies in the automated CTG analysis have focused on the detection of morphological features such as baseline, accelerations, and decelerations in accordance with the guidelines [10, 11]. These basic morphological features have been extended with other features obtained from different domains in the most of applications in automated CTG analysis [12]. Moreover, the recent automated CTG analysis models require using advanced methods and algorithms performing the nonlinear analysis [13]. Approximation entropy (ApEn), Sample Entropy (SampEn), Shannon Entropy as the new predictors are associated with the activities of the fetus during the intra-uterine life [14]. Apart from the mentioned studies, image-based time-frequency features, sometimes called texture features, have been used to improve the classification success of fetal hypoxia detection [15, 16]. In addition to feature extraction stage identifying entire of FHR time-series in a lower dimension, machine learning techniques constitute a significant part of the computerized CTG analysis. Therefore, the classifiers also covering the optimization of the network topology such artificial neural network (ANN) [17], support vector machine (SVM) [18], extreme learning machine (ELM) [19], radial basis function network (RBFN) [20], and random forest (RF) [21] have been utilized to determine abnormal FHR samples and to improve the generalization performance of the classifiers.

Recently, convolutional neural networks have become attractive due to their performance on image classification [22]. In this study, we concern with the deep convolutional neural network (DCNN) as a tool to distinguish the fetuses with hypoxia from normal, and the colored spectrogram images provided by Short Time Fourier Transform (STFT) are the basic sources for feeding DCNN. The main aim of this study is to ensure a novel model which has a robust diagnostic ability.

2 Methods

The proposed fetal hypoxia detection system consists of three stages: signal prepro-cessing, obtaining spectrogram images using STFT and DCNN based classification. An overall schema of the proposed model is illustrated in Fig. 1.

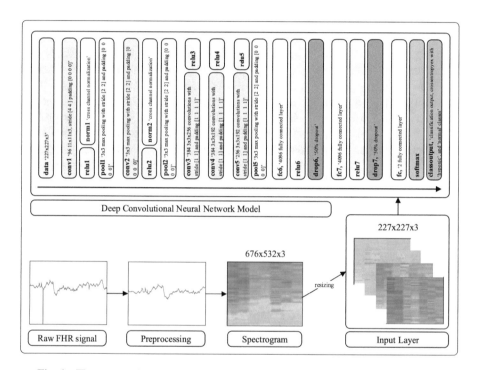

Fig. 1. The proposed hypoxia detection schema based on the pre-trained model [23].

2.1 Open Access Intrapartum CTU-UHB Database

The CTU-UHB is an open database which ensures the set of raw 552 intrapartum CTG recordings. This set is a subset of 9164 intrapartum recordings collected between 2010 and 2012. The recordings were selected carefully considering several characteristic criteria such as singleton pregnancies, gestational age > 36 weeks, and availability of biochemical parameters of the umbilical arterial blood sample. The recordings sampled at 4 Hz were collected via STAN S21/S31 and Avalon FM40/50 EFM devices, and all signals were stored OB TraceVue® system in an electronic form. The visual interpre-tations of the experts and also biochemical markers such umbilical artery pH are pre-sented in the study to separate retrospectively the signals as normal and abnormal [24].

In this case, the umbilical artery pH value was set to 7.05 and this value was utilized as a borderline to mark hypoxic samples. The samples which have pH value equal or lower to 7.05 were agreed as hypoxic and the rest of them were normal. As a result of this selection, we achieved 44 hypoxic and 508 normal CTG recordings for the experimental study and we focused on the last 15 min of the signals.

2.2   Preprocessing

Preprocessing is an essential step almost all biomedical signal processing application in order to improve the signal quality as well as to remove artifacts caused by maternal and fetal movements, equipment, labor-based stress conditions [25].

A basic preprocessing schema is operated in this study. Firstly, the raw CTG including FHR and UC signals are obtained from CTU-UHB database, and only FHR signals are passed to the further steps, in other words, UC signals are ignored. Then the long gaps which are greater than 15 s are isolated from the FHR time-series [26]. Then, the small gaps are detected by an algorithm again and cubic Hermite spline interpolation technique provided by MATLAB is utilized to fix the missing beats problem [27]. In the next step, a standard median filter is employed to soften lightly FHR time series without allowing the loss of variability. In addition, we consider the outliers values in both locally and globally. To this end, we use outliers detection algorithm proposed by Romano [28]. Due to the using of nonlinear techniques, we detrend the signals in order to preserve the dynamics in the nature of FHR signals in the last step of the preprocessing. A sample FHR time-series is displayed in Fig. 2 considering before and after the preprocessing schema. Figure 2 consists of small squares and big rectangles. The small squares correspond to 10 bpm on the vertical axis and 30 s on horizontal axis whereas big rectangles correspond to 30 bpm in the vertical axis and 3 min on the horizontal axis, respectively [9].

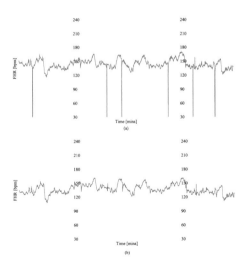

Fig. 2. FHR time series (a) before preprocessing (b) after preprocessing

2.3   Short Time Fourier Transform

STFT is a helpful tool for time-frequency representation for non-stationary signals such as CTG [29]. For a discrete-time signal $x(n)$, the STFT is expressed as follow:

$$X_w(nL, \omega) = \sum_{m=-\infty}^{\infty} x(m)w(nL - m)e^{-j\omega m} \qquad (1)$$

Herein the subscript w in $X_w(nL, \omega)$ indicates the analysis window, $w(n)$. L denotes the separation in time between adjacent short-time sections and it is set using positive integers. For a certain value of n, $X_w(nL, \omega)$ shows the Fourier transform with respect to m of short-time section of $f_n(m) = x(m)w(nL - m)$. In this manner, the large-scale signal is divided into equal shorter parts. After each shift of L samples, the STFT is obtained by using the sliding window [29]. The spectrogram is a graphical representation of the magnitude of the discrete STFT in log scale and it is defined as

$$S(n, k) = log|X(n, k)|^2 \text{ and } X(n, k) = X(n, \omega)|_{\omega=\frac{2\pi}{N}k} \qquad (2)$$

In this study, the colored spectrogram images are the basic information resource to feed DCNN. In order to provide enough data to the DCNN, we take into account the four different frequency intervals: very low frequency (VLF, 0–0.03 Hz), low frequency (LF, 0.03–0.15 Hz), middle-frequency (MF, 0.15–0.50 Hz) and high frequency (HF, 0.50–1 Hz) [30]. The specified frequency intervals are associated with some physiological events. In this context, LF points the fetal maturation, MF carries the signs of fetal movements and maternal respiratory, and HF is influenced essentially parasympathetic nervous system fluctuations [31]. This case is illustrated in Fig. 3. As shown in Fig. 3, after the normalized spectrogram is obtained, sub-frequency intervals are considered and these sub-spectrograms are also used to train the DCNN.

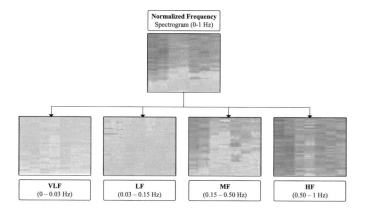

Fig. 3. The spectrogram images

2.4   Deep Convolutional Neural Network

DCNN which is intended to process data that come in the form of multiple arrays has become quite attractive models day to day because of their capabilities in processing images, videos, speech, and audio [32]. It can learn features from data and consolidates the spatial information of a given input image between the pixels and the nodes of the network. DCNN architecture commonly includes convolution, pooling, and fully connected layers. In the network architecture, the convolution layers are followed frequently pooling layers. Recently, different types of layers such as ReLU, normalization, and dropout have been embedded the network to make training faster and to decrease computational cost and to prevent overfitting [23]. The features are obtained in the convolution and followed layers such as pooling whereas the classification task is realized in the fully connected layer.

A classification task using deep learning model can be performed using different approach either training from scratch, transfer learning or feature extraction. Training from scratch is a less common approach since it requires a huge amount of data as well as a long training period from days to weeks. In most of the applications, transfer learning approach has been preferred since it needs only fine-tuning process based on studied data and a pre-trained model such as GoogLeNet or AlexNet [33]. In this study, we employed AlexNet that was trained using 1.2 million images into 1000 classes. The architecture of the network is illustrated in Fig. 1. As seen in Fig. 1, it contains 5 convolution layers, some of which are followed by ReLU, norm and pooling layers. It also has 3 fully-connected layers. The weights of the network are updated in agreement with stochastic gradient descent with momentum (SGDM) as shown following:

$$v_{i+1} := 0.9 * v_i - 0.0005 * \in * w_i - \in * \left\langle \frac{\partial L}{\partial w} | w_i \right\rangle_{D_i} \tag{3}$$

$$w_{i+1} = w_i + v_{i+1} \tag{4}$$

herein i indicates iteration index, v shows the momentum, $\in$ symbolizes the learning rate, and $|\left\langle \frac{\partial L}{\partial w} |_{wi} \right\rangle_{D_i}$ is the average over the ith batch D_i of the objective with respect to w, assessed at w_i [23]. By the way, the momentum and a weight decay were set to 0.9 and 0.0005, respectively.

3   Results

The main aim of this study is to ensure a novel model based on DCNN for classifying FHR recordings regardless of clinical information. To this end, the experimental study was performed on open access intrapartum CTU-UHB database consisting of 552 CTG recordings with additional clinical information [24]. The recordings were divided into two classes considering umbilical artery pH value which was set to 7.05. As a result, the recordings, which have pH value greater than 7.05, were agreed as normal (508 items) and the rest of them (44) abnormal (hypoxic).

In network training phase, we used only spectrogram images provided by STFT instead of clinical information or morphological features since DCNN requires images for feeding the network [22]. This transformation approach ensured the colored time-frequency information embedded in an image with $676 \times 532 \times 3$ resolution (cf. Fig. 3). The specified four frequency intervals were taken into account for identifying the signals. Also, we defined a preprocessing function before feeding DCNN in order to resize the images from $676 \times 532 \times 3$ to $227 \times 227 \times 3$ resolution. As mentioned before, we utilized transfer learning approach for training the network. For this particular purpose, the adjusted fine-tuning parameters are indicated in Table 1. The network has 5 convolution layers and some of which are followed pooling, ReLU and normalization layers. Also, the values of the parameters for filtering, padding, and strides are indicated in Fig. 1.

Table 1. The fine-tuning parameters of DCNN

Parameter	Value
Mini batch size	10
Maximum epoch	4
Initial learning rate	0.0001
Validation frequency	5

Table 2. Performance metrics with short descriptions

Metric	Formulation	Short description
Accuracy (*Acc*)	$\frac{TP+TN}{TP+FP+FN+TN}$	*Acc* gives the general performance of the network
Sensitivity (*Se*)	$\frac{TP}{TP+FN}$	*Se* explains the success of the classifier on positive samples
Specificity (*Sp*)	$\frac{TN}{TN+FP}$	*Sp* indicates the success of the classifier on negative samples
Quality Index (*QI*)	$\sqrt{Se.Sp}$	*QI* means the geometric mean of *Se* and *Sp*

The network was trained 10 different times, and the performances of each testing were reported in Table 3 respectively considering the several performance metrics indicated in Table 2. These metrics are derived from confusion matrix including 4 prognostic indices which are True Positive (TP), True Negative (TN), False Positive (FP) and False Negative (FN). TP and TN correspond to the number of hypoxic and normal samples identified correctly whereas FP and FN correspond to the number of hypoxic and normal samples identified incorrectly. Also, receiver operating characteristic (ROC) curve and the area under this curve (*AUC*) were taken into account in the evaluation. Lastly, TT in Table 2 stands for training time.

According to experimental results, the average *Acc*, *Se*, and *Sp* were achieved as 93.32%, 56.15%, and 96.51%, respectively. It should be noted that *Sp* was superior to *Se* due to the imbalanced data distribution among the classes. Such the case, the *QI* and *AUC* become more useful. In this scope, the average *QI* and *AUC* were obtained as

72.84% and 0.84, respectively. Consequently, we achieved encouraging results regarding the using DCNN in the detection of fetal hypoxia.

Table 3. Performance results of the proposed model

No.	Acc (%)	Se (%)	Sp (%)	QI (%)	AUC	TT (s)
1	92.12	53.84	95.39	71.67	0.81	38
2	96.34	53.84	100	73.38	0.84	48
3	96.97	69.23	99.34	82.93	0.93	45
4	96.36	69.23	98.68	82.65	0.86	52
5	93.33	53.84	96.71	72.16	0.83	37
6	90.90	53.84	94.07	71.17	0.76	35
7	92.72	15.38	99.34	39.09	0.78	38
8	95.75	53.84	99.34	73.13	0.86	36
9	81.81	69.23	82.89	75.75	0.82	18
10	96.97	69.23	99.34	82.93	0.91	80
Avg.	**93.32**	**56.15**	**96.51**	**72.48**	**0.84**	**42.7**

4   Discussion

When literature is examined, it is seen that the main trend in automated CTG analysis relies on a set of processes that constitutes preprocessing [28], feature transform (feature extraction [9–12] and feature selection [26, 34]), and classification [19–21]. For this reason, it can be noticed that using the convolutional neural network (CNN) for determining fetal adverse outcomes is not very common. As an exception, Bursa et al. [22] carried out an experimental study based on CNN. They used CTU-UHB database and preferred to use continuous wavelet transform (CWT) with different scales for representing the signal in two dimensions. Also, the gray-scale images were utilized in the study to feed the network. Furthermore, they utilized Tensorflow framework. As another leading difference is that the researchers performed an experimental study on both FHR and UC signals. As a result, they reported 94.1% accuracy for the network. In summary, DCNN is found as a useful tool for detecting fetal hypoxia. Moreover, we believe that different transformation approaches such as CWT or STFT with the various parameter settings can ensure more efficient results. To achieve more efficient results, CNN based classifier should be used on larger databases.

References

1. Murray, H.: Antenatal foetal heart monitoring. Best Pract. Res. Clin. Obstet. Gynaecol. **38**, 2–11 (2017)
2. Brown, R., Wijekoon, J.H.B., Fernando, A., Johnstone, E.D., Heazell, A.E.P.: Continuous objective recording of fetal heart rate and fetal movements could reliably identify fetal compromise, which could reduce stillbirth rates by facilitating timely management. Med. Hypotheses **83**, 410–417 (2014)

3. van Geijn, H.P.: 2 Developments in CTG analysis. Baillieres Clin. Obstet. Gynaecol. **10**, 185–209 (1996)
4. Ayres-de-Campos, D., Spong, C.Y., Chandraharan, E.: FIGO consensus guidelines on intrapartum fetal monitoring: Cardiotocography. Int. J. Gynecol. Obstet. **131**, 13–24 (2015)
5. Tongsong, T., Iamthongin, A., Wanapirak, C., Piyamongkol, W., Sirichotiyakul, S., Boonyanurak, P., Tatiyapornkul, T., Neelasri, C.: Accuracy of fetal heart-rate variability interpretation by obstetricians using the criteria of the National Institute of Child Health and Human Development compared with computer-aided interpretation. J. Obstet. Gynaecol. Res. **31**, 68–71 (2005)
6. Czabanski, R., Jezewski, J., Matonia, A., Jezewski, M.: Computerized analysis of fetal heart rate signals as the predictor of neonatal acidemia. Expert Syst. Appl. **39**, 11846–11860 (2012)
7. Garabedian, C., Butruille, L., Drumez, E., Schreiber, E.S., Bartolo, S., Bleu, G., Mesdag, V., Deruelle, P., De Jonckheere, J., Houfflin-Debarge, V.: Inter-observer reliability of 4 fetal heart rate classifications. J. Gynecol. Obstet. Hum. Reprod. **46**, 131–135 (2017)
8. Palomäki, O., Luukkaala, T., Luoto, R., Tuimala, R.: Intrapartum cardiotocography: the dilemma of interpretational variation. J. Perinat. Med. **34**, 298–302 (2006)
9. Cömert, Z., Kocamaz, A.F.: Novel software for comprehensive analysis of cardiotocography signals CTG-OAS. In: KARCI, A. (ed.) International Conference on Artificial Intelligence and Data Processing (IDAP17), pp. 1–6. IEEE, Malatya (2017)
10. Bernardes, J., Ayres-de-Campos, D., Costa-Pereira, A., Pereira-Leite, L., Garrido, A.: Objective computerized fetal heart rate analysis. Int. J. Gynecol. Obstet. **62**, 141–147 (1998)
11. Warrick, P., Hamilton, E., Macieszczak, M.: Neural network based detection of fetal heart rate patterns. In: IEEE International Joint Conference on Neural Networks, pp. 2400–2405 (2005)
12. Cömert, Z., Kocamaz, A.F.: Evaluation of fetal distress diagnosis during delivery stages based on linear and nonlinear features of fetal heart rate for neural network community. Int. J. Comput. Appl. **156**, 26–31 (2016)
13. Magenes, G., Pedrinazzi, L., Signorini, M.G.: Identification of fetal sufferance antepartum through a multiparametric analysis and a support vector machine. In: 26th Annual International Conference of the IEEE Engineering in Medicine and Biology Society, pp. 462–465 (2004)
14. Monteiro-Santos, J., Gonçalves, H., Bernardes, J., Antunes, L., Nozari, M., Costa-Santos, C.: Entropy and compression capture different complexity features: the case of fetal heart rate. Entropy **19**, 688 (2017)
15. Cömert, Z., Kocamaz, A.F.: Cardiotocography analysis based on segmentation-based fractal texture decomposition and extreme learning machine. In: 25th Signal Processing and Communications Applications Conference (SIU), pp. 1–4 (2017)
16. Cömert, Z., Kocamaz, A.F.: A study based on gray level co-occurrence matrix and neural network community for determination of hypoxic fetuses. In: International Artificial Intelligence and Data Processing Symposium (IDAP), pp. 569–573. TR (2016)
17. Cömert, Z., Kocamaz, A.F.: A study of artificial neural network training algorithms for classification of cardiotocography signals. Bitlis Eren Univ. J. Sci. Technol. **7**, 93–103 (2017)
18. Spilka, J., Frecon, J., Leonarduzzi, R., Pustelnik, N., Abry, P., Doret, M.: Sparse support vector machine for intrapartum fetal heart rate classification. IEEE J. Biomed. Health Inform. **21**(3), 664–671 (2016)
19. Cömert, Z., Kocamaz, A.F., Gungor, S.: Cardiotocography signals with artificial neural network and extreme learning machine. In: 24th Signal Processing and Communication Application Conference (SIU) (2016)

20. Sahin, H., Subasi, A.: Classification of the cardiotocogram data for anticipation of fetal risks using machine learning techniques. Appl. Soft Comput. **33**, 231–238 (2015)
21. Cömert, Z., Kocamaz, A.F.: Comparison of machine learning techniques for fetal heart rate classification. Acta Phys. Pol. A **132**, 451–454 (2017)
22. Bursa, M., Lhotska, L.: The use of convolutional neural networks in biomedical data processing. In: Bursa, M., Holzinger, A., Renda, M.E., Khuri, S. (eds.) Proceedings of Information Technology in Bio- and Medical Informatics: 8th International Conference, ITBAM 2017, Lyon, France, 28–31 August 2017, pp. 100–119. Springer, Cham (2017)
23. Krizhevsky, A., Sutskever, I., Hinton, G.E.: ImageNet classification with deep convolutional neural networks. In: Pereira, F., Burges, C.J.C., Bottou, L., Weinberger, K.Q. (eds.) Proceedings of the 25th International Conference on Neural Information Processing Systems, vol. 1, pp. 1097–1105. Curran Associates, Inc., USA (2012)
24. Chudáček, V., Spilka, J., Burša, M., Janků, P., Hruban, L., Huptych, M., Lhotská, L.: Open access intrapartum CTG database. BMC Pregnancy Childbirth **14**, 16 (2014)
25. Cesarelli, M., Romano, M., Bifulco, P., Fedele, F., Bracale, M.: An algorithm for the recovery of fetal heart rate series from CTG data. Comput. Biol. Med. **37**, 663–669 (2007)
26. Spilka, J., Georgoulas, G., Karvelis, P., Oikonomou, V.P., Chudáček, V., Stylios, C., Lhotská, L., Jankru, P.: Automatic evaluation of FHR recordings from CTU-UHB CTG database. In: Bursa, M., Khuri, S., Renda, M.E. (eds.) Proceedings of Information Technology in Bio- and Medical Informatics: 4th International Conference, ITBAM 2013, Prague, Czech Republic, 28 August 2013, pp. 47–61. Springer, Heidelberg (2013)
27. Kahaner, D., Moler, C., Nash, S.: Numerical Methods and Software. Prentice-Hall Inc., Upper Saddle River (1989)
28. Romano, M., Faiella, G., Bifulco, P., D'Addio, G., Clemente, F., Cesarelli, M.: Outliers detection and processing in CTG monitoring. In: Roa Romero, L.M. (ed.) XIII Mediterranean Conference on Medical and Biological Engineering and Computing 2013, MEDICON 2013, Seville, Spain, 25–28 September 2013, pp. 651–654. Springer, Cham (2014)
29. Nawab, S., Quatieri, T., Lim, J.: Signal reconstruction from short-time Fourier transform magnitude. IEEE Trans. Acoust. **31**, 986–998 (1983)
30. Romano, M., Iuppariello, L., Ponsiglione, A.M., Improta, G., Bifulco, P., Cesarelli, M.: Frequency and time domain analysis of foetal heart rate variability with traditional indexes: a critical survey. Comput. Math Methods Med. **2016**, 1–12 (2016)
31. Groome, L.J., Mooney, D.M., Bentz, L.S., Singh, K.P.: Spectral analysis of heart rate variability during quiet sleep in normal human fetuses between 36 and 40 weeks of gestation. Early Hum. Dev. **38**, 1–9 (1994)
32. LeCun, Y., Bengio, Y., Hinton, G.: Deep learning. Nature **521**, 436–444 (2015)
33. Yu, Y., Lin, H., Meng, J., Wei, X., Guo, H., Zhao, Z.: Deep transfer learning for modality classification of medical images. Information **8**(3), 91 (2017)
34. Subha, V., Murugan, D.: Genetic Algorithm based feature subset selection for fetal state classification. J. Commun. Technol. Electron. Comput. Sci. **2**, 13–17 (2015)

Analysis of Causes for Differences in Modeling Results of Multi-hop Wireless Networks Using Various Network Simulators

I. O. Datyev[1], A. A. Pavlov[1(✉)], M. V. Ashkadov[1],
and M. G. Shishaev[1,2]

[1] Institute for Informatics and Mathematical Modelling of the Kola Science
Center RAS, 184209 Apatity, Russia
{datyev,pavlov,ashkadov,shishaev}@iimm.ru
[2] Murmansk Arctic State University, 183720 Murmansk, Russia

Abstract. Simulation is the main way to test multi-step wireless networks. Creating a simulation model of a multi-hop wireless network is a laborious task associated with the use of specialized software tools, called network simulators. In this paper, the modern experience of modeling multi-hop wireless networks is considered, and the main problems are formulated. One of the main problems is the impossibility of a comparative analysis of the simulation experiments results conducted by various researchers. We attempted to classify the reasons for the differences in the results of simulation of multi-hop wireless networks using various network simulators. The proposed classification is designed to help researchers improve the quality of multi-hop wireless networks modelling and improve interaction among researchers. In addition, the paper suggests preventive ways to minimize differences in the results of simulation experiments.

Keywords: Multi-hop wireless networks · Simulation · Network simulators

1 Introduction

Multi-hop wireless networks (MWN) are radio networks, whose nodes can act as repeaters, which increases the network coverage area. Currently, the development of such networks based on mobile devices is an urgent task. The scope of multi-hop wireless networks includes entertaining, smart home, military, and others. The main advantages of such networks are speed of deployment and lack of the need for a developed telecommunication infrastructure. The main problems of multi-hop wireless networks based on mobile devices are complexity of data routing, due to the nodes mobility, and limited battery life of mobile nodes. Various solutions of these problems are offered by a scientists and developers around the world.

An important practical task is evaluation of effectiveness of different MWN under identical operating conditions, or reverse is to determine the area of the most effective use of a single MWN. The proposed MWN usually tested by using specialized software-network simulators. Therefore, initial data for comparing or positioning a new MWN among analogues is previously created models and/or results of simulation experiments.

© Springer International Publishing AG, part of Springer Nature 2019
R. Silhavy (Ed.): CSOC 2018, AISC 763, pp. 249–258, 2019.
https://doi.org/10.1007/978-3-319-91186-1_26

The results of experiments obtained with the help of various simulators cannot be compared directly, at least, because of differences in composition of MWN model parameters. Impossibility of direct comparison of results entails the need to recreate an imitation experiment using other simulators or MWN model. However, even when recreating an experiment, researchers face many difficulties, due to differences in implementation of various network simulators and insufficiently detailed documentation of other researchers.

This work is devoted to revealing and classification of causes for differences in simulation results conducted with various network simulators.

2 Related Works

Testing the effectiveness of different solutions (routing metrics, algorithms, protocols, etc.) developed for multi-hop wireless networks is possible in several ways: (1) real-life experiments in real conditions, (2) creation of testbeds, and (3) the use of network simulators. The first and second ways are resource intensive, therefore, network simulators are used as the main tool. Network simulator is a specialized software tool for simulating computer networks. There are works devoted to comparison of various network simulators.

Many works are devoted to review of existing network simulators [1, 2]. Typically, such works cover all popular simulators. But in our opinion, they lack an important detail: there is no comparison of the simulation results of identical models.

In [3], an overview of the NS-2, NS-3, OMNET++, and GloMoSiM network simulators is presented. The purpose of this paper is to evaluate the following performance parameters: CPU utilization, memory usage, computational time, and scalability by simulating a MANET routing protocol. The aim of the work is to help the scientific community determine optimal network simulator for MANET research.

In this work [4] there have been presented a comparative study between results obtained from a real testbed and three usual network simulators (NS-2, Qualnet and OPNET). The main goal of this study is to evaluate the relevance of these simulators in indoor and outdoor environments. According to this study, authors can point out that for the simulators, the choice of the PHY layer characteristics is predominant, particularly the propagation model and the associated parameters.

In paper [5] there have been presented the results of the simulation of the flooding algorithm using three popular mobile ad-hoc network simulators and author's MAN-SIM simulator.

To determine the state of MANET simulation studies, authors [6] surveyed the 2000–2005 proceedings of the ACM International Symposium on Mobile Ad Hoc Networking and Computing (MobiHoc). From this survey, they found significant shortfalls. They then summarize common simulation study pitfalls found in this survey. Finally, authors discuss the tools available that aid the development of rigorous simulation studies. Authors offer these results to the community with the hope of improving the credibility of MANET simulation-based studies.

In this paper [7], two network simulators were compared with a network testbed. The accuracy of NS-2 and Modeler from OPNET was compared using CBR data traffic and an FTP session. Several scenarios were evaluated and regenerated in the simulation tools and the network testbed. The results provide interesting guidelines to network researchers in the selection of network simulation tools. From the researcher's point of view, NS-2 provides very similar results compared to OPNET Modeler, but the "freeware" version of NS-2 makes it more attractive to a researcher. However, the complete set of OPNET Modeler modules provides more features than NS-2, and it therefore will be more attractive to network operators.

As a rule, the purpose of the studies is to confirm some hypothesis or to identify the drawbacks of certain network simulators. In the works reviewed, the authors do not attempt to systematize the causes for differences in simulation results. However, such systematization, in our opinion, could improve the quality of the MWN simulation.

3 Our Work

We tried to classify the causes for the differences in the results of simulation experiments (Fig. 1).

According to the proposed classification, these causes can be divided into two main groups:

1. Causes beyond the control of researcher:

   - Features of network simulator implementation. Like any other software product, each network simulator is unique. It is influenced by many factors, such as different programming languages, different sets of implemented technologies that affect the results of experiments.
   - Closed source code. Some network simulators have closed source code and therefore it is not possible to view or modify the implementation of embedded technologies used in the network simulator.

2. Causes under the control of researcher:

   - Hardware and software configuration of the model. Network device parameters: physical and link layer standards has a great influence on simulation results.
   - Operating conditions. The simulation parameters can be greatly influenced by the parameters of the simulation model used: nodes mobility model, environment model (if provided by the capabilities of the network simulator).

Based on this classification, it can be concluded that it is not possible to create two identical models using different network simulators because of many causes beyond the control of the user. But using our classification and further advice, in some cases, it is possible to increase the accuracy of the comparison of the results of simulation experiments.

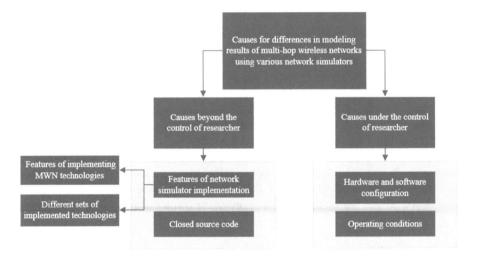

Fig. 1. Classification of causes for differences in simulation experiments

3.1   Causes Beyond the Control of Researcher

Currently, one of the most popular, developing and available network simulators are NS-3 [8], Riverbed modeler [9], Omnet++ [10]. Each of the presented network simulators has its own set of variable parameters. Difficulties in comparative analysis appear during the development of a similar model, where it is not possible to specify one of the important parameters that is not implemented in another network simulator (Table 1). It should also be considered that missing parameter can greatly influence the results of entire simulation experiment.

Table 1. Comparison of some parameters availability of different network simulators

	NS-3	Riverbed modeler	Omnet++, INET
Version	3.25	17.5 Academic Edition	INET 3.5.0 for OMNeT++ 5.1
Data Rate (bps)	+	+	+
Transmitter Power	+	+	+
AODV, OLSR, DSR routing protocols	+	+	+
Phy mode	+	+	+
Wi-Fi standard	+	+	−
Signal Propagation model	+	−	−
Packet Size	+	+	+
Bandwidth (MHz)	−	+	+

We conducted simulation experiments to determine the effect of difference in network simulators parameter sets on simulation results.

Simulation models of multi-hop wireless network had the most approximate sets of similar parameters and their values in different network simulators (Table 2).

Table 2. Values of model parameters of different network simulators

	OMNeT++, INET	Riverbed Modeler	NS-3
Routing protocol	AODV	AODV	AODV
Transmitter Power	10 mW	0.01 W	10 dBm
Data Rate[a]	256000 bps, 512000 bps	125 packets per second, 250 packets per second	256000 bps, 512000 bps
Data packet size	256 bytes, 512 bytes	2048 bits, 4096 bytes	256 bytes, 512 bytes
Phy mode	DsssMode1Mbps	Direct Sequence 1Mbps	DsssRate1Mbps
Wi-Fi standard	802.11 g	802.11 g	802.11 g
Signal Propagation model	Not defined	Not defined	Friis
Duration of simulation (seconds)	100	100	100

[a] The data rate in each simulator is set in its own way. We gave the values of the parameters according to the simulator used.

The experiments were conducted with different network topologies, reflecting the specific difficulties associated with routing in a multi-hop wireless network.

In the "matrix" model, the network nodes are placed similarly to the matrix elements at 100 meters from each other (both horizontally and vertically). With this topology, each node has no more than eight links. The information source is located on the right lower vertex of the matrix, and the receiver is in the upper left. This topology allows to get the maximum number of routes between the corresponding pair, as well as the largest number of transit nodes.

The "line" model was used to evaluate the effectiveness of the dynamic routing protocols for long route lengths. All nodes are placed in one line at 100 meters from each other. The data is transferred between two end nodes of the line.

Thus, the results of the NS-3 and Riverbed Modeler simulations using the AODV protocol differ not so much in comparison with Omnet++ (Table 3).

This may be since the implementation of the AODV protocol used in NS-3 and Riverbed Modeler does not differ much in the set of parameters from each other, in contrast to the implementation of AODV in Omnet++.

Table 3. Simulation results for a packet size of 256-bit and bitrate of 256,000 bps

Minimum delay (sec)			
	4 × 4 ("matrix" model)	10 nodes ("line" model)	8 × 8 ("matrix" model)
NS-3	0.0027	0.0027	0.0058
Riverbed	0.0034	0.0034	0.019
Omnet++	0.00012	0.00037	0.00037
Average delay (sec)			
	4 × 4 ("matrix" model)	10 nodes ("line" model)	8 × 8 ("matrix" model)
NS-3	0.0028	0.0028	0.044
Riverbed	0.0036	0.0037	0.06
Omnet++	0.00014	0.0013	0.00068
Maximum delay (sec)			
	4 × 4 ("matrix: model)	10 nodes ("line" model)	8 × 8 ("matrix" model)
NS-3	0.046	0.0065	3.214
Riverbed	0.026	0.0074	3.896
Omnet++	0.0041	0.33	0.013

Based on this, it can be concluded that a comparison of simulation results between the NS-3 and the Riverbed Modeler is possible. However, it must be considered that the models in these network simulators have certain differences, which in other cases can have a much greater impact and simulation results will differ significantly. Therefore, before creating simulation models, it is necessary to study the features of implementing MWN technologies in the network simulators planned for use.

Standards related to MWN described in the RFC [11]. For example, the AODV routing protocol is presented in RFC3561 [12]. Developers of MWN algorithms implemented in network simulators, as a rule, take the basis of RFC. Most documents are changed over time, for example, there are 13 versions of AODV protocol RFC. For this reason, different versions of seemingly the same protocol can be used in different network simulators.

Therefore, during the experiments it is important to pay attention to what RFCs were taken as a basis.

For example, implementation of DSR protocol in NS-3 and Riverbed Modeler network simulators have serious differences.

In NS-3, the authors describe their implementation [13], which indicates that it does not fully comply with RFC: «The model is not fully compliant with RFC 4728 [14]. As an example, DSR fixed size header has been extended and it is four octets longer then the RFC specification. Consequently, the DSR headers can not be correctly decoded by Wireshark».

Differences between the implementation of DSR protocol in NS3 and Riverbed Modeler can also be seen in the protocol parameters that are available for modification (Table 4, Fig. 2).

Table 4. Available parameters for the DSR protocol of NS-3

Parameter	Description	Default
MaxSendBuffLen	Maximum number of packets that can be stored in send buffer	64
MaxSendBuffTime	Maximum time packets can be queued in the send buffer	Seconds (30)
MaxMaintLen	Maximum number of packets that can be stored in maintenance buffer	50
MaxMaintTime	Maximum time packets can be queued in maintenance buffer	Seconds (30)
MaxCacheLen	Maximum number of route entries that can be stored in route cache	64
RouteCacheTimeout	Maximum time the route cache can be queued in route cache	Seconds (300)
RreqRetries	Maximum number of retransmissions for request discovery of a route	16
CacheType	Use Link Cache or use Path Cache	LinkCache
LinkAcknowledgment	Enable Link layer acknowledgment mechanism	True

The possibilities for changing the parameters of the DSR protocol in the Riverbed Modeler are much greater than in the NS-3. Therefore, it can be assumed that the implementation of DSR in Riverbed Modeler has been elaborated in more detail. The same can be attributed to many other modules of these network simulators.

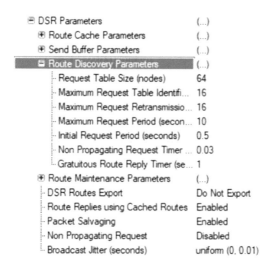

Fig. 2. Available parameters for the DSR protocol in the network simulator Riverbed modeler

Table 5. Simulation results for a packet size of 256 bits and bitrate of 512,000 bps (DSR protocol)

Minimum delay (sec)			
	4 × 4 ("matrix" model)	10 nodes ("line" model)	8 × 8 ("matrix" model)
NS-3	0,0029	0,0029	0,0062
Riverbed Modeller	0,0032	0,0031	0,019
Average delay (sec)			
	4 × 4 ("matrix" model)	10 nodes ("line" model)	8 × 8 ("matrix" model)
NS-3	0,00292	0,00292	0,0048
Riverbed Modeller	0,00328	0,0344	0,315
Maximum delay (sec)			
	4 × 4 ("matrix" model)	10 nodes ("line" model)	8 × 8 ("matrix" model)
NS-3	0,003	0,0176	0,294
Riverbed Modeller	0,0036	0,0479	1,129

Based on the conducted experiments it should be noted that in some cases, for example using DSR protocol, the results obtained using of NS-3 and Riverbed Modeler (Table 5) are very different and can not be used for direct comparison.

3.2   Causes Under the Control of Researcher

An important condition for improving the similarity of simulation results using different network simulators is the attention paid to tuning the simulation model parameters. Each seemingly insignificant parameter has a great influence on the results of the model.

Table 6. Parameters values of the simulation model for NS-3

Parameter name	Value
Routing protocol	AODV
Bitrate (bps)	512000
phyMode	DsssRate11Mbps
TransmissionPower (dBm)	9
PropagationLoss, PropagationDelay	FriisPropagationLossModel, ConstantSpeedPropagationDelayModel
nodeSpeed (m/s)	13
PacketSize (byte)	128, 256, 512
MobilityModel	RandomWaypointMobilityModel
Wi-Fi	WIFI_PHY_STANDARD_80211b

Here is a small example that confirms that you need to pay attention to each parameter when building a model. We used one model (Table 6) and conducted several experiments in which the value of only one of the parameters, the size of the transmitted packet, was changed.

The results of the simulation show (Table 7) that a change in at least one parameter can have a big effect on the result of an imitation experiment. A more detailed study of the individual parameters effect on the modeling results was carried out in work [15].

Table 7. Simulation results using NS-3

Packet size (byte)	Mean delay (msec)	Packet loss ratio (%)
128	53.4	10.9
256	19.9	6.6
512	27.8	7.4

Based on these results, it can be concluded that researchers of multi-hop wireless networks need to pay more attention to the development and publication of simulation models. In our opinion, preventive measures to minimize discrepancies in the results of simulation experiments are reduced to two main groups:

- Researchers should publish as much as possible detailed characteristics of models in their publications so that other researchers can reproduce these models as accurately as possible.
- Researchers, who recreate the simulation experiment, pay more attention to tuning the parameters during the preparation of this experiment.

At the same time, the implementation of these measures can be complicated by various factors. For example, often the maximum volume of publication is limited, so researchers difficult to publish a detailed list of parameters for recreating an exact copy of the simulation model. Another reason may be the reluctance of researchers to disclose some features of their models.

However, often during the development of a model for conducting a simulation experiment, researchers can simply underestimate the importance of each individual parameter influence on the simulation results overall.

4   Conclusion

Simulation is one of the main ways to test the effectiveness of multi-hop wireless networks. Currently, there is many network simulators. Each of them is unique software product. Therefore, a direct comparison of simulation results obtained by different researchers using different network simulators is almost impossible. In this paper, we summarized the experience of other researchers, and attempted to classify the causes for simulation results difference of multi-hop wireless networks using various network simulators. The proposed classification can be used to analyze the causes for the difference in the results of simulation experiments in order to minimize modifications of simulation models to enable comparison of these results. The main feature of this classification is the division of causes those under the control of researcher and beyond the control of researcher. Accordingly, we can minimize the influence of the first class of causes, and we can only take into account the second class of causes. In addition, this classification contributes to the development of preventive measures to minimize the discrepancies in the results of the experiments. As a preventive measure, we propose to carefully consider the issue of choosing a network simulator, to study the features of implementing multi-hop wireless network technology, which is planned to be modeled, in a specific network simulator. In addition, an important aspect is to provide other researchers with the fullest possible specification of the simulation experiment for more effective interaction among researchers.

References

1. Lakshmanarao, K., VonodKumar, C.R., Kanakavardhini, K.: Survey on simulation tools for wireless networks. IJERT **2**(10), 608–612 (2013)
2. Abu Salem, A.O., Awwad, H.: Mobile ad-hoc network simulators, a survey and comparisons. Int. J. P2P Netw. Trends Technol. (IJPTT) **9**, 12–17 (2014)
3. ur Rehman Khana, A., Bilalb, S.M., Othmana, M.: A Performance Comparison of Network Simulators for Wireless Networks. https://arxiv.org/ftp/arxiv/papers/1307/1307.4129.pdf. Accessed 11 Jan 2018
4. Rachedi, A., Lohier, S., Cherrier, S., Salhi, I.: Wireless Network Simulators Relevance Compared to a Real Testbed in Outdoor and Indoor Environments. https://hal.archives-ouvertes.fr/hal-00620343/document. Accessed 11 Jan 2018
5. Zoican, R., Galaţchi, D.: A Comparison of the Present MANET Simulators. TELE-COMUNICAŢII. 1/2008, pp. 84–90 (2008)
6. Kurkowski, S., Camp, T., Colagrosso, M.: MANET simulation studies: the incredibles. Mob. Comput. Commun. Rev. **1**(2) (2018). http://citeseerx.ist.psu.edu/viewdoc/download?doi=10.1.1.329.1133&rep=rep1&type=pdf. Accessed 11 Jan 2018
7. Lucio, G.F., Paredes-Farrera, M., Jammeh, E., Fleury, M., Reed, M.J.: OPNET modeler and Ns-2: comparing the accuracy of network simulators for packet-level analysis using a network Testbed. http://citeseerx.ist.psu.edu/viewdoc/download?doi=10.1.1.64.7286&rep=rep1&type=pdf. Accessed 11 Jan 2018
8. NS-3. https://www.nsnam.org/. Accessed 11 Jan 2018
9. Riverbed modeler. https://www.riverbed.com/gb/products/steelcentral/steelcentral-riverbed-modeler.html. Accessed Jan 11 2018
10. Omnet++. https://www.omnetpp.org/. Accessed 11 Jan 2018
11. Request for Comments (RFC). https://www.ietf.org/rfc.html. Accessed 11 Jan 2018
12. RFC3561. https://www.ietf.org/rfc/rfc3561.txt. Accessed 11 Jan 2018
13. NS-3, DSR. https://www.nsnam.org/docs/models/html/dsr.html. Accessed 11 Jan 2018
14. RFC 4728. https://tools.ietf.org/html/rfc4728. Accessed 11 Jan 2018
15. Datyev, I.O., Pavlov, A.A., Shishaev, M.G.: The parameters list for multihop wireless networks cross-layer routing metric. In: Cybernetics and Mathematics Applications in Intelligent Systems: Proceedings of the 6th Computer Science On-line Conference 2017 (CSOC2017), Prague, Czech Republic, 26–29 April 2017, vol. 2, pp. 150–160 (2017)

Pattern Lock Evaluation Framework for Mobile Devices: Memorizability and Timing Issues

Agnieszka Bier, Adrian Kapczyński, and Zdzisław Sroczyński[✉]

Institute of Mathematics, Silesian University of Technology,
Kaszubska 23, 44-100 Gliwice, Poland
{agnieszka.bier,adrian.kapczynski,zdzislaw.sroczynski}@polsl.pl

Abstract. The paper concerns the influence of memory, forgetting and timing issues on the security of mobile applications. The designed framework system was used to further elaboration of the automatic measure, which estimates the quality (strength) of unlock gesture patterns. The data analysis described in detail presents the relations between human subjective ratings of patterns' complexity and memorizability levels in regards to computed values of quality measure and real-life time spans needed to enter and re-enter the pattern.

Keywords: Pattern strength measure · Human memory
Authentication · Mobile access control

1 Introduction

The meaning of the mobile security grows rapidly because of widespread implementations of the mobile technologies in almost every aspect of personal and corporate computing. The access to the data held by mobile devices should be authorized as often and as strictly as possible. On the other hand, the unlock procedure for a mobile device should provide fast and comfortable access to the desired functionality.

There are some disadvantages of common alphanumeric passwords known from desktop personal computers, because small virtual keyboards rendered at the touch screen of the mobile devices are vulnerable to imprecise input. This way the time needed to enter the password increases, along with the number of failed attempts. Hence, the most popular unlock procedures used for mobile devices usually incorporate entering an unlock pattern, as this method is significantly less sensitive to the inaccuracy of input precision. In the following we analyze the timing and memorizability issues of pattern locks.

2 Related Work

There are many alternative methods for the unlock procedure, which try to fit the needs of mobile device's user: numeric and alphanumeric PIN/passwords [1],

© Springer International Publishing AG, part of Springer Nature 2019
R. Silhavy (Ed.): CSOC 2018, AISC 763, pp. 259–268, 2019.
https://doi.org/10.1007/978-3-319-91186-1_27

gestures on grid or picture [2,3], behavioral [4–7] including motion patterns [8,9] and keystroke dynamics [10], biometrical with commonly implemented finger-prints [11,12], face/image recognition [13] and novel 3D face recognition [14,15].

The usability of the unlock process which implements one of the particular methods can significantly differ, depending on peoples' age, possible disability, computer skills and education [16,17]. One of the key factors describing the comfort of usage for unlock methods are memorizability and speed of the input. The secret provided to authorize the user should be easy to memorize and remember, being still complex enough to prevent counterfeiting or guessing. Moreover, the time needed to enter the secret should be limited to prevent time waste while continuous use of the mobile device.

Timing and memorizability issues are especially important for pattern unlock method. The user should connect a pattern of control points from the rectangular grid shown at the touch screen. Obviously, the longer pattern ensures the greater security. On the other hand, human memory has surprising limits [18] and without any additional context long patterns may become hard to recall [19]. Long patterns can also take extra time to enter, therefore some users consider them impractical. The memory capacity differs for the people depending on age, education/trainning, overall psychophysical fitness, and elderly or handicapped people need a special assistance in many areas of everyday activity [20]. Thorough general study of human memory, including research experiments, theories and applications is given in [21].

Memory problems mentioned above are certainly the reasons, why grid size used in pattern unlock is generally 3×3 in most applications [22–24], although 4×4 were examined as well [25]. The questionnaire-based evaluation of the quality of given unlock pattern can reflect subjective human impressions about the security level it assures. But there is also a need for automatic evaluation, hence the research and analysis of grid-pattern unlock procedure have given some quality measures [24,26–28].

The measure which corresponds with human subjective classification should incorporate the set of quality factors forming the overall security of the given pattern. Popular quality factors are: pattern length [2], pattern complexity (the number of direction changes) [24], the number of repeated nodes and the number of connections of specific direction [29], symmetry of the pattern [2,16], the probability distribution of starting node in [22,25,26], tendency of choosing patterns resembling letters and other common shapes [30], and memorizability [31].

Summarizing, the limited human immediate memory span and the impact of forgetting on the effectiveness of mobile security is definitely worth elaboration. These factors were essential for our investigations described in next sections. In the following experiments regarding timing and memorizability of pattern unlock procedure we have used the measure and testing framework introduced in [27] and extended in [28].

3 Framework Specification

In order to conduct the experiments related with pattern lock evaluation, a dedicated framework was designed and implemented. First, two schemes were introduced: the numbering scheme of the nodes and the directions scheme. The numbering scheme assumes there are 9 nodes, which are numbered with digits from 0 to 8. The directions scheme assumes, that each direction is encoded with a letter. The nodes numbering scheme and the directions encoding scheme were presented on Fig. 1.

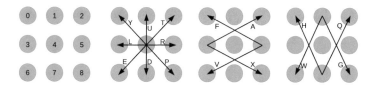

Fig. 1. Nodes numbering scheme and the encoding of directions.

At the first glance, the directions encoding scheme might be considered as redundant, however it is helpful for an efficient computation of the complexity measure for the patterns. Moreover, it can be used in case of modified grids, e.g. consisting of larger number of points (e.g. 4 × 4 [25]) or having the points arranged in a different way.

Entered patterns are stored in internet-based repository, which was implemented as the RESTful webservice using cloud-hosted database. It has been developed with assumption of security, stability and scalability for further experiments requiring larger storage space. Each record in the database carries the following information: lists of nodes and directions in given pattern, complexity rating, memorizability rating, the mode (which can be: new, for new patterns or eval, for evaluation patterns), timestamp, the version of the operating system, time of the first attempt needed to draw the pattern, time of the second attempt needed to draw the pattern (as repetition) and the user's ID called "nick", which is optional.

The RESTful service uses JSON notation, which allows using different client applications. The API specification introduces the methods to manipulate the patterns (see Table 1).

The mobile test application is based on multiplatform FMX framework and embedded cross-compilers available in RAD Studio [32]. This development toolchain maintains single source code, which is shared between different platforms, however the user interaction dynamically introduces the proper GUI for the given operating system. Thanks to that the user experience during lock patterns input in our experiments is very close to native GUI and UX in popular mobile operating systems, in contrast to the other pen-and-paper laboratory experiments [30].

Table 1. REST service API.

Input	Method	Parameters	Result
URL/patterns	POST	Nick, ratecomplex, ratememory, mode, time1, time2, params	Complexity measure for the given pattern, inserts the pattern into the repository
URL/patterns	GET	-	Nodes and encoded directions for the randomly chosen pattern from the repository
URL/measure	GET	Nodes, directions, a (optional), b (optional)	Complexity measure for the pattern
URL/measures	GET	-	Array with all stored patterns and their complexity measures

This is the reason why it was possible to perform the research for different, not compatible operating systems, without the necessity of development of similar software solutions dedicated for given operating system.

Despite the fact, that the pattern lock mechanism is used commercially only in Android, we have provided the research software running on Apple iOS operating system and Microsoft Windows operating system. This way it is possible to detect even the slightest differences in the pattern lock usage dependant on the origin of the user. The research application does not imply the strict (required by Android OS) lock regulations and length limitations, as we wanted to enable users to provide even the most complicated and long pattern lock combinations.

The research environment consists of the test application and the web service. The test application mimics interface implemented in mobile operating systems with all the details: icons, tabs placement and virtual keyboard and, making it as close as possible to the real interface. This strongly distinguishes our research application from the others, where simplified mockups were used.

The user interface of the test application running in different operating systems was presented below (Fig. 2).

The test application works in two modes. The first mode is designed for provision of the new pattern lock and for it's evaluation. The second mode is designed for evaluation of the pattern downloaded from the repository.

In the first mode, the user can input the pattern of his own choice. The length of the pattern is not limited and the nodes can be repeated, however the connections cannot overlap the nodes. This is different from the standard Android unlock procedure and gives the potential to explore some novel, lock patterns.

The system forces the user to repeat the pattern before rating it and sending to the repository. This should prevent from entering extremely long or random patterns, actually impossible to memorize. The user can enter long patterns with repeating parts, which can reduce the risk of the smudge analysis [33,34] or thermal analysis [35].

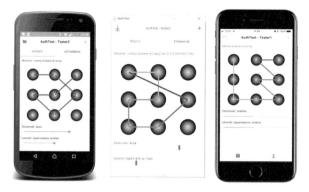

Fig. 2. The GUI of the multiplatform test application running on different operating systems.

In the next step the user can evaluate the difficulty level for the provided pattern and specify the memorizability for it. The encoded pattern supplemented with these ratings can be written to the repository.

In the second mode the lock pattern is downloaded from the repository. Next, the user can rate its complexity and memorizability and resend the results to the repository. The user is not obliged to rate every proposed pattern lock, instead, the user can continuously request for new pattern to be downloaded. The lock pattern in the second mode is animated, which illustrates the connections between the modes. This helps the user to understand the way the pattern was created and also to determine its complexity or the easiness to memorize it.

4   Pattern Strength Measure

The measure introduced in [27] and elaborated in [28] was developed to provide an automatic estimate of the unlock pattern strength. In the following we use the notation in which the pattern P will be viewed as a tuple $(P_0; d_1, d_2, d_3 \cdots d_k)$, where $P_0 \in \{0, 1, 2, 3, 4, 5, 6, 7, 8\}$ is the starting node and $d_1, d_2, d_3 \cdots d_k$ is the sequence of consecutive directions – see formula 1 and Fig. 1 for the reference.

$$d_i \in dir = \{L, U, R, D, Y, T, E, P, H, Q, W, G, X, V, F, A\} \tag{1}$$

Statistic distribution for the choice of the particular P_0 node and consecutive directions determine the complexity of the pattern. Let $P' = (P_0; d'_1, d'_2, d'_3 \cdots d'_q)$ denote a pattern P with consecutive repeating directions reduced.

The following quality measure (formula 2) for the unlock patterns on 3×3 grid was introduced to reflect the pattern characteristics by assigning properly defined weights to particular node and direction choices [27]:

$$m(P) = m(P_0; d_1, d_2, d_3 \cdots d_k) = (1 - p(P_0))(1 - \alpha^k) \frac{\sum_{i=1}^{q} w(d'_i)}{3k} \left(\frac{1}{2} + \frac{2^{N_r} - 1}{2^{N_r+1}} \right) \tag{2}$$

where: $\alpha \in [0.75, 0.99]$ is a parameter adjusting measure sensitivity to pattern length, p is the probability distribution of choosing P_0 as the starting node, $q = length(P')$, N_r is the number of nodes visited more than one time and w is the weight function of directions.

The goal of above proposal was to establish the relation between the unlock pattern strength measure and the subjective human perception of a lock pattern. In our calculations $\alpha = 0.8$, and the starting node probability distribution and the direction weight function were assumed as in [27] and are presented in the Tables 2a and 2b below.

Table 2. Direction weight function (a), starting node probability distribution (b).

a.

d	L	U	R	D	Y	T	E	P	H	Q	W	G	X	V	F	A
$w(d)$	1	1	1	1	2	2	2	2	3	3	3	3	3	3	3	3

b.

P_0	1	2	3	4	5	6	7	8	9
$p(P_0)$	0.39	0.1	0.12	0.06	0.04	0.06	0.075	0.055	0.1

5   Experimental Results

The experiment was carried out in November 2016 at the Faculty of Applied Mathematics at Silesian University of Technology (SUT, Poland). The participating students and lecturers were asked to evaluate the memorizability and complexity of various lock patterns basing on their own perception, reenter a given pattern lock or enter a new one to the database. The system acquired data on the activities performed by respondents, such as time needed to input a new pattern or reenter an already existing one, the type of the device used or gender of the respondent.

A dataset of nearly 200 records on timing issues was analyzed in terms of interdependencies between the input time of a pattern and its complexity and memorizability rates.

On Fig. 3a, 3b, and 4a a general input time statistics are presented with respect to the three main pattern features: complexity rate, memorizability rate and total length. The plots show the respective dependencies of the difference of time t_1 needed to enter the pattern for the first time and the time t_2 needed to reenter it. Within the range of rates up to 3 and length up to 10, a natural increase tendency is observed in $t_1 - t_2$ along with the pattern complexity and length increase. Similarly, $t_1 - t_2$ decreases with the increase of memorizability rate. However, for more complex patterns these tendencies turn into their converse; more complex and less memorizable patterns required less time to be entered/reentered. To explore this phenomena, the direct relation of the human perception ratings and average of t_1 has been investigated. On Fig. 4b the natural tendencies are more evident.

In addition to the simple statistics, a synthesis of the interdependencies between time t_1, t_2 and $|t_1 - t_2|$ and both human ratings are presented on Fig. 5a, 5b and 6a respectively. These comprehensive data show that the time

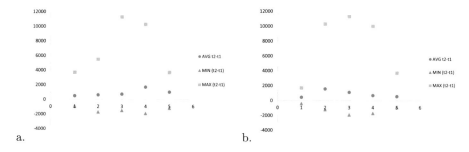

Fig. 3. Time statistics vs complexity rate (a), time statistics vs memorizability rate (b).

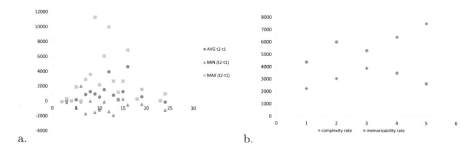

Fig. 4. Time statistics vs pattern length (a), time statistics vs human ratings (b).

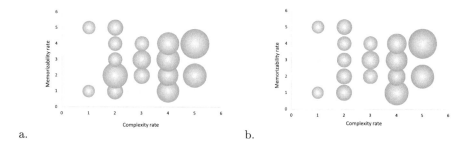

Fig. 5. Average pattern input time vs memorizability and complexity rate (a), Average pattern repetition time vs memorizability and complexity rate (b).

needed for pattern input is more sensitive to the complexity rating, than to the memorizability one, however there is no evident tendency in neither of the statistics. There are few possible explanations of this phenomenon. First, the memorizability and complexity rates are subjective evaluations dependent on an individual's perception. This subjectivity may disturb the correlation between the evaluations and independent objective time metrics, since patterns evaluated as more complex may require less time to be entered.

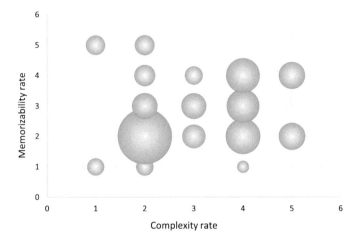

Fig. 6. Average time difference vs memorizability and complexity rate.

Another significant bias in the dependencies may be introduced by the diversity of the group of human testers/evaluators. More experienced pattern lock users usually are more fluent in pattern entering than the less experienced ones. Hence they may input longer, more complex patterns in a shorter time and rate them relatively low. These hypotheses could be verified in another experiment, with a well defined group of testers/users.

6   Summary

In this paper we have pointed some conclusions from the memorizability and timing issues research in the multiplatform framework designed for evaluation of pattern locks strength. In general, the correlation of pattern input time and evaluated complexity rate seems stronger than with memorizability rate.

The observed phenomena helped to raise further research questions for next investigations in this area. The diversity of users' reactions and subjective ratings of memorizability factors can become essential for the development of strengthened and comfortable mobile security solutions in the future.

References

1. Von Zezschwitz, E., Dunphy, P., De Luca, A.: Patterns in the wild: a field study of the usability of pattern and pin-based authentication on mobile devices. In: Proceedings of the 15th International Conference on Human-Computer Interaction with Mobile Devices and Services, pp. 261–270. ACM (2013)
2. De Luca, A., Hang, A., Brudy, F., Lindner, C., Hussmann, H.: Touch me once and i know it's you!: implicit authentication based on touch screen patterns. In: Proceedings of the SIGCHI Conference on Human Factors in Computing Systems, pp. 987–996. ACM (2012)

3. Meng, Y., Wong, D.S., Schlegel, R.: et al.: Touch gestures based biometric authentication scheme for touchscreen mobile phones. In: International Conference on Information Security and Cryptology, pp. 331–350. Springer (2012)
4. Kapczynski, A., Kasprowski, P., Kuzniacki, P.: User authentication based on behavioral patterns. Int. J. Comput. **6**(1), 75–79 (2014)
5. Kapczynski, A., Sroczynski, Z.: Behavioral HCI-based user authentication. In: Rostanski, M., Pikiewicz, P., Buchwald, P. (eds.) 10th International Conference Proceedings of Internet in the Information Society 2015. Academy of Business in Dabrowa Gornicza Press (2015)
6. Lee, J.D., Im, H.J., Kang, W.M., Park, J.H.: Ubi-rke: a rhythm key based encryption scheme for ubiquitous devices. Math. Prob. Eng. **2014** (2014)
7. Zargarzadeh, M., Maghooli, K.: A behavioral biometric authentication system based on memory game. Biosci. Biotechnol. Res. Asia **10**(2), 781–787 (2013)
8. Neverova, N., Wolf, C., Lacey, G., Fridman, L., Chandra, D., Barbello, B., Taylor, G.: Learning human identity from motion patterns. IEEE Access **4**, 1810–1820 (2016)
9. Buriro, A., Crispo, B., DelFrari, F., Wrona, K.: Hold and sign: a novel behavioral biometrics for smartphone user authentication. In: 2016 IEEE Security and Privacy Workshops (SPW), pp. 276–285. IEEE (2016)
10. Primo, A.: Keystroke-based continuous authentication while listening to music on your smart-phone. In: The 8th IEEE Annual Ubiquitous Computing, Electronics & Mobile Communication Conference, New York City, NY (2017)
11. Cao, K., Jain, A.K.: Hacking mobile phones using 2D printed fingerprints. Technical report, MSU Technical report, MSU-CSE-16-2 (2016)
12. Szczepanik, M., Jóźwiak, I.J., Jamka, T., Stasiński, K.: Security lock system for mobile devices based on fingerprint recognition algorithm. In: Information Systems Architecture and Technology: Proceedings of 36th International Conference on Information Systems Architecture and Technology–ISAT 2015–Part III, pp. 25–35. Springer (2016)
13. Cejudo-Torres-Orozco, M., Garcia-Rios, E., Escamillahernandez, E., Nakano-Miyatake, M., Perez-Meana, H.: Counterfeit image detection in face recognition systems using stereo vision and optical flow methods. In: MCASE (2014)
14. Smith, K.A., Zhou, L., Watzlaf, V.J.: User authentication in smartphones for tele-health. Int. J. Telerehabilitation **9**(2), 3 (2017)
15. Wojewidka, J.: Why the mobile biometrics surge demands true liveness. Biometric Technol. Today **2017**(10), 8–11 (2017)
16. Andriotis, P., Tryfonas, T., Oikonomou, G., Yildiz, C.: A pilot study on the security of pattern screen-lock methods and soft side channel attacks. In: Proceedings of the Sixth ACM Conference on Security and Privacy in Wireless and Mobile Networks, pp. 1–6. ACM (2013)
17. Aviv, A.J., Fichter, D.: Understanding visual perceptions of usability and security of android's graphical password pattern. In: Proceedings of the 30th Annual Computer Security Applications Conference, pp. 286–295. ACM (2014)
18. Miller, G.A.: The magical number seven, plus or minus two: some limits on our capacity for processing information. Psychol. Rev. **63**(2), 81 (1956)
19. Cowan, N.: The magical mystery four: how is working memory capacity limited, and why? Curr. Dir. Psychol. Sci. **19**(1), 51–57 (2010)
20. Połap, D., Woźniak, M.: Introduction to the model of the active assistance system for elder and disabled people. In: International Conference on Information and Software Technologies, pp. 392–403. Springer (2016)

21. Jagodzińska, M.: Psychology of the memory: research, theories, applications. In: Polish:Psychologia pamieci: badania, teorie, zastosowania, Helion (2008)
22. Goodin, D., Loge, M.: New data uncovers the surprising predictability of android lock patterns; tell me who you are, and i will tell you your lock pattern (2015). http://arstechnica.com/security/2015/08/new-data-uncovers-the-surprising-predictability-of-android-lock-patterns/. Accessed 23 Feb 2017
23. Siadati, H., Gupta, P., Smith, S., Memon, N., Ahamad, M.: Fortifying android patterns using persuasive security framework. In: UBICOMM 2015, p. 81 (2015)
24. Song, Y., Cho, G., Oh, S., Kim, H., Huh, J.H.: On the effectiveness of pattern lock strength meters: measuring the strength of real world pattern locks. In: Proceedings of the 33rd Annual ACM Conference on Human Factors in Computing Systems, pp. 2343–2352. ACM (2015)
25. Budzitowski, D., Aviv, A.J., Kuber, R.: Do bigger grid sizes mean better passwords? 3×3 vs. 4×4 grid sizes for android unlock patterns. In: Symposium On Usable Privacy and Security (SOUPS) (2015)
26. Uellenbeck, S., Dürmuth, M., Wolf, C., Holz, T.: Quantifying the security of graphical passwords: the case of android unlock patterns. In: Proceedings of the 2013 ACM SIGSAC Conference on Computer & Communications Security, pp. 161–172. ACM (2013)
27. Bier, A., Sroczynski, Z.: Evaluation of pattern lock codes strength for increased security in mobile applications. In: Rostanski, M., Pikiewicz, P., Buchwald, P., Maczka, K. (eds.) Proceedings of the 11th Scientific Conference Internet in the Information Society 2016. Academy of Business in Dabrowa Gornicza Press (2016)
28. Bier, A., Kapczyński, A., Sroczyński, Z.: Pattern lock evaluation framework for mobile devices: human perception of the pattern strength measure. In: International Conference on Man–Machine Interactions, pp. 33–42. Springer (2017)
29. Sun, C., Wang, Y., Zheng, J.: Dissecting pattern unlock: the effect of pattern strength meter on pattern selection. J. Inf. Secur. Appl. **19**(4), 308–320 (2014)
30. Aviv, A.J., Prak, J.L.: Comparisons of data collection methods for android graphical pattern unlock. In: Symposium On Usable Privacy and Security (SOUPS) (2015)
31. Egelman, S., Jain, S., Portnoff, R.S., Liao, K., Consolvo, S., Wagner, D.: Are you ready to lock?. In: Proceedings of the 2014 ACM SIGSAC Conference on Computer and Communications Security, pp. 750–761. ACM (2014)
32. Sroczynski, Z.: Human-computer interaction on mobile devices with the FM application platform. In: Rostanski, M., Pikiewicz, P. (eds.) Internet in the Information Society. Insights on the Information Systems, Structures and Applications. Academy of Business in Dabrowa Gornicza Press (2014)
33. Aviv, A.J., Gibson, K.L., Mossop, E., Blaze, M., Smith, J.M.: Smudge attacks on smartphone touch screens. Woot **10**, 1–7 (2010)
34. Kwon, T., Na, S.: Tinylock: affordable defense against smudge attacks on smartphone pattern lock systems. Comput. Secur. **42**, 137–150 (2014)
35. Abdelrahman, Y., Khamis, M., Schneegass, S., Alt, F.: Stay cool! understanding thermal attacks on mobile-based user authentication. In: Proceedings of the 2017 CHI Conference on Human Factors in Computing Systems, pp. 3751–3763. ACM (2017)

On Finding Model Smells Based on Code Smells

Erki Eessaar[(✉)] [ID] and Ege Käosaar

Department of Software Science, Tallinn University of Technology, Tallinn, Estonia
erki.eessaar@ttu.ee, ege.kaosaar@gmail.com

Abstract. A smell in an artifact is a sign that the artifact may have a technical debt, meaning that it may contain the results of one or more sub-optimal design decisions. The debt makes it more difficult to understand, maintain, extend, and reuse the artifact. Technical debt could appear in any technical artifact. Currently the debt in source code is in the center of attention. Modeling is an important system development activity. Models can have technical debt and a sign of it is the presence of one or more model smells. This paper introduces a catalog of 46 model smells (https://github.com/erki77/model-smells) that has been created based on a set of code smells. The cataloged smells are general but their examples are given based on system analysis models. The main takeaway is that most of the considered code smells are instances of generic problems that can also occur in models.

Keywords: Technical debt · Model smell · Code smell · System analysis
CASE

1 Introduction

Modeling is an important system development activity. Regardless of whether we sketch models just for communication and for presenting ideas explicitly or whether we want to use the full power of model-driven development, it is useful to achieve and maintain a good quality of models. Not only should the models say truth about the modeled system, they should also do it in a right manner. Their structure, style, and level of detail should facilitate communication and understanding. It should not impede system validation and verification based on models but instead make it easier, regardless of whether it is done by humans or by software. It should make it as easy as possible for modelers to change models, reuse models, and use the models as an input for a software that evaluates the models or generates new artifacts based on these.

There is quite a lot of interest towards the concept of technical debt, smells, which are the signs of the presence of the debt, and refactoring, which is the process of getting rid of the debt. Although technical debt can appear in any technical artifact, currently the biggest attention is towards the debt, smells, and refactorings in the source code, particularly in the source code (aka *code*) of object-oriented languages [1, 2]. The concepts of debt, smell, and refactoring are applicable to the models too. Most of the research about model smells concentrates to the application design models and many of the presented smells are also code smells. For instance, most of the model smells in the

© Springer International Publishing AG, part of Springer Nature 2019
R. Silhavy (Ed.): CSOC 2018, AISC 763, pp. 269–281, 2019.
https://doi.org/10.1007/978-3-319-91186-1_28

Table 6 of a literature review [3] are also code smells or system architecture smells. There is no clear line between design and code smells because design models are high-level descriptions of solutions that the source code implements. Tufano et al. [4] show, based on the investigation of the change histories of more than 200 open source projects, that most of the smells are introduced to the source code already at its creation time, not as the result of later modifications. The earlier we discover a problem, the easier and less costly it is to fix it. Thus, it is advantageous to be able to detect problems already in design models in order to avoid their propagation to the code. Even if some model smell (like "No Incoming Transitions" [3] of state machines) has no direct counterparts in the code smells it may be a manifestation of a general problem that can appear in both models and code. In case of the particular smell an artifact contains an unused part that appears in the programming domain as the smell "Dead Function" [1].

Jaakkola et al. [5] estimate that errors that have been made in analyzing software requirements increase costs by the multiplying factor three in the each following system development phase. Thus, the effort that is needed to correct the analysis mistakes in the design phase is three times, in the implementation phase nine times, and in the system testing phase 27 times higher than correcting these already during the system requirements analysis. Therefore, it is very important to test and inspect the ideas of the system from early on by using its models. If the models have problems of structure, presentation, or style, then it makes it more difficult to use the models for the purpose. Thus, maintaining and increasing the quality of analysis models is an important topic.

A strategy of doing it is searching occurrences of smells from the models and refactoring the models to remove their underlying problems. The body of literature of smells of analysis models is not very extensive. The existing catalogs concentrate to specific model types (see Related Works). At the same time there are extensive catalogs of code smells. One such is a work of Martin [1] that we used as the input to our work. Most of the code smells are general, meaning that they are applicable in case of different programming languages. Their representation contains references to code fragments (and thus to specific programming languages) as examples. It gave us an idea that perhaps it is possible to use a catalog of code smells as a basis in order to work out a catalog of smells for analysis models. Our hypothesis before we started the endeavor was that although code and model smells are about different domains there is a set of core problems that are common to both domains and can manifest themselves in the artifacts of both domains. This would be similar to how the principle of separation of concerns manifests in the database domain through the normalization theory and in the system functionality domain through the theory of normalized systems [6]. Thus, if we use a catalog of code smells as an input and are able to identify the core problem behind each smell, then we can use the information to write down corresponding general smells for the modeling domain and use different types of analysis models, perhaps created in different languages, as examples of the model smells. The possibility of such connection is predicted by Mohagheghi et al. [7], who state, without further thorough elaboration, that many practices regarding maintenance of code also apply to models.

The method of research is *design science* [8]. The driving business need is to have models with good quality to be able to discover system problems early on. A goal of the work is to create a new design artifact – a catalog of model smells that could characterize analysis models and could be used as an educational material and as a checklist for improving the quality of models. To create the artifact, we have to know the state of the art of the field. Thus, Sect. 2 introduces some relevant related work. Section 3 explains the method of creating the catalog. Section 4 presents the key statistics of the catalog together with the reference to its location. To evaluate the catalog, we use a combination of analytical, descriptive, and experimental methods. The results are outlined in the Sect. 5. Moreover, Sect. 5 envisions how a future CASE system may help us to reduce the number of smells in models and discusses threats to the validity of the research. Finally, we conclude and point to the future work with the current topic.

2   Related Works

Extensive literature review together with the definition of six model quality goals by Mohagheghi et al. [7] is a good example of the depth and breadth of the research about model quality. One strategy of achieving a better quality is trying to follow best practices and avoiding bad practices, which can be represented as patterns/heuristics and anti-patterns/smells, respectively. There exist some catalogs of model smells. Arendt and Taentzer [9] present a catalog of model smells about class diagrams (12 smells), use case diagrams (three smells), and state machines (two smells). Many of the class diagram smells have direct counterparts in documented code smells (see [1, 2]). El-Attar and Miller [10] present eight antipatterns of use case models and refer to the location with 18 more. Each antipattern is a structured description of a recurring problem together with suggestions of how to fix it. The relationship between smells and antipatterns is that each smell could be written down in a structured manner as an antipattern. Actually in case of some other model smell catalogs ([9, 11]) the presentation is also well-structured and contains information about refactoring. In case of such catalogs the terms "antipattern" and "smell" could be used interchangeably. Similarly, the smells in our catalog use structured presentation (see Sect. 3). Weber et al. [11] present eight process model smells. Gerlitz et al. [12] introduce 21 model smells for MATLAB/Simulink, which is a software for model-driven development in the industrial domain. McDonald and Matts [13] present seven smells that can occur in conceptual data models or domain class models. As a part of extensive literature review, Misbhauddin and Alshayeb [3] give a list of 27 model smells that can appear in UML class, state, sequence, or component diagrams.

Opposite to smells and antipatterns are best practices, which could be written down in a less structured manner as heuristics [14] or in a more structured manner as patterns [15]. These offer information about what and how to refactor in models in order to get rid of smells. Because our general smells can incorporate smells of different model types as examples, the refactorings of specific model types can also serve as examples.

3   The Method

The catalog of model smells was mainly created by translating the code smells from the book about clean code [1]. The only exception is the smell G0 ("Not Remembering the

History") that suggests keeping model history and that was not created by translating the code smells of Martin [1]. It was written because it is a reason of other model smells – *having metadata in models in a wrong place* and *having expired model elements*.

The translation was conducted as a mental exercise by the authors to see the extent of similarities of flaws that modelers and programmers can have in their work results. The translation does not mean translating the smells word by word but rather translating the idea. In mathematical terms, if we exclude G0, then a translation function *f_translate* from a set *CodeSmells* to a set *ModelSmells* is surjective [16], meaning that for every element *m* in the codomain *ModelSmells* of *f_translate* there is at least one element *c* in the domain *CodeSmells* of *f_translate* such that f_translate(c) = m. There are elements of *CodeSmells* that do not have a counterpart in the *ModelSmells*. The function *f_translate* may map more than one element of *CodeSmells* to the same element of *ModelSmells*. Each code smell is mapped to at most one model smell.

Gerlitz et al. [12] mention code smells as a source of inspiration. Our approach is similar to the work of Weber et al. [11] in the sense that they too used a list of code smells as the starting point. A difference is that they filtered input smells whereas we used all the smells from the input catalog. Weber et al. [11] used literature study to confirm that each candidate smell represents a true problem whereas we tried to find examples based on our experience with modeling. The work [11] added candidate model smells based on the literature study, i.e., code smells were not the sole input. We added only one model smell that does not have a corresponding code smell in the input catalog. Not remembering the history would be also a problem of source code.

Many of the smells in our catalog are general enough to be applicable to any kind of software models. However, while writing the smell examples in the catalog, we had specifically in mind a set of eight model types that one can use during system analysis [17]. These model types include textual models (contracts of database operations), visual (diagrammatic) models (activity, state machine, and system sequence models), models that contain both visual and textual component (use case, conceptual data, and domain class models), and models that could be presented either visually or textually (subsystems). The smells are not UML-specific, although all the visual components of the models can be created in UML. All models can contain comments. Thus, for each smell, we had to think as to whether it could affect a comment of any model.

Compared to the free-form presentation of code smells [1], we use a more structured presentation that is similar to the presentation of antipatterns [10].

Just like in [1], the smells have been divided into classes but the classes are different. Each smell belongs to exactly one class. The catalog has four classes of smells.

- G: General smells of models and diagrams.
- S: Smells about the style of models and diagrams.
- N: Smells about the naming of models, diagrams, and their elements.
- T: Smells about testing models and testing modeled systems based on models.

Depending on the used modeling language, diagrams might not be classified as model elements. For instance, UML 2.5 defines diagrams as "graphical representations of parts of the UML model." [18] According to the UML definition a diagram contains graphical elements that represent elements in a UML model. Thus, our catalog classifies smells according to their scope – smells that apply only to model elements (for instance, "Expired Model Element"), only to diagrams (for instance, "Too Big Diagrams"), or both (for instance, "Metadata about Models in a Wrong Place"). We see here an analogy with the database world. Diagrams and model elements have the same relationship as database views and base data structures in a database. Database views are database objects and in some cases (for instance, naming and avoiding exact duplicates) should follow the same design rules as base data structures. On the other hand, there are also differences. For instance, in case of base data structures of operational databases, we should try to reduce data redundancy within and across the data structures as much as possible (see the database normalization theory and the orthogonal design principle). On the other hand, values of views are automatically calculated based on the values of base data structures and having some data redundancy within or across views is not a problem because it is a *controlled redundancy*. Just like each database view is a part of a database, each diagram should be considered a part of the model that it helps us to represent. To summarize, the first version of smells is influenced by UML, considers diagrams as parts of models but does not count diagrams as model elements.

We created the catalog in an iterative manner. Firstly, we replaced the highest-level code related terminology with the model-related terminology and removed code-specific sentences like code examples. For instance, we replaced words "code" and "source file" by "model" and "model file", respectively. In case of more code-specific sentences, we tried to come up with a more general sentence and then started to search examples of it from the modeling domain. For instance, the description of the code smell "Feature Envy" contains the statement "The methods of a class should be interested in the variables and functions of the class they belong to, and not the variables and functions of other classes." [1] Initially, we replaced it with the statement "The element in a container should be interested in the elements of the container they belong to, and not the elements of other containers." By searching examples, we found, for instance, that the element could be a database operation and the container is in this case the register that contains the operation as a part of its public interface.

As we proceeded with the catalog creation, we used the code and model smells as a checklist to improve the catalog. For instance, the number of model smells is smaller than the number of code smells (see Table 1) because in case of some code smells the initial translation resulted with too similar model smells. To avoid *redundancy*, we merged these into a single smell. For preventing *inconsistencies of representation*, we worked out a uniform structure for all the model smells. An *inconsistency of representation* in the input code smells is that the names of some (like "Obsolete Comment") refer to problems in code but the names of others (like "Function Names Should Say What They Do") refer to actions that should be taken to solve problems. Each smell describes a problem and if its name does not express the problem, then *the name does not express its nature*. In our catalog, we selected to all the smells a name that describes a problem (for instance, "Metadata about Models in a Wrong Place"). The same approach is also used in the names of the smells in [3, 9–13]. Relationships between smells are important because they point to the problems that could co-occur with a smell. Martin [1] states in the code smell "Base Classes Depending on Their Derivatives", which is about *mixing different abstraction and decomposition* levels that "In general, base classes should know nothing about their derivatives." It gave us a guideline to refer from the more specific model smells to the more general model smells but not vice versa in order to decouple the smells and allow the use of more general smells without necessarily having to consult with the more specific smells. The more specific smell might describe a special case of the more general smell or describe a specific reason or result of having the more general smell in models. In order to avoid *illogical order*, we tried to present more general smells before more specific smells.

Table 1. Model smells that have the biggest number of corresponding code smells

Model smell	Code smells that were the input for the creation
N1: The Name Does Not Express Its Nature	G16: Obscured Intent G20: Function Names Should Say What They Do N1: Choose Descriptive Names N4: Unambiguous Names N7: Names Should Describe Side-Effects
G6: Redundancy	G5: Duplication G25: Replace Magic Numbers with Named Constants G28: Encapsulate Conditionals G35: Keep Configurable Data at High Levels

4   The Catalog

Due to the space restrictions, we present in this paper statistics and conclusions based on the catalog but not the actual smells. The full catalog is available at https://github.com/erki77/model-smells.

The input catalog [1] contains 66 code smells. The created catalog of model smells consists of 46 model smells, 45 of which have been written down by translating the code smells. Only three code smells ("Output Arguments", "Avoid Long Import Lists by Using Wildcards", and "Don't Inherit Constants") do not have a corresponding model smell. The reason is that the code smells are too code-specific. Table 1 names the model smells that were created based on the biggest number of code smells. If one is able to create one model smell based on more than one code smells, then it shows that the code smells describe quite similar problems that one can generalize. At the same time there are 36 model smells that have exactly one corresponding code smell.

Most of the model smells (28) belong to the general class. There are four model smells about style, five about naming, and nine about testing.

Most of the model smells are generic. We estimate that 18 defined smells (39%) can occur in the comments of any model type.

We estimate that 26 defined smells (57%) can occur in all the eight target analysis model types. If we take into account comments, then the number of such smells is 30 (65%). The most referenced smell within the catalog is "Redundancy" (seven references from other model smells), followed by "Imprecise model" and "Inconsistencies of Representation" (both six references), "Not Understanding the Modeled Process", and "Not Following the Popular Modeling Style Conventions" (both five references), i.e., these general smells have the most associated smells that give additional details.

For the 29 (63%) model smells in our catalog, we did not find a corresponding smell in the existing catalogs of model smells [3, 9–13]. There is no corresponding smells for all the nine smells about testing. All the corresponding smells in other sources are more specific and can be used as examples in case of our smells.

5 Discussion

Removing the smells form analysis models improves the chances to discover mistakes based on the models. While we did not experiment with it for the paper, the first author of the paper makes this observation based on more than 15 years of experience of supervising and evaluating almost 2000 student projects as well as graduation theses. On the other hand, avoiding and removing the smells in case of analysis models is not a guarantee that technical debt will not appear in some other form in the artifacts of the later system development phases. For instance, Izurieta et al. [19] comment that mapping models to code might introduce new technical debt. Analysis helps us to understand the problem and design leads to a solution, which is finally implemented. Avoiding and getting rid of smells in analysis models helps us to solve the right problem but unfortunately it does not remove the possibility of doing it in a bad way.

Another value of defining model smells is that these can be used to discuss the quality of a particular set of models. The names of the smells would be a part of the vocabulary about the topic. If these models have been generated by a tool, then it also provides a framework for discussing the quality of work results of the tool and thus also for suggesting improvements for the tool. For instance, Thakur and Gupta [20] use UML model smells in order to characterize (in the particular case to criticize) the quality of some analysis models that were generated by a tool.

Similarly, if the analyzed models are the result of an educational process, then information about the presence of smells in these would give lecturers feedback about what to improve in the teaching process. We conducted a small experiment by manually reviewing 30 accepted course projects that contain a system analysis component. The reviewer (the second author) did not participate in the original evaluation of the projects. Out of the 24 different types of detected problems the most frequent were:

- "Illogical Order" (of use cases) (20 projects),
- "Imprecise Model" and "Not Following the Popular Modeling Style Conventions" (both 18 projects),
- "Long Lines on a Diagram" (16 projects),
- "Stating the Obvious" (14 projects),
- "Inconsistencies of Representation" (13 projects),
- "The Name Does Not Express Its Nature" (12 projects).

The median number of different types of smells (regardless of how many times they appeared) in a project was six.

The first author of the paper teaches university courses about database design and development. Some of the participating students and their curriculum are more oriented towards programming and some are more oriented towards modeling. Students with the programming background sometimes express views that improving the quality of models is a waste of time. Explaining to them the importance in terms of the removal of code smells helps them to better understand the importance and usefulness of the process. Similarly, for the students who work more with models it is useful to see that the practices that they follow in case of models are equally important in case of coding. Seeing the similarities may help them to reduce potential uneasiness towards programming subjects.

5.1 Common Problems of Models and Code

The results of creating model smells by translating code smells show that there is a set of common problems in both domains. Examples of such general problems are redundancy, inconsistencies of representation, mixing different abstraction and decomposition levels, high coupling, losing focus, and not using the power of existing tools that are supposed to automate the work.

Next, we give an example of the shared problems based on concrete artifacts. The use case diagram A in Fig. 1 illustrates a set of model smells. The diagram B is the result of removing the smells, i.e., refactoring. The diagrams are followed by a query with multiple code smells (query A), and finally by a refactored query (query B).

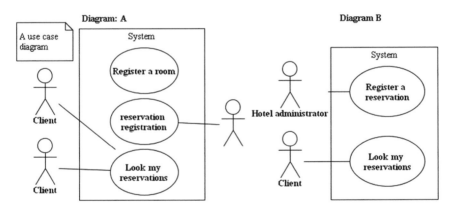

Fig. 1. A use case diagram with smells (diagram A) and after refactoring (diagram B)

```
--Query A
select R.hotel_nr, R.room_nr, R.hotel_nr, Ucase(comment)
from Reservation AS R INNER join hotel
ON Hotel.hotel_nr=R.hotel_nr; /*Find reservations.*/
--Query B
SELECT room_nr, hotel_nr, Ucase(comment) AS comment
FROM Reservation;
```

In the next description, we use the letter C to refer to code smells in [1] and the letter M to refer to model smells in our catalog. Both the diagram A and query A have a comment that states the obvious (M: "Stating the Obvious"; C: "Redundant Comment"). The diagram A duplicates the actor Client and the query A has multiple columns with the same name and content (*hotel_nr*) in the result (M: "Redundancy"; C: "Duplication"). The diagram A has an unnamed actor whereas in the query A the expression Ucase (comment) produces a column that name is not given by the query writer, i.e., the name is implementation-dependent (M: "The Name Does Not Express Its Nature"; C: "Choose Descriptive Names"). The diagram A presents use case "Register a room" that is not a part of the overall theme of the diagram (management of registrations) (M: "No Clear Focus"). Similarly, the query A joins table *Hotel* that is not needed for the correct result (C: "Artificial Coupling"). The name of use case "reservation registration" does not begin with a strong verb (M: "Not Following the Popular Naming Conventions") and some keywords in the query A are not in uppercase (C: "Follow Standard Conventions"). The names of use cases in the diagram A have different naming style, including the use of uppercase letters (M: "Inconsistencies of Representation"). Similarly, in the query A some keywords are in uppercase and some are not (C: "Inconsistency"). Moreover, in some places a table name starts with an uppercase letter and in some place with a lowercase letter.

Atkinson and Draheim [21] describe the notion of *deep standardization* that means finding and fixing problems shared by multiple technologies, development processes, tools, or different types of artifacts. Establishing problems that are shared by models and code is an act of deep standardization.

5.2   A Vision for Reducing the Number of Smells in Models

Realization of the vision of the cloud-based and view-based CASE system (called CASE 2.0) [21] that offers different software engineering artifacts as views on top of the single underlying model (SUM) would mitigate a lot of problems that are expressed in the model smells. Next, we use *italics* to refer to the corresponding model smells in our catalog. Having a set of tools operating against the single underlying source of information would reduce *redundancy* and *the time that is needed to generate a new artifact from a model*. Both problems are partly caused by having multiple CASE projects that contain models of the same modeling project. The union of different types of models could present information redundantly but it is not a problem because the SUM would be the "single, unambiguous, authoritative representation" [22] of information within the CASE system and other models are derived from it. As such it is an example of application of the Dont Repeat Yourself design principle [22]. *Collecting and storing metadata* and *keeping the historic versions of models* would be centralized and done in a uniform manner based on the SUM. These processes could be transparent to the modelers, i.e., the system does it automatically. Each model or a source code artifact is a view value that is calculated based on the SUM. If one wants to look information about the current understanding of the system, then the queries behind the views ensure that *model elements that represent expired system elements* are not in the results. The burden of *not mixing different levels of abstractions* in a model, providing models with *clear focus*, presenting all the information *in an expected part of a model*, *showing dependencies* in all the places of models where these have to be visible, presenting model elements and diagrams in a *logical order*, and having *consistent representations* moves from modelers to tool providers who implement the SUM and the views that present information that is captured in the SUM. The same is true for multitude of *model style* issues. It would be possible to define and enforce *naming conventions* centrally, in the cloud, at the level of the SUM. *Unauthorized representations* would not be a problem because at the beginning of the project the representations will be agreed and the CASE system is instructed to present models that follow the agreement. View mechanism would make it possible to switch between *multiple software languages* because for each such language there is a distinct set of views. Thus, each model consumer can use his/her preferred languages. Checking as to whether some system element is not needed by any other element is possible across different system development artifacts (models, code, tests, etc.). For instance, in a current state of the art development environment there could be a database stored procedure that implements a database operation, which is described by using a contract. The procedure is invoked by an application that implements a functional subsystem. In the models of the system the database operation is not used by any functional subsystem. Just checking the model would result in removing the contract of the database operation. Having models and code artifacts as view values that are calculated based on the SUM would avoid it.

5.3   Threats to Validity

A threat to validity is that only two people created the catalog. Other people might have mapped the input code smells differently and produced a different set of model smells. This is partly a question about validity, meaning that all the model smells should express problems that are real and relevant in the modeling domain. To evaluate validity, we tried to map the model smells that are the result of our work (we denote the set as S) with the analysis and design model smells in the catalogs [9–13] (61 in total). We denote the set of these model smells as V. There is a correspondence between s∈S and v∈V if v can be used as an example of s. We were unable to map only one of the smells in V with a smell in S – "Specialization Aggregation" [9]. It shows that the smells in S cover quite well the reported problems in models. The smells of V were mapped only with the 16 smells in S. However, because we were able to give our own examples for all the smells in S, we do not think that the result invalidates S.

A threat to validity is that only one catalog of code smells was used as the basis for the process of writing down model smells. This is a question about completeness, meaning that the set of model smells should *ideally* cover all the problems that are real and relevant in the modeling domain. There are more documented code smells than presented in [1]. Misbhauddin and Alshayeb [3] conducted a systematical literature review about model refactoring that included 63 primary studies. They present a set C of 27 models smells that "are the union of all smells covered by all surveyed articles." Most of these model smells are also either code smells or system architecture smells. The model smells are defined in terms of specific diagrams – mostly class diagrams but in case of some smells also state, sequence, or component diagrams. To evaluate completeness, we tried to map the model smells that are the result of our work (we denote the set as S) with the model smells in C. There is a correspondence between s∈S and c∈C if c can be used as an example of s. We were unable to map only three of the smells in C with a smell in S. These smells are "Unnecessary Behavioral Complexity", "Structural Complexity", and "Specialization Aggregation". These smells are about excessive complexity – the first two are very generic and the third is an example of unnecessary complexity in UML class diagrams.

The smells like "No Clear Focus", "One Process with Multiple Tasks", and "Unnecessary Decomposition" in S indicate that the model is more complex than it should be. Nevertheless, there is not a general smell in S that warns against too complex models. Models are simplifications, models should help modelers to convey ideas, and for this purpose overly complex models are counterproductive. On the other hand, in case of using model-driven development one should represent all the necessary details in models in order to be able to generate source code from the models or use the models as an input to a system that interprets and executes these. Thus, the models cannot be overly simplified. We agree with [12] that smells are hints rather than guarantees of problems. In this case, whether a complexity of models is a problem depends on the task where the models are needed. To conclude, the results of mapping S and C show that our set of model smells is quite complete but we do not claim absolute completeness.

A threat to validity would be not paying attention to all the necessary quality aspects. What types of quality should be achieved in case of describing smells? We think that a quality framework of conceptual modeling [23] gives a good answer. We have already covered validity and completeness that are the only two semantic quality goals in the framework. The framework also prescribes two other types of quality – syntactic quality and pragmatic quality. The only syntactic quality goal is syntactic correctness [23] of an artifact. In case of structured textual documents like smells the syntactic quality means both syntactic correctness of text and following of the predefined structure of the artifact. We used speller in the writing process. We proposed a structure of smell documentation and followed it. The only pragmatic goal is comprehensibility [23] of an artifact. For the better comprehension, we structured all the smells in a similar manner, placed more general smells before more specific smells, and tried to come up with concrete examples based on different model types. Additional means would be a part of future work like turning the catalog into a searchable website and automating the detection of smells in models.

As an experiment, we tried to map all the UML patterns [15] (48 in total) (set P) with the smells in our catalog S. There is a link between s∈S and p∈P if p can be used to refactor a model in order to get rid of or mitigate the underlying problem of s. We were unable to map only three UML patterns ("Rotated Text", "Configuration Management Model", and "Model for Maintenance") to the model smells of our catalog. We found at least one refactoring UML pattern for 21 (46%) of model smells in S.

The generality of smells in S could be seen as a weakness, because for each smell there is not a single algorithm how to detect the smell in models and how to remove it. These algorithms would be specific to model types. We speculate that if we group together under one umbrella a set of specific model or code smells that already have automatic detection algorithm and a set of model smells that do not yet have it, then the algorithms for the first can be used to work out the algorithms for the second.

The main threat to the validity of the mapping experiments is that the mapping is based only on the opinion of one or two people. The mapping results are available in the catalog where the smells contain cross–references as well as references to the previously documented smells and patterns.

6   Conclusion

The paper presents the results of a work to produce a catalog of general model smells by translating the ideas behind a catalog of code smells [1]. The created catalog consist of 46 smells. Due to the space restrictions the paper does not contain the catalog but the full catalog is available at https://github.com/erki77/model-smells.

The translation process meant finding the core idea of each code smell and deciding as to whether the problems that fall under the same idea can appear in models as well. The resulting model smells are general and we were able to formulate examples of the smells in terms of analysis models. The main lesson of the work is that there is a set of generic problems that can manifest itself both in the domain of models as well as in the domain of source code. It is important to share this information in the educational process to strengthen the understanding that modeling and coding have things in common. A cloud-based and view-based software engineering environment (CASE 2.0) would mitigate many of the problems described in the model smells.

The future work should search ways of automating the detection and removal of model smells. Perhaps it is possible to modify the detection algorithms of code smells to work with model smells. Most certainly the code smells give inspiration as to what to check in case of models. Martin [1] acknowledges that his set of smells is incomplete. Thus, one can work with other catalogs of code smells to increase the completeness of the model smell catalog. We agree with [1] that total completeness is unrealistic.

References

1. Martin, R.C.: Clean Code. A Handbook of Agile Software Craftsmanship. Pearson Education, Upper Saddle River (2009)
2. Shvets, A.: Refactoring.Guru. https://refactoring.guru/
3. Misbhauddin, M., Alshayeb, M.: UML model refactoring: a systematic literature review. Empir. Softw. Eng. **20**, 206–251 (2015). https://doi.org/10.1007/s10664-013-9283-7
4. Tufano, M., Palomba, F., Bavota, G., Oliveto, R., Di Penta, M., De Lucia, A., Poshyvanyk, D.: When and why your code starts to smell bad. In: Proceedings of the 37th International Conference on Software Engineering, vol. 1, pp. 403–414. IEEE Press, Piscataway (2015). https://doi.org/10.1109/icse.2015.59
5. Jaakkola, H., Henno, J., Welzer-Družovec, T., Thalheim, B., Mäkelä, J.: Why information systems modelling is difficult. In: 5th Workshop of Software Quality, Analysis, Monitoring, Improvement, and Applications, pp. 29–39. University of Novi Sad (2016)
6. Eessaar, E.: The database normalization theory and the theory of normalized systems: finding a common ground. Balt. J. Mod. Comput. **4**, 5 (2016)

7. Mohagheghi, P., Dehlen, V., Neple, T.: Definitions and approaches to model quality in model-based software development–a review of literature. Inform. Softw. Tech. **51**, 1646–1669 (2009). https://doi.org/10.1016/j.infsof.2009.04.004

8. Hevner, A.R., March, S.T., Park, J., Ram, S.: Design science in information systems research. MIS Quart. **28**, 75–105 (2004)

9. Arendt, T., Taentzer, G.: UML model smells and model refactorings in early software development phases. Universitat Marburg (2010)

10. El-Attar, M., Miller, J.: Improving the quality of use case models using antipatterns. Softw. Syst. Model. **9**, 141–160 (2010). https://doi.org/10.1007/s10270-009-0112-9

11. Weber, B., Reichert, M., Mendling, J., Reijers, H.A.: Refactoring large process model repositories. Comput. Ind. **62**, 467–486 (2011). https://doi.org/10.1016/j.compind.2010.12.012

12. Gerlitz, T., Tran, Q.M., Dziobek, C.: Detection and handling of model smells for MATLAB/Simulink models. In: International Workshop on Modelling in Automotive Software Engineering, pp. 13–22 (2015)

13. McDonald, K., Matts, C.: The seven information smells of domain modelling. https://www.infoq.com/articles/seven-modelling-smells

14. Berenbach, B.: The evaluation of large, complex UML analysis and design models. In: Proceedings of the 26th International Conference on Software Engineering, pp. 232–241. IEEE Computer Society, Washington (2004). https://doi.org/10.1109/icse.2004.1317445

15. Evitts, P.: A UML Pattern Language. Macmillan Technical, Indianapolis (2000)

16. Surjective function. https://en.wikipedia.org/wiki/Surjective_function

17. Eessaar, E.: A set of practices for the development of data-centric information systems. In: José Escalona, M., Aragón, G., Linger, H., Lang, M., Barry, C., Schneider, C. (eds.) Information System Development, pp. 73–84. Springer, Cham (2014). https://doi.org/10.1007/978-3-319-07215-9_6

18. OMG® Unified Modeling Language® (OMG UML®) Version 2.5.1

19. Izurieta, C., Rojas, G., Griffith, I.: Preemptive management of model driven technical debt for improving software quality. In: Proceedings of the 11th International ACM SIGSOFT Conference on Quality of Software Architectures, pp. 31–36. ACM, New York (2015). https://doi.org/10.1145/2737182.2737193

20. Thakur, J.S., Gupta, A.: AnModeler: a tool for generating domain models from textual specifications. In: Proceedings of the 31st IEEE/ACM International Conference on Automated Software Engineering, pp. 828–833. ACM, New York (2016). https://doi.org/10.1145/2970276.2970289

21. Atkinson, C., Draheim, D.: Cloud-aided software engineering: evolving viable software systems through a web of views. In: Mahmood, Z., Saeed, S. (eds.) Software Engineering Frameworks for the Cloud Computing Paradigm. Computer Communications and Networks, pp. 255–281. Springer, London (2013). https://doi.org/10.1007/978-1-4471-5031-2_12

22. Dont Repeat Yourself. http://wiki.c2.com/?DontRepeatYourself

23. Lindland, O.I., Sindre, G., Solvberg, A.: Understanding quality in conceptual modeling. IEEE Softw. **11**, 42–49 (1994). https://doi.org/10.1109/52.268955

Synthesis of Intellectual Tools to Support Models Translation for Mobile Robotic Platforms

Maxim Polenov[✉], Artem Kurmaleev, and Sergey Gushanskiy

Department of Computer Engineering, Southern Federal University,
Taganrog, Russia
{mypolenov,smgushanskiy}@sfedu.ru,
art.kurmaleev@gmail.com

Abstract. This paper is about the models translation support for mobile robotic platform simulation. Application of expert system and multilanguage translation tools were proposed to hasten and simplify the process of models' code generation in the required format at mobile robotic platform development. Usage of intelligent support and the created translating module was explained and illustrated on the example of models conversion into the format required for Robot Operating System. The Developed application was examined and the results were reviewed.

Keywords: Model's translation · Multitranslator
Distributed Storage of Models · Expert system · Mobile robotic platform
Robot Operating System

1 Introduction

Nowadays in the presence of lots of different modeling tools many researchers and developers of complex technical systems face the need of recreation or reusing of the same models for different tools depends on their modeling aims [1]. To solve this problem the Multitranslator [1] was created. The Multitranslator is the tool for the development of translating modules. They together provide automatization of multilanguage models translation from one modeling language into another and support conversion of models for different formats and modeling tools. In addition, sometime after the standalone version of the Multitranslator was released [2], it became certain that it was necessary to upgrade Multitranslator, in order to:

1. Utilize the simplifier user-version for translations processing only;
2. Avoid investing into a user's PC for reducing time costs of translations;
3. Store all the models and translating modules in a safe place without any need for backup for each PC that Multitranslator was installed at.

To accomplish these requirements the Distributed Storage of Models [3] was developed. The Distributed Storage of Models is based on a distributed architecture [4]

R. Silhavy (Ed.): CSOC 2018, AISC 763, pp. 282–291, 2019.
https://doi.org/10.1007/978-3-319-91186-1_29

and developed as a client-server application. Moreover, the reusable models data base approach was applied on the server side [5].

For a better understanding of the next section of the paper the basic principles of models' translation by Multitranslator need to be explained. One of the main components of Multitranslator is the translating module. The translating module is required for each pair of languages that need in translation processing. It consists of two parts of instructions. One is for parsing of an input language and another is for generating the structures of an output language. However, in certain cases translation cannot be completed if there are insufficient input data or uncertainty of solutions [2]. This uncertainty can happen for example when too many outcomes are possible to occur while parsing the source code of some models.

To resolve this problem, the use of expert system was proposed [6, 7]. It has already allowed to reduce time costs on the examined tasks. Currently our team is working at mobile robotic platform simulation. In addition, there is a set of models that were created for MATLAB [8] during the previous researches. These models could be utilized if they would have been translated into Robot Operating System [9] format. Therefore, models translation was needed while working on mobile robotic platform simulation. Nevertheless, to apply multilanguage models translation with intellectual support for current task our system needed some adjustments and improvements. They are described in the next sections.

2   Integration with Mobile Robotic Platform Simulation System

First, to explain what part multilanguage translation is taking in mobile robotic platform simulation, it's necessary to make a short review of its specifics.

Nowadays there are a lot of robots and mobile robotic platforms. Many of those platforms are based on Robot Operating System (ROS) that utilizes C++ [10] and Python [11] languages. The ROS is used in the current researches, so a brief examination of specifics has to be made.

2.1   Robot Operating System

The structure of ROS core can be represented as a few nodes that are connected with each other as a graph. And to complete the picture the main parts of the ROS [12] have to be considered:

1. Nodes – is a process, which performs some forms of computation. Nodes communicate with each other using streaming topics, RPC services and the Parameter Server (described below). They are used to communicate with the main hardware and transmit the data into a ROS understandable format;
2. Messages – are used by nodes to communicate, behaves as a data structure that contains typed fields;
3. Topics – is as "graph edge" that transfer the messages between nodes;

4. Services – are used when publish/subscribe model is not appropriate for the task. It provides request/reply model and it is defined by two messages (request and reply);
5. Master – processes naming and registration services to nodes. It tracks publishers and subscribers to the topics and services;
6. Parameter Server – behaves as a shared, multivariate dictionary that is given access through network API's. Nodes use it to store and get the parameters at runtime.

ROS is designed to use a modular architecture using independent nodes. The format of message used between a pair of nodes is predefined and called "Bag" due to its "bag" extension.

The structure of node communication is considered below.

1. The output of a node is "published" in a specific type of message;
2. A node "subscribes" to a topic when it wants to receive data from it. In this case, the node that outputs data is called a "publisher". And the node which receives data is called a "subscriber";
3. When a publisher and a subscriber use the same topic, data is transferred. Hence, topics behave as the virtual data bus in a ROS system.

There can be multiple publishers and subscribers to a topic. The ROS master keeps track of all the publishers and subscribers and provides registration service to all the nodes in a ROS system.

An example of such a structure is shown in Fig. 1.

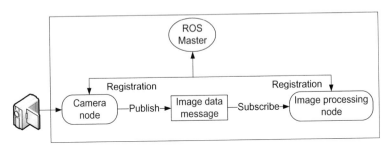

Fig. 1. An Example of the ROS nodes structure

In the figure above, the following happen:

1. "Camera" node receives data from the camera device;
2. "Camera" node publishes image data;
3. "Image processing" node receives data through subscription;
4. "Image processing" node processes the data.

From the figure one, modular architecture simplifies models development since each module can be debugged separately. For example, the "Camera" node, which publishes camera data, was debugged so it is able to run then constantly. The "Image processing" node, which uses data from the camera, can be disabled or enabled when no actions to initial data publisher are required to be taken. This can be a problem if you are making another application with different OpenCV [13] implementations. Different nodes can use other implementation of OpenCV. Therefore, this will reduce time costs if the mobile robotic system model becomes more complex and interconnected.

2.2 Explanation of Translation

Due to compatibility and format problems, it was decided to switch to Octave [14] that is largely compatible with MATLAB syntax. Therefore, reuse of those models without creating them repeatedly can be a big deal for future work.

However, to implement intelligent support with the multilanguage translation system here the appropriate translating module that will convert MATLAB-models into C++ is required and another knowledge base for the expert system [7] will be needed.

To achieve this aim, recently developed translating module [15] that converts MATLAB models into C++ language can be used. So now, it has to be only adopted and it will support the models translation process with the new base of expert system oriented towards their description in ROS.

3 Synthesis of Models Translation Support

3.1 Translating Module Analysis

The recursive rule structure of the Multitranslator's translating module (TM) that allows converting models from MATLAB language into C++ is listed below.

- MFile – links to "Scenario" and "MFile" rules;
- Scenario – links to STMS rule. Parses "clear", "all", ";" elements;
- STMS – links to "STM" and "STMS" rules;
- STM – links to "Variabiles", "Conditional brach", "Cyclic operators" and "C`all library functions" rules;
- Function – links to "List ParamFunc", "VarList" and "Block" rules. Parses "=", "(", ")" elements;
- List ParamFunc – links to "List ParamFunc" rule. Parses ",", STD_ID elements;
- Block – links to "STMS" rule. Parses "end" element.

To understand the process better it's necessary to detail the elements listed above:

- "MFile" element also can include all the required preload code fragments to integrate them into the project;
- "STMS", "STM" describe parsing of variables, operators, expressions and declarations;
- "Function", "ListParamFunc" – all together they describe function parsing for its body and parameters.

However, it was taken into consideration that the class structure for knowledge base should be made in the same way in general. This doesn't include "MFile" and "Scenario" elements, since it is not necessary to make them as classes in COOL [16], used for the expert system.

3.2   Expert System Adjustments

The expert system allows improving models translation system by processing exceptional cases and inputting data uncertainty. So implementing such support will make possible avoiding the need of expert's help.

Since the corresponding TM was analyzed it's possible now to create appropriate knowledge base for the expert system. The expert system was implemented earlier [7] so the points required to determine needs for the current task are mentioned below:

- "INPUT" and "DATA" classes remain the same;
- Class and code specifier's names remain unchanged since MATLAB has the same types of operands except the "prefix type specifier" and type specifier, since language format doesn't require those (conditional operator, loop operator, logic operator, mathematical function);
- There are still 3 levels of rules (expert, request and management rules).

The filler of class and code specifiers is completely remade. Even if the types of operands are the same, operands themselves are different and syntax is different as well. For example, "logic operator" type specifier will contain the following changes:

```
([logic-operator] of DATA-TYPE
(namedata <= >= == != || && < >)
(namespec Logic operator))
```

Further, involving experts of MATLAB language is planned to continue completing of knowledge base to maximize the cover of exceptional cases that can happen in models translation.

4   Developing of Intellectual Support

4.1   Approaching Data Transfer Organization for CLIPS Based Expert System

When it was decided to use the expert system for an intellectual support of models translation, an alpha version of the expert system was developed. It allowed to process the basic functions of the Multitranslator. However, there was a problem of data transfer between CLIPS [17] based expert system, Multitranslator and the user. The user is on the client side of the complex when CLIPS is on the server and data have to be transferred automatically. This went into the following goals to achieve:

1. Making both side data transfer between the Multitranslator and CLIPS expert system;
2. Simplifying debugging process for us. It's not necessary to put the expert system into CLIPS environment repeatedly to get debugging log;
3. Possibility for the user to get answers from the expert system if it is necessary;
4. Develop an intellectual editor for the user. It will allow editing and adding rules for expert system if user has appropriate knowledge.

CLIPS environment works with the user in a dialogue mode, so the CLIPS API have to be used to achieve these goals.

4.2   Developing Interaction with CLIPS API

To transfer data between CLIPS and external applications global variables are needed. The global variables are CLIPS objects that save data when used not in the defining constructions of CLIPS language (any "def" construction).

Global variables in CLIPS can be defined with "defglobal" operator, for example:

```
(defglobal ?*name*=expression)
```

The definition of global variable can be changed anywhere in code by using "bind" operator. It is used in the same construction as "defglobal".

Changing the global variables is allowed from external applications using CLIPS API by the next functions:

```
GetDefglobalValue("name",&pointer)
SetDefglobalValue("name",&pointer)
```

CLIPS API also allows to edit core memory depending on the user answers by using "FunctionCall" operand:

```
FunctionCall("function", "arguments", &pointer)
```

It calls the function with "function" name transferring "arguments" as the parameters. All the functions of the expert system are based on CLIPS language. CLIPS API loads expert system and all the internal memory upon initializing.

Therefore, the data transfer between CLIPS and external application are achieved through reading and writing global variables by calling CLIPS API core functions.

4.3   Synthesis of Debugging Interface

To review the achieving of goals the user interface on server side on Fig. 2 have to be considered. There are two dropdown menus in MDI type interface. There are two possible ways to process debugging completely, here they are:

1. Click "Data" dropdown menu and input the testing data into "Data" window and press "ok". It will save the data and it will automatically run your data through the expert system and write the results in logs. After the processing of expert system result log will automatically open for review. This excludes transferring data from Multitranslator when debugging expert system is required;
2. When Distributed Storage of Models transfers data from Multitranslator to expert system then data are being written in "Data" window for debugging. Afterwards press "Expertise" at "Menu" manually and "Result" window will be opened automatically for review. In this case, after the review of debugging information is finished press "Transfer" manually to see how the expert system processing results will affect the translation process in the end.

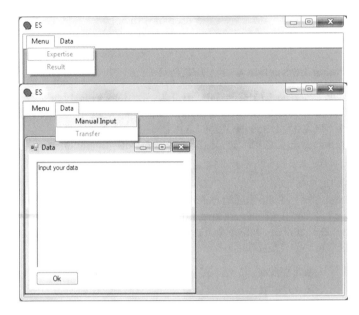

Fig. 2. Server side user interface

It is possible to close child windows if it is necessary and so they can be opened again, what is the only purpose for "Result" row presently.

It allows to develop the expert system more precisely and much faster as well as debug the whole work of the complex in systematic method.

5 Model's Translation Test

After a few experiments, the one was to review in this paper. Input model code example for Octave:

```
width = 5;
heigth = 5;
K = zeros(heigth+1, width+1);
for y = 1:heigth
    for x = 1:width
        % calculate element (row,col)
        K(y+1,x+1) = 0;
        for yy = 1:y
        for xx = 1:x
            K(y+1,x+1) = K(y+1,x+1) + I(yy,xx);
        end
        end
    end
end
```

After building and running translation module in Multitranslator the model file is chosen as input parameter. After parsing success message is printed and preset output file is filled with following C++ code of ROS format:

```cpp
#include <message_filters/subscriber.h>
#include <message_filters/time_synchronizer.h>
#include <sensor_msgs/Image.h>
#include <sensor_msgs/CameraInfo.h>
using namespace sensor_msgs;
using namespace message_filters;
void callback(const ImageConstPtr& image, const CameraIn-
foConstPtr& cam_info)
{
  int K[heigth+1][width+1] = {0};
  for (int y = 0; y < image.heigth; y++)
  {
    for (int x = 0; x < image.width; x++)
    {
      // calculate element (row,col)
      K[y+1][x+1] = 0;
      for (int yy = 0; yy < y+1; yy++)
      {
        for (int xx = 0; xx < x+1; xx++)
        {
          K[y+1][x+1] = K[y+1][x+1] + image.data[yy][xx];
        }
      }
    }
  }
}
int main(int argc, char** argv)
{
  ros::init(argc, argv, "vision_node");
  ros::NodeHandle nh;
  message_filters::Subscriber<Image> image_sub(nh, "im-
age", 1);
  message_filters::Subscriber<CameraInfo> info_sub(nh,
"camera_info", 1);

  TimeSynchronizer<Image, CameraInfo> sync(image_sub, in-
fo_sub, 10);
  sync.registerCallback(boost::bind(&callback, _1, _2));
  ros::spin();
  return 0;
}
```

From the example of translation above, the test was successful and the results are correct. Translating of node models to integrate into work of mobile robotic platform simulation system shouldn't call any exceptions.

6   Conclusion

The proposed approach of intelligent support of models translation for mobile robotic platform simulation system development that is partially based on ROS and the Multitranslator translating module has been synthesized and new knowledge base for the expert system has been developed. Data exchange interaction between Multi-translator and CLIPS core has been proposed. It allows to configure user interface and develop intelligent editor in the future. This approach utilizes the Multitranslator within the Distributed Storage of Models as translation tool, extends its functionality and improves universality of suggested solutions.

Expert system debugging process through the server side of user interface has been developed and explained in details to review the improvements of suggested tools. The whole tools debugging process has been explained in systematic method that allowed simplifying the development of expert system and reduced time costs of the system model building. Application of this approach will extend the possibilities of created tools even further allowing not to just avoid the expert's help, but also to extend the functionality of the such tools for each user on its own.

Acknowledgments. This work was carried out within the State Task of the Ministry of Education and Science of the Russian Federation (Project part No. 2.3928.2017/4.6) in Southern Federal University.

References

1. Chernukhin, Y., Guzik, V., Polenov, M.: Multilanguage Translation for Virtual Modeling Environments. Publishing house of Southern Scientific Center of Russian Academy of Sciences, Rostov-on-Don (2009). (in Russian)
2. Chernukhin, Y., Guzik, V., Polenov, M.: Multilanguage translation usage in toolkit of modeling systems. WIT Trans. Inf. Commun. Technol. **58**, 397–404 (2014)
3. Polenov, M., Guzik, V., Gushanskiy, S., Kurmaleev, A.: Development of the translation tools for distributed storage of models. In: Proceedings of 9th International Conference on Application of Information and Communication Technologies (AICT 2015), pp. 30–34. IEEE Press (2015)
4. Tanenbaum, E., Van Sten, M.: Distributed Systems: Principles and Paradigms, 2nd edn. Prentice-Hall, Upper Saddle River (2006)
5. Robinson, S., Nance, R.E., Paul, R.J., et al.: Simulation model reuse: definitions, benefits and obstacles. Simul. Model. Pract. Theory **12**, 479–494 (2004)
6. Polenov, M., Guzik, V., Gushansky, S., Kurmaleev, A.: Intellectualization of the models translation tools for distributed storage of models. In: Informatics, Geoinformatics and Remote Sensing (Proceedings of 16-th International Multidisciplinary Scientific Geoconference (SGEM 2016)), vol. 1, pp. 255–262. STEF92 Technology (2016)

7. Polenov, M., Gushanskiy, S., Kurmaleev, A.: Synthesis of expert system for distributed storage of models. In: Software Engineering Trends and Techniques in Intelligent Systems. Advances in Intelligent Systems and Computing, vol. 575, pp. 220–228. Springer (2017)
8. Matlab. https://www.mathworks.com/products/matlab.html
9. Robot Operating System (ROS). http://www.ros.org/
10. C++ language. http://www.cplusplus.com/
11. Python language. https://www.python.org/
12. ROS, Nodes explanation. http://wiki.ros.org/Nodes
13. OpenCV, MatLab application. https://www.mathworks.com/discovery/matlab-opencv.html
14. Octave. https://www.gnu.org/software/octave/
15. Polenov, M., Lapshin, V.: Development of mechanisms for translating software models presented in the matlab environment format. In: Innovative Technologies and Didactics in Teaching (ITDT-2017), Collected Papers of International Conference, Berlin, Conference Print, pp. 199–208 (2017)
16. CLIPS Object Oriented Language (COOL). https://www.csie.ntu.edu.tw/~sylee/courses/clips/bpg/node9.html
17. CLIPS. A Tool for Building Expert Systems. http://www.clipsrules.net/

The Tool for the Innovation Activity Ontology Creation and Visualization

Sergey V. Kuleshov, Alexandra A. Zaytseva[✉], and Alexey J. Aksenov

St.-Petersburg Institution for Informatics and Automation of RAS, St.-Petersburg, Russia
{kuleshov,a_aksenov,cher}@iias.spb.su

Abstract. In this paper the problem of automatic application of the semantic analysis methods to documents on financial and economic topics in order to visualize the semantic environment map of innovation activity is discussed. The tool for the innovation activity ontology creation and visualization based on associative ontology approach is proposed.

Keywords: Ontology · Innovation activity · Ontology model visualization
Corpus of text · Associative ontology

1 Introduction

In the tasks of economics phenomena analysis including estimation of innovation potential and prediction of its changing tendencies the one of the first stages is the bibliography search and on topic data accumulation. Traditionally this stage is performed manually on the economic texts and factual materials, economic and business news, aggregated from information agencies.

The problem of automation of this stage is an important task in the modern conditions of growing quantities of textual information. For the forming of the model of economy growing potential in the relation with innovation activity as the basic growth driver it is necessary to implement modern methods of automated ontology creation from corpus of texts.

The obtained data could be used to reveal casual relationship between given macro economical indicators and to develop empirical predictive model of economy growth in Russian Federation. The data collection can be implemented iteratively simultaneously with ontology and model structure correction [1].

The procedure of data collection consists of automated download of documents from available network resources, assessment of topic relevance, thematic clustering, tracking the dynamics of the topic growth [2, 3].

The financial and economic texts have following syntactic features: syntactic completeness of statements, frequent use of patterns, narrow thesaurus, the expanded system of connecting elements (unions, union words) [4–7].

© Springer International Publishing AG, part of Springer Nature 2019
R. Silhavy (Ed.): CSOC 2018, AISC 763, pp. 292–301, 2019.
https://doi.org/10.1007/978-3-319-91186-1_30

2   The Associative-Ontological Approach to Graph Creation

One of the perspective methods for ontology formation is based on usage of a corpus of texts collected over given period of time.

In general, the methodology for text corpuses ontology creation and semantic environment visualization suggests the system oriented on storage and joint processing of thematic ontologies which supposes structural decomposition of texts and creation of notion graphs (semantic environment) [8–10].

The majority of current systems for intellectual texts analysis such as [11] utilize detection of syntactic and semantic definitions based on preliminary given classifiers. Also the number of approaches are exist based on preliminary created ontology matching [1–3, 12–15].

In contrast to mentioned above the proposed method based on dynamic associative ontology creation and usage of associative relations graph as a graph of notions formed directly on thematic ontology texts [8]. The results of structural text decomposition are represented as JSON expressions. The semantic concepts are being stored according to value of their relations in texts corpora.

The structure of method of ontology creation and visualization is shown on Fig. 1.

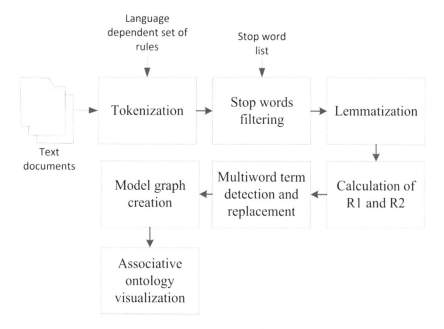

Fig. 1.   The structure of method of ontology creation and visualization

Formally the proposed method can be described as follows.

Let's assume the word **w** as a lexeme consisting of the sequence of literals which are the belong to the set of natural language alphabet symbols. The terminal symbol (the symbol of the end of the sentence) will be considered as the elements of the set

{".", "?", "!"}. The sentence S is a syntactic construct that is a set of words $[w_1, w_2, \dots]$ with terminal symbols at the end.

Let's assume the text as a set of syntactically related words

$$w_i \in \{W\} \subset \{\omega\}, \tag{1}$$

where w_i is a word, W—set of words, ω—is the set of valid words in the language, $i \in \mathbb{N}$. A valid set of words in the language $\{\omega\}$ is defined by some thesaurus.

Each word in the text is being lemmatized – normalized using the function m of the morphological analysis. In this context the normalization means the obtaining the base form of the word (called 'lemma'):

$$w_i \xrightarrow{m} \overline{w_i}, w_i \in \{W\}, i \in \mathbb{N}, \tag{2}$$

where $\{W\}$ is the set of word forms, $\overline{w_i}$ is the base form of the word. The words belonging to the set of stop-words are ignored.

For each word in the base form the frequency of occurrence is calculated which is then used in the ontology graph constructing and visualization. For all words, relations R1 and R2 are constructed according to the following rule:

$$(\overline{w_i}, \overline{w_j}) \in R1, \ i, j \in \mathbb{N}, \tag{3}$$

if w_i, w_j are the consecutive words in the sentence (S),

$$\forall w_i \in S \ \& \ \forall w_i \in S \ \Rightarrow (\overline{w_i}, \overline{w_j}) \in R2. \tag{4}$$

The relation R1 can be used as one of the criteria for collocation detection in further analysis. If frequency of occurrence $\xi_{R1}(\overline{w_i}, \overline{w_j}) > \varepsilon_1$, the words w_i, w_j are considered to be a multi-word term. After the detection of all the multi-word terms (collocations), the processing is repeated, but each of the multi-word terms is considered as single term.

The relation R2 defines the associations in this method. Associations are used as a way of displaying potential real-world relations obtained as a result of text analysis. For each element of relations R1 and R2 the frequency of occurrence (ξ_{R1} and ξ_{R2}) is calculated.

If $\xi_{R2}(\overline{w_i}, \overline{w_j}) > \varepsilon_2$, then the words w_i и w_j are supposed to be semantically related.

$$\xi_{R2}(\overline{w_i}, \overline{w_j}) > \varepsilon_2 \Rightarrow (\overline{w_i}, \overline{w_j}) \in E_2, \overline{w_i} \in V_2 \ \& \ \overline{w_j} \in V_2, \tag{5}$$

where ε_1 and ε_2 – are the threshold parameters. The resulting semantic environment (model of associative ontology) can be represented as undirected weighted graph $G = \{E_2, V_2\}$ the edges of which are the elements $(\overline{w_i}, \overline{w_j}) \in E_2$, and vertices are the words form set V_2.

The ε_2 parameter can be dynamically changed to control the size of the semantic environment. One example of the application of semantic environment management can be an interactive study of the subject area by the user.

3   The Development of System for the Ontology Creation and Visualization

To apply semantic analysis to documents on financial and economic topics, innovation policy, expert comments on banking and investment instruments a set of documents was collected using automatic document collection instruments (crawlers, parsers and search engines) and backstage researches [5].

A software module is being developed that creates a list of text documents from the analyzed resources and stores them into the database for further analysis.

When developing the service, the following technologies are used: HTML parsing technology and web pages download automation.

The developed software allows analyzing the documents over various periods of time (from days to years) independently and/or in historical context of ontology changes.

The data from proposed service is assigned for the analysts in the economic and innovation potential thematic fields.

Figure 2 shows the web service architecture for proposed system.

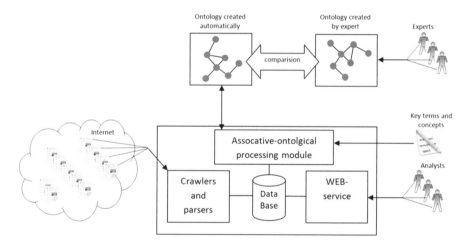

Fig. 2. Web service architecture

The Experts using experience and knowledge in the subject area create an ontology containing entity and relations between them. Simultaneously, the software system using the associative-ontological text processing module to analysis of Internet resources automatically creates an ontology based on the predefined key concepts, discovering their relations and, if necessary, introducing new objects that are not found in the initial key concept list. An automatically created ontology usually turns out to be less qualitative in the sense of detecting regularities and generalization of complex concepts but comprises relevant and up to date semantic environment map. The comparison of automatic and manual created ontologies allows experts to create the unified methodology for prediction of innovative activity based on integration of formalized and non-formalized approaches which in turn increases the degree of reliability and foundation for top level economic management decisions.

The core of proposed system is designed as web-service supporting REST-requests. The resulted graph of ontology model is being returned in a form of following JSON-expression:

```
{ "links":[
{"id1":17,"id2":19},
{"id1":17,"id2":67},
...
], "objects":[
{"level":0,"id":17,"text":"INNOVATION"},
{"level":0,"id":18,"text":"HUMAN CAPITAL"},
{"level":0,"id":19,"text":"BANK"},
...
, ] }
```

The model is stored in the form of an undirected weighted graph, the vertices (elements) of which are the objects, subjects, concepts or actions, and the edges are dependencies between elements. Figure 3 shows an example of ontology graph created on text [16].

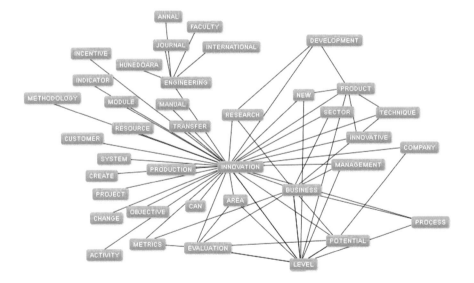

Fig. 3. Example of ontology automatically formed in the innovation potential field [16]

4 Results and Discussion

At the time being the Internet contains enormous volume of textual information on various topics. Computers can easily operate with such text volumes but computer algorithms are not always able to make correct decision about the significance of the revealed connections between facts or events – the common task for experts. Unfortunately it is difficult for a person to perceive and analyze even a 1000 objects simultaneously located on the screen in a relatively short time. The developed system solves the problem of visual representation of the terms of the subject domain in a form that is easily perceived by the expert to reveal hidden patterns and cause-and-effect relationships of phenomena.

For the evaluation of proposed system performance the texts on economics were collected using semantic search system [1] and filtered by relation to "innovation activity" topic.

The texts passed through filtering system formed the corpus of text database allowing for further ontology creation and visualization. Figure 4 shows the distribution of corpus text volume of economic news in Russia by the years. It is possible to build ontology for the texts published on specific periods of time. Obtained ontologies then can be matched and compared. For better observation by experts the ontology detailing can be changed by modifying filtering threshold.

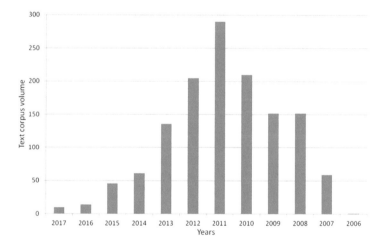

Fig. 4. The distribution of corpus text volume of economic news in Russia by the years

The Fig. 5, 6 and 7 show the ontologies created by proposed system for corpuses of Russian economic news texts published in different years and filtered by "innovation activity" and "innovation potential" core terms.

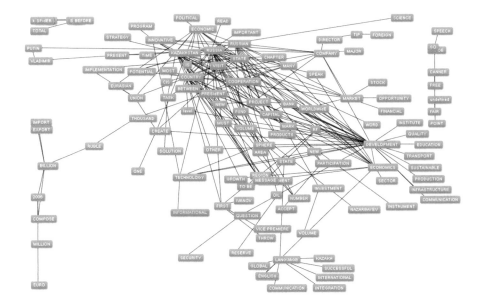

Fig. 5. The graph of the ontology created for corpus of Russian economic news texts published in 2007

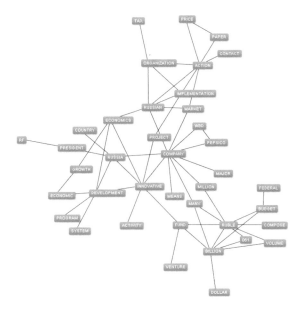

Fig. 6. The graph of the ontology created for corpus of Russian economic news texts published in 2011

Fig. 7. The graph of the ontology created for corpus of Russian economic news texts published in 2015

Application of the developed system allowed to reveal the fact that the appearance of term "innovation activity" in Russian economic news texts was on it's peak in 2007-2011 and since then was in decline. Additionally we were able to determine what the term was associated with in those years. Since 2015 the innovations were mainly associated with the activity of pharmaceutical companies, and in 2011 it was closely related to geopolitical processes in the world. Thus a convenient tool for automatic ontology creation in given thematic field and visualization of the hidden relations between the phenomena reflected in these ontologies in the form of terms, which makes similar dependencies easily observable for analysts.

5   Conclusion

The paper describes approaches to creation of the ontologies for the innovation activity model formalization. It is shown that the application of the concept of automatic ontology creation based on associative ontology approach [3, 17] provides visualization of the semantic environment map for the key terms in the field of innovation activity.

The system aggregates the expert opinions obtained from open Internet resources by automatic processing of textual data sets with the methods of semantic analysis and visualization of the key terms semantic environment. The application of such visualization provides the possibility of establishing new significant non-evident relations between the terms in thematic field. Which provides usability of proposed tool in semantic analysis of documents on financial and economic topics and thus further increasing reliability, providing foundation for top level economic management decisions.

The application of proposed system for automatic ontology visual maps creation in the economic and financial field will improve the process of empirical model construction for innovating economic growth potential.

Acknowledgments. This research is supported by the Russian Foundation for Basic Research, project N 16-29-12965\17.

References

1. Kuleshov, S.V., Zaytseva, A.A., Markov, S.V.: Associative-ontological approach to natural language texts processing. J. Intellect. Technol. Transp. **4**, 40–45 (2015). (in Russian)
2. Zaytseva, A.A., Kuleshov, S.V., Mikhailov, S.N.: The method for the text quality estimation in the task of analytical monitoring of information resources. J. SPIIRAS Proc. **37**(6), 144–155 (2014). https://doi.org/10.15622/sp.37.9. (in Russian)
3. Mikhailov, S.N., Malashenko, O.I., Zaytseva, A.A.: The method for the infology analysis of patients complaints semantic content in order to organize the electronic appointments. J. SPIIRAS Proc. **42**(5), 140–154 (2015). https://doi.org/10.15622/sp.42.7. (in Russian)
4. TECHNOPOLIS GROUP & MIOIR: Evaluation of Innovation Activities. Guidance on methods and practices. Study funded by the European Commission, Directorate for Regional Policy (2012). http://ec.europa.eu/regional_policy/information/evaluations/guidance_en.cfm#1
5. Korableva, O., Razumova, I., Kalimullina, O.: Research of innovation cycles and the peculiarities associated with the innovations life cycle stages. In: Proceedings of the 29th International Business Information Management Association Conference – Education Excellence and Innovation Management through Vision 2020: From Regional Development Sustainability to Global Economic Growth, pp. 1853–1862 (2017)
6. Segev, E.: Google and the Digital Divide: The Biases of Online Knowledge, 171 p. Chandos Publishing, Oxford (2010). ISBN 978-1-84334-565-7
7. Introna, L.D., Nissenbaum, H.: Shaping the web: why the politics of search engines matters. J. Inf. Soc. **16**(3), 169–185 (2000). https://doi.org/10.1080/01972240050133634
8. Kuleshov, S.V.: The development of automatic semantic analysis system and visual dynamic glossaryies. Ph.D. (Tech) thesis, Saint-Petersburg (2005). (in Russian)
9. Alexandrov, V.V., Kuleshov, S.V.: Semiological information systems – analytical self-referring. In: Materials of X International Conference and Russian Scientific School. INNOVATICA-2005, vol. 6, pp. 9–14. Moskva. Radio i svjaz' (2005). (in Russian)
10. Kuznecova, J.M., Osipov, G.S., Chudova, N.V.: Intellectual analysis of scientific publications and the current state of science. J. Large-scale Syst. Control **44**, 106–138 (2013). (in Russian)
11. Smirnov, A.V., Pashkin, M., Chilov, N., Levashova, T.: Agent-based support of mass customization for corporate knowledge management. J. Eng. Appl. Artif. Intell. **16**(4), 349–364 (2003)
12. Smirnov, A., Levashova, T., Shilov, N.: Patterns for context-based knowledge fusion in decision support systems. J. Inf. Fusion **21**, 114–129 (2015)

13. Raufi, B., Ismaili, F., Ajdari, J., Zenuni, X.: Knowledgebase harvesting for user-adaptive systems through focused crawling and semantic web. In: ACM International Conference Proceeding Series, vol. 1164, pp. 323–330. Association for Computing Machinery (2016)
14. Kim, H., Kang, S., Oh, S.: Ontology-based quantitative similarity metric for event matching in publish/subscribe system. J. Neurocomput. **152**, 77–84 (2015)
15. Khan, S., Safyan, M.: Semantic matching in hierarchical ontologies. J. King Saud Univ. Comput. Inf. Sci. **26**(3), 247–257 (2014)
16. Sabadka, D. Innovation potential metrics. J. Ann. Fac. Eng. Hunedoara Int. J. Eng. (2012). http://annals.fih.upt.ro/pdf-full/2012/ANNALS-2012-3-79.pdf
17. Kuleshov, S.V., Yusupov, R.M.: Is softwarization the way to import substitution? J. SPIIRAS Proc. **46**(3), 5–13 (2016). https://doi.org/10.15622/sp.46.1. (in Russian)

Modeling and Performance Analysis of Priority Queuing Systems

Dariusz Strzęciwilk[1(✉)] and Włodek M. Zuberk[2]

[1] Department of Applied Informatics, University of Life Sciences, Nowoursynowska Street 159,
02-787 Warsaw, Poland
dariusz_strzeciwilk@sggw.pl
[2] Department of Computer Science, Memorial University, St. John's, NL A1B 3X5, Canada
wlodek@mun.ca

Abstract. The paper presents the results of modeling and analysis of data performance on systems that support QoS (*Quality of Service*). In order to evaluate the performance of the modeled systems used were TPN (*Timed Petri Nets*). Studied were mechanisms of traffic shaping systems based on PQS (*Priority Queuing System*). Tested was the impact of the mechanism of generating traffic using TPN. Moreover, discussed were the basic mechanisms and queuing systems occurring in QoS structures. It is shown that models can be effectively used in the modeling and analysis of the performance of computer systems.

Keywords: Priority queuing system · Petri nets · Performance analysis
Modeling · QoS data

1 Introduction

Queuing theory is one of the heavily used tools in modeling and testing the quality of transmission in computer networks [1–3]. This theory uses a mathematical apparatus associated with the theory of stochastic processes, in particular Markov processes [4]. By queuing system herein is meant a system that on the one hand receives notification requiring maintenance, on the other hand, there are the so-called maintenance devices, designed to meet the needs of these applications. If the process of receiving applications exceeds the capabilities of their immediate service, a queue is created. A queuing system may be characterized by regulations of queues, i.e., the way one determines the order of service applications in the system [5]. The most common queuing systems are FIFO (*First In First Out*), LIFO (*Last In First Out*), SIRO (*Select In Random Order*), PQ (*Priority Queuing*). The basic mechanism that supports the transfer of packages is FIFO scheduling that is easy to implement and treats all packets equally. FIFO scheduling is not suitable to provide for a good quality of service transmission, as when the packets come from different traffic flows, one of them can easily disrupt the flow of the other remaining streams. Packet processing in the order of flue means that an aggressive stream can appropriate the higher capacity of a router queue. This can result in poor transmission causing, for example, sudden increase in delays of transmitting packets. Developed was a lot of packet scheduling algorithms that have better insulation between

© Springer International Publishing AG, part of Springer Nature 2019
R. Silhavy (Ed.): CSOC 2018, AISC 763, pp. 302–310, 2019.
https://doi.org/10.1007/978-3-319-91186-1_31

the streams [6]. In the case of the priority scheduling algorithms, some application can be handled before others, regardless of when they occurred in the system. Priority queuing systems form a large class of queuing systems where the incoming requests are to be distinguished by their importance [7]. A typical example of the use of such algorithms are routers, to which are flowing subsequent packets. Core routers classify incoming packets to the specified classes of traffic, and then handle packets belonging to the aggregated streams. Packets are handled in accordance with an implemented queuing mechanism and specific support and traffic shaping policies in order to provide services with the agreed QoS [8]. QoS is one of the most important challenges arising during the design and maintenance of both modern computer networks and next generation networks [9]. Guaranteeing adequate quality of service is of particular importance in the case of real-time applications such as Voice over IP [10] and video - IPTV [11]. These services are particularly sensitive to delay and require a guaranteed bandwidth [12]. To provide the desired QoS packages for the entire route from the sender to the recipient has been the subject of research for many years [13–15]. Research in this area can be divided into two groups. In the analytical methods the authors sought solutions of algebraic or differential equations which bind together the probability of events in the system. In the simulation methods were used most often implantation queuing algorithms, which were then subjected to statistical analysis. Based on the analysis of available research, it can be stated that the analytical methods most commonly include relatively simple queuing systems, which require implementation of many of the assumptions of the stochastic nature of the traffic flow. Complex systems are very difficult in the analysis and their functioning can be effectively examined by simulation methods. It should be noted that, to construct a sufficiently accurate model is not simple, and the waiting time for results could be discouragingly long. Hence, the aim of this study was to use models of Petri nets to assess the efficiency and to study the effectiveness of queuing mechanisms of PQS. Such an assessment may also be useful in the design and analysis of data in computer networks, distributed systems and multiprocessor systems. Constructed queuing models allowed estimation of significant features and parameters of the system under test.

2   Petri Nets and Network Models

Petri nets are a graphical and mathematical tool used in many fields. They are seen as a mathematical tool for modeling of concurrent systems [16, 17]. Although there are many varieties of Petri nets [18, 19] their common feature is the structure based on a bipartite directed graph, i.e. graph with two types of vertices, alternately connected by facing edges (or arcs). These two types of vertices represent, in general terms, conditions and events occurring in a modeled system, but each event can occur only when fulfilled are all the conditions associated with it. Formally Petri net is defined as a system $N = (P, T, A)$ composed of a finite set of p-elements P (representing conditions), a finite set of t-element T (representing events) and set A of arcs connecting the p-components with T-elements and t-elements with p-elements, $A \subseteq P \times T \cup T \times P$, set A is called the relation of parity. P-elements are connect by arcs directed to t-element and are called its input elements, while

p-elements connected by arcs facing away from T-element are called its output elements. The overlapping of events is represented here by the so-called tokens assigned to the p-network elements, typically p-element, with which is associated at least one marker indicates that the condition represented by the element is fulfilled. Location of markers in the p-components can be described by marking function, $m: P \rightarrow \{0, 1, 2 \ldots\}$ or presented as a vector specifying the number of tags assigned to the further p elements of network $m = \left[m(p_1), m(p_2), \ldots\right]$. The network N together with the (initial) marking function m_0 is called the labeled M network, $M = (N, m_0) = (P, T, A, m_0)$. To evaluate the performance of the modeled system, i.e. to determine how quickly certain events may follow each other, in a Petri networks one must also take into account the duration of the modeled events [20]. Extension of networks by definitions of time allows for their use in modeling of real-time systems [21]. Formally temporal models are an extension of token models, with the additional elements of the description defining the times of overlapping of events and the probability or frequency of occurrence of random events. The temporal T network is thus defined as the system $T = (M, c, f)$ where: M is a labeled network, $M = (P, T, A, m_0)$, c is a function of the resolution of conflicts $c: T \rightarrow [0, 1]$, which for each class of decision gives the probability of particular events belonging to this class, and for other conflict events gives their relative frequency used for random conflict resolution, and f - defines the times of occurrence of the event, $f: T \rightarrow \mathbb{R}_+$, where $\mathbb{R}_+$- denotes the set of non-negative real numbers. The duration time of events can be deterministic, specified by the $f(t)$ value or stochastic described by a corresponding function of probability of density with the $f(t)$ parameter. In the case of distributions described by more parameters the value of the f function should be treated as vectors of the appropriate values. Performance evaluation of the model using a temporal network is well described in the literature [17, 22].

3   The Model

Several versions of Petri nets have been used as models of systems which exhibit concurrent and parallel activities. Stochastic Petri nets and timed Petri nets have many similarities but deal with temporal properties of models in a different way, so sometimes the similarities may be misleading. The basic model used here is known as an inhibitor Petri net N, which is a bipartite directed graph $N = (P, T, A, H)$ where P and T are two disjoint sets of vertices $P \cap T = \emptyset$, called places and transitions, respectively, A is a set of directed arcs connecting places with transitions and transitions with places, $A \subseteq P \times T \cup T \times P$, and H is a set of inhibitor arcs connecting places with transitions, $H \subset P \times T$. Normally, $A \cap H = \emptyset$. For performance analysis of Petri net models, temporal characteristics of occurring transitions must be taken into account. This can be done in several ways assigning occurrence timed to places, or to transitions, or even to arcs of the Petri net models. In timed nets, the occurrence times of some transitions may be equal to zero, which means that such occurrences are instantaneous, all such transitions are called immediate, while the others are called timed. It should be noted that such a convention effectively introduces the priority of immediate transitions over the timed ones, so the conflicts of immediate and timed transitions are not allowed in timed nets. To evaluate the performance of the modeled systems used was a Petri net model shown schematically in Fig. 1. Based on the prepared models studied were mechanisms of traffic shaping in systems based on priority queues. In the studied models, have been made assumptions that the PQ model will consist of three priority queues (places p_1, p_2, p_3). The three places, p_1, p_2 and p_3, are queuing for class-1, class-2 and class-3 packets, respectively. It is assumed that all queues have infinite capacities. This will allow the display of traffic data with high, medium and low priority. In the studied models, data has been denoted as class-1 (high priority), class-2 (medium priority), class-3 (low priority). Timed transitions t_1, t_2 and t_3 represent a transmission channel for class-1 packets (t_1), class-2 packets (t_2) and class-3 packets (t_3). In the studied model, for simplicity, has been assumed that the transmission time is deterministic and is 1 time unit (for all three classes). In the studied model, place p_5 is shared by the transitions t_1, t_2 and t_3. In order to ensure that only one packet is transmitted at the moment, place p_4 has been marked with a single token (i.e., $m_0(p_5) = 1$). Furthermore, inhibitor arc (p_1, t_2)

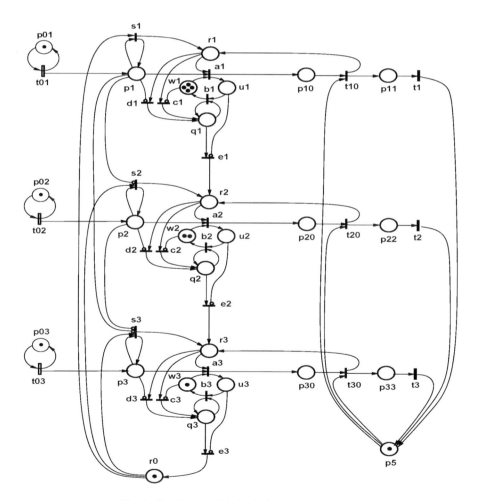

Fig. 1. Petri net model of priority queueing system

does not allow transmission of data with the transit t_2 when the package is ready to be transmitted from class-1. Similarly, inhibitor arc (p_1, t_3 and p_2, t_3) allows transmissions only when there is no data type class-1 and class-2 waiting for transmission. A characteristic feature of priority queues is the fact that if any packets are in the queue with a higher priority packets waiting in queues with lower priority cannot be sent. Data transfer only the queues with higher priority can lead to "starvation" of the data classified into lower priority queues. The phenomenon of starvation data (*starvation problem*) can be eliminated by introducing configurable queuing CQ (*Custom Queuing*). This mechanism allows to handle the queues on the basis of round-robin by downloading the first x bytes from the first queue, then y bytes from the second queue and z bytes from the third queue. This design prevents the extinction of less privileged streams. In the studied models, assumptions have been made that the model X421 will support successively 4 packages of type class-1, then 2 packages of type class-2, and finally 1 package of type class-3. Based on so prepared models were studied mechanisms of traffic shaping in systems based on Priority Queuing with the priorities of 4-2-1 and 4-3-1. As a source of data generation in all three queues, used were M-timed Petri nets (Markovian nets).

4   The Results of Research and Discussion

In order to compare the performance of systems made was a comparative analysis of parameters such as traffic intensity, average waiting time, throughput and utilization. Based on the simulations prepared were charts showing the behaviour of the studied systems. Detailed analysis concerned the behaviour of the system on the verge of its stability.

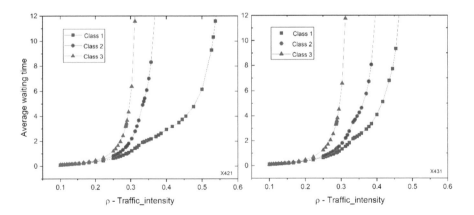

Fig. 2.   Average waiting times for X421 model (left) and X431 model (right)

Research has shown that for the same number of data generated and the channel utilization, waiting times are different (see Fig. 2). In both models, it was found that the average waiting times characteristic in the range for $\rho < 0.20$ is linear for all classes in the both studied models. However, after exceeding this value, in models occur dynamic changes. In order to illustrate the dynamics of the changes models are also shown on a logarithmic scale (Fig. 3). The waiting time for data transmitted in the class-3 initially increases exponentially, but after exceeding the value $\rho = 0.25$ this time increases faster than exponentially. A similar effect was observed in the model X431 (Fig. 3).

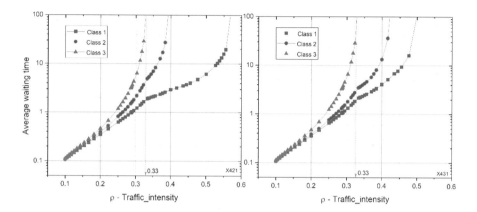

Fig. 3. Average waiting times for X421 model (left) and X431 model (right) on a logarithmic scale

Moreover, it was found that the changes in the course of the characteristics of waiting times for data class-2 and class-3 occur when the limit values of ρ are exceeded. For both models X421 and X431 this value is $\rho = 0.33$ (Fig. 3). Studies on queue length have shown that when the traffic intensity approaches 0.30, the most affected class is a class-3, the lowest priority class. Moreover, studies on queue length in a function of traffic intensity revealed that in the model X421 queue length increases dramatically for $\rho_3 > 0.30$ (class-3), $\rho_2 > 0.35$ (class-2), $\rho_3 > 0.50$ (class-1) (Fig. 4). Similar behavior was observed for the X431 model in the case of values $\rho_3 > 0.30$, $\rho_2 > 0.38$ and $\rho_1 > 0.45$ (Fig. 4). In a next step were examined the two model parameters throughput and utilization. Studies have shown that in the X421 model, the maximum throughput for the data class-1 is 0.57 and remains at this level of value $\rho_1 > = 0.58$. In the case the data type class-2 maximum throughput occurs for $\rho_2 = 0.4$. Increasing the value ρ_2 above 0.4 caused that in the tested model were observed "blocking effects" created by priority queuing ρ_1 (class-1). Throughput class-2 stabilizes at the level of 0.28, when ρ_2 assumes a value of 0.58. As in the case of data type class-2, also in the case of data type class-3 observed were "blocking effects". However, in the case of the data class-3 this effect occurs in two stages. The maximum throughput for the data class-3 is 0.33 at $\rho_2 = 0.33$. After reaching the maximum, data throughput class-3 is rapidly declining. This is caused by "blocking effects" created by priority queuing ρ_1 (class-1) and priority queuing ρ_2 (class-2). In the range to the throughput decreases to the value of 0.20. After crossing $\rho_2 = 0.40$ decrease rate of throughput decreases and stabilizes at a level of 0.14 when the value $\rho_2 = 0.60$. Above the $\rho_1 = \rho_2 = \rho_3 = 0.60$ the running characteristics for all classes is uniform and rectilinear. As expected, the characteristics of throughput models X421 and X431 has a similar character. The maximum throughput in the X431 model for class-1 is 0.50 for $\rho_1 = 0.50$, for class-2 is 0.40 for $\rho_1 = 0.40$ and for class-3 is 0.33 for $\rho_3 = 0.33$ (Fig. 5). Almost identical course shows the characteristics of utilization obtained for models X421 and X431.

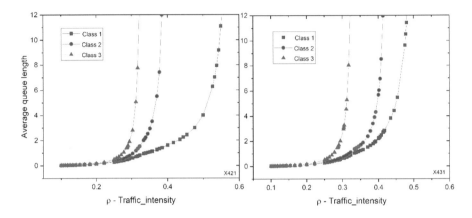

Fig. 4. Average queue length for X421 model (left) and X431 model (right)

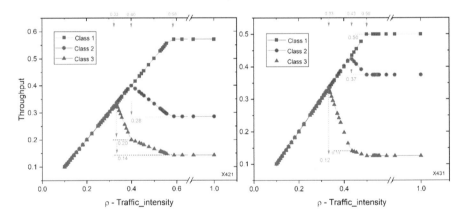

Fig. 5. Throughput for X421 model (left) and X431 model (right)

5   Summary

Through telecommunications networks, including the Internet pass huge amounts of information. We may relatively easily calculate the transfer time from node to node, knowing the distance between the nodes and the transmission speed of the link. Unknown is, however, time and nature of expectancy in the node. On the one hand, it is important to make the best use of network resources, and on the other an increase in network usage worsens the quality of service (growing queue of nodes, increasing like-lihood of overflow). Hence the planning and development of computer networks can be supported using mathematical modeling and computer simulation. The study demonstrated that the use of models of temporal Petri nets can be used to evaluate the performance and effectiveness of the Priority Queuing Systems. It should be noted that the models can often be very complex, and analyze them without a flexible and robust software tools is almost impossible. Although there are many tools for modeling and data analysis with Petri nets, only a few of them can be used in the study of complex systems. It was shown that hierarchical modeling in which we consider some parts of the model at a very detailed level and the other portions on a more general level can be easily implemented based on the Petri net models. Model or its parts can easily undergo further and detailed analysis.

References

1. Buchholz, P., Kriege, J., Felko, I.: Input modeling with phase-type distributions and Markov models: theory and applications. Springer (2014)
2. Czachórski, T., Domański, A., Domańska, J., Rataj, A.: A study of IP router queues with the use of Markov models. In: International Conference on Computer Networks. Springer International Publishing (2016)
3. Donthi, R., Renikunta, R., Dasari, R., Perati, M.: Study of delay and loss behavior of internet switch-markovian modelling using Circulant Markov Modulated Poisson Process (CMMPP). Appl. Math. **5**(3), 512–519 (2014)
4. Puterman, M., Markov, L.: Decision Processes: Discrete Stochastic Dynamic Programming. Wiley, New York (2014)
5. Bose, S.K.: An Introduction to Queueing Systems. Springer Science & Business Media, New York (2013)
6. Bhatti, S.N., Crowcroft, J.: QoS-sensitive flows: issues in IP packet handling. IEEE Internet Comput. **4**(4), 48–57 (2000)
7. Mishkoy, G., et al.: Priority queueing systems with switchover times: generalized models for QoS and CoS network technologies and analysis. In: The XIV Conference on Applied and Industrial Mathematics, Satellite Conference of ICM, Chisinau (2006)
8. Aurrecoechea, C., Campbell, A.T., Hauw, L.: A survey of QoS architectures. Multimedia Syst. **6**(3), 138–151 (1998)
9. Carmona-Murillo, J., et al.: QoS in next generation mobile networks: an analytical study. In: Resource Management in Mobile Computing Environments, pp. 25–41. Springer International Publishing (2014)
10. Strzęciwilk, D.: Examination of transmission quality in the IP multi-protocol label switching corporate networks. Int. J. Electron. Telecommun. **58**(3), 267–272 (2012)
11. Kim. H.J., Choi, S.G.: A study on a QoS/QoE correlation model for QoE evaluation on IPTV service. In: The 12th International Conference on Advanced Communication Technology (ICACT), vol. 2. IEEE (2010)
12. Li, M.: Queueing Analysis of Unicast IPTV with User Mobility and Adaptive Modulation and Coding in Wireless Cellular Networks. arXiv preprint arXiv:1511.01794 (2015)
13. Zhang, Q., et al.: Early drop scheme for providing absolute QoS differentiation in optical burst-switched networks. In: Workshop on High Performance Switching and Routing, HPSR. IEEE (2003)
14. Zuberek, W., Strzeciwilk, D.: Modeling Quality of Service Techniques for Packet–Switched Networks. Dependability Engineering (2018). ISBN 978-953-51-5592-8
15. Tarasiuk, H., et al.: Performance evaluation of signaling in the IP QoS system. J. Telecommun. Inf. Technol., 12–20 (2011)

16. Menth, M., Briscoe, B., Tsou. T.: Precongestion notification: new QoS support for differentiated services IP networks. IEEE Commun. Mag. **50**(3), 94–103 (2012)
17. Zuberek, W.M.: Timed Petri nets definitions, properties, and applications. Microelectron. Reliab. **31**(4), 627–644 (1991)
18. Gianfranco. B.: Introduction to stochastic Petri nets. Lectures on Formal Methods and Performance Analysis, pp. 84–155. Springer, Heidelberg (2001)
19. Roux, O.H., Déplanche, A.M.: A t-time Petri net extension for real time-task scheduling modeling. Eur. J. Autom. **36**(7), 973–987 (2002)
20. Coolahan, J.E., Roussopoulos, N.: Timing requirements for time-driven systems using augmented Petri nets. IEEE Trans. Softw. Eng. **5**, 603–616 (1983)
21. Cheng, A.: Real-Time Systems: Scheduling, Analysis, and Verification. Wiley (2003)
22. Strzeciwilk, D., Zuberek. W.M.: Modeling and performance analysis of QoS data. In: Photonics Applications in Astronomy, Communications, Industry, and High-Energy Physics Experiments 2016. International Society for Optics and Photonics (2016)

Method of User Interface Design Based on Semantic Approach

Svetlana A. Belikova, Yury I. Rogozov[✉], and Alexandr S. Sviridov

Institute of Computer Technology and Information Security, Southern Federal University,
Taganrog, Russian Federation
{belousova,yrogozov,asviridov}@sfedu.ru

Abstract. Eliminate the reason and achieve the preservation of the meaning of user actions in the domain, and fully meet the user requirements in the interface of the information system is possible if to create a methodological framework (a general methodological description) of user interfaces development that will be a common sign form of models organization and will be understandable to all participants of the interface development process, including the user. It is proposed to represent the user actions in the domain in the form of mechanisms of action and on the bases of them to build a process of interface design. The paper presents the method of user interface design based on semantic approach and demonstrates a short example of its usage.

Keywords: User interface design · Semantic approach · User activity
Table model · Design space · Mechanism of action

1 Introduction

Information systems (IS) are an integral part of people's lives in the modern world. The user interface is the central element of any modern software system, since users interact with the system through an interface that is the embodiment of the functionality of the IS for the user. The purpose of any IS is to automate some operations that have been performed by specialists manually. Since the appearance of the first software systems, specialists have been working on solving the problem of constructing an effective interface and have been trying to achieve efficiency in various ways, since the speed of the tasks performing by the user and user's overall satisfaction with the system as a whole depend on this efficiency. The structure and functionality of the interface are determined by developer based on the study of the subject area in which the system will be operated, and the requirements of users are also taken into account. Users impose various and numerous requirements based on their own preferences, and also on the basis of purpose of the system, the conditions and the specifics of the subject area of its use. These factors, together with systems development technologies, require the creation of high-quality interfaces for various platforms. As a result, the design, development, modification and maintenance of the user interface are extremely complex and time-consuming, since the process of developing the user interface is poorly formalized and is performed empirically by specialists. Since no less than half of the time required for

© Springer International Publishing AG, part of Springer Nature 2019
R. Silhavy (Ed.): CSOC 2018, AISC 763, pp. 311–318, 2019.
https://doi.org/10.1007/978-3-319-91186-1_32

the development of the entire system is spent on the development of the interface, statistical data obtained with respect to software systems can be attributed to the development of the interface as part of the system: according to the studies on the analysis of IT performance projects, conducted by the company Standish Group in 2016, only 29% of the projects are successful, 52% are problematic, and 19% are failures [1].

Proceeding from the foregoing, research aimed at solving the problems of reducing the complexity of designing, prototyping, implementing and maintaining the user interface, are relevant at present time.

On the other hand, an equally important area of research is the involvement of the user in the process of interface development. The reason for the above statistics is also the semantic gap that occurs when the user and the developer interact at the domain analysis stage. Unfortunately, at present there are no methods of how to relate the domain and interface, and build the latter without losing the sense of the user's activity, which is very important, because in the user's view the interface is a tool for organizing its professional functions in the system, which will undertake part of these functions.

Eliminate the reason and achieve the preservation of the meaning of user actions in the domain, and fully meet the user requirements in the interface of the information system is possible if to create a methodological framework (a general methodological description) of user interfaces development that will be a common sign form of models organization and will be understandable to all participants of the interface development process, including the user. Moreover, this sign form will allow the end user himself to be the developer of the interface and to describe the subject area in the language of this sign form. A sign form will be a set of mechanisms of user actions. Filling the contents of the structure of mechanisms will be the result of the joint activity of all participants in the interface development process, and therefore it will be understandable and equally interpreted by all participants, which will lead to solving the problem – reducing the semantic gap.

In this case, the behavior and structure of the interface constructed in this way can be "read" by the engineer, programmer, customer, and end user.

Using a single sign form significantly simplifies the process of modifying the interface, because in case of interface modification, the developer or the user himself can make adjustments to the content of the action mechanisms.

2   The Prerequisites for Using the User Activity as the Basis for Interface Development

Object-oriented approach models the object, but not from the point of view of how the object is created, but what kind of behavior it has, what functions it performs, and what data it requires; functions and data of the object are combined. This is a prerequisite for the appearance of many methods of interface construction, including data-based ones.

The system, and as a consequence, the interface is treated as a collection of objects and connections between them. This is a consequence of the fact that developers build systems based on the subject domain, and transfer from it to the technical system the objects and their properties. Then the whole essence of the system is reduced to a set of interacting objects, while the logic, rules, meaning of their interaction remain for the user behind the scenes.

The object-oriented paradigm is oriented to the human perception of the world, objects really exist and possess properties. Therefore, technical systems are built on the basis of an object-oriented approach. But if you look at the objects of the real world in different way, you can see that the objects exist to operate on them: guided by the goal, the person performs the actions, operating with these objects. (For example, there is an object – student. From the teacher's point of view, the student exists only in the context of the actions performed on him – he teaches the student, puts him grades, gives him homework, etc.) By the same principle, it is necessary to build an interface: user, as a performer of activity, needs to see in the interface form only those elements that will allow him to perform the necessary actions and obtain the result of these actions. Therefore, the interface structure should be based on the structure of the user's actions in the domain. And due to the fact that most interfaces are built on the basis of object-oriented approach, there are a lot of problems, the main one of which is that the user does not understand the logic of interaction with the interface.

In the proposed approach, on the contrary, the object is modeled from the point of view of its creation in the context of the action performed for this purpose.

3   Mechanism of Action, as a Way of Representing User Activity

At present, two groups of methods are used to visualize the activity. The first group is graphical ways of visualizing the process of activity itself, for example, storyboarding [9], various models of visualization of business processes. But this category does not allow to fully reflect the data the person operates with. In addition, such methods do not allow a person who reads them to unequivocally interpret the implied sense, since there is no clear methodology for creating and describing such models. In the other group, there is only an attempt to reflect the process, the main aim is to reflect the structure of the data used in this process. Such methods are, for example, infographics, various methods of data representation, ways of visualization of ontologies [8]. But the activity is much more than the data.

Different methods of task analyses including activity theory [7] were developed and promoted by famous researchers (such as A. Leontev, S. Rubinshtein, Y. Engestrom) in their works. In this paper, it will be used the structure of activity proposed by Yu. Rogozov in [5] and other papers.

The proposed semantic approach assumes the joint work of all participants in the interface design process, based on the results of which a model will be constructed containing all the information necessary for interface design: starting with the description of user tasks and actions and ending with a description of the system elements that implement these actions.

Such structure of interaction between the user and developers will allow to create a model understandable to all stakeholders in which a high-level description of the user's activity according to certain rules is decomposed by the system analyst, allowing to get a representation of this activity in terms of the system, which in turn is decomposed by the developer in order to obtain specific system objects, that are used for the implementation of the user's activity in the system.

All activities will be reduced to a single structure. Mechanism of action is the presentation form of information about the domain and the system objects [10]. In comparison with other forms of representation of user tasks [2], the mechanism of action, among other things, has the main advantage – it has a simple form and structure that is understandable for user who does not possess any knowledge and skills in developing the interface and software in general. The action mechanism filled with content is a multidimensional table model, where each element can be represented as an embedded mechanism for the creation of this element. Because the table model is the result of the interaction between the user and the developer, the contents of this table should be clear to both. The way of building the tables is as follows: top-level tables are decomposed until the information contained in the tables is sufficient for the design of the information system interface.

4   State of the Art

The proposed method can be attributed to model-oriented methods of user interface development (such as, for example [4]), where the model of the user's activity in the subject area is used.

Task models are used in existing methods of interface design. There are two main types of task models. The first type is oriented to show a sequence of tasks. The second is to show the flows of the data used in the tasks. Existing task models are used both to describe the current task state entirely, and to describe the process of accomplishing the task in the future, taking into account the interface for the design of which this task model is needed.

But while all other methods use task models to show low-level aspects of interface which has to be designed, the model in form of mechanism of action allows to show the sequences of user tasks in problem domain, to show data flows that are used in these tasks, also the structure of mechanism of action is clear to end user so he is able to be involved in the interface design process from the very beginning. The mentioned aspects are advantages of using the mechanism of action in the interface design process and make the proposed method connected with the notion of Participatory Design [3].

5   User Interface Design Method

The proposed method is a methodological framework for user interfaces development (storage technology combined with execution), the content of which can be objectified in the form of mechanisms of different levels. The methodological framework defines the interface structure. What is described by the participants at high level of abstraction is further objectified to obtain the interface. This approach has a number of advantages over other methods of user interface design. First, there is no coding step in the usual sense. High-level specification of the user activity in the domain in the form of a set of action mechanisms is gradually transformed into an interface. Secondly, in case of need to make changes, the developer of the interface (or the user himself) returns to the specification, which has a higher level of generality (in comparison with the program in the usual programming language). Thus, the testing and maintenance stages are simplified. Thirdly, users, who do not have experience in software development and programming, can easily master such a high level language of specification. The proposed method consists of four main stages.

1. Selection of user functions, which should be modeled in the interface.
   1.1 Identify a set of related user functions in the domain that he performs as part of his problem domain.
   1.2 Present each function, detached in 1.1, in the form of action mechanism of its formation. (Representing a function as an action mechanism will allow to determine how the user performs this function in the domain and what data uses).
   1.3 Select from this set of functions, represented as actions mechanisms, those which can be delegated to the system and, therefore, which should be represented (modeled) in the interface.
   1.4 According to the way of how the user will perform these functions in the interface, to converse the user action mechanisms performed in the subject area of 1.3 to the action mechanisms of performing these functions in the interface. Conversion can be performed in one of the following ways:
      (i) By replacing the contents of one of the elements of the mechanism,
      (ii) By decomposing the mechanism into several interrelated mechanisms of action.
2. Creation of mechanisms library by means of which it is possible to model the actions of item 1 and present them in the interface.
   2.1 Create a classification of user actions in the interface according to the type (meaning) of the performed function.
   2.2 Create a classification of typical interface elements by the meaning of the function performed.
   2.3 Establish the correspondence between two classifications.
   2.4 Given the established correspondence, each element of the user's activity classification should be presented in the interface in the form of action mechanism for selecting a typical interface element, which allows to perform this action in the interface.
3. Interface design: the formation of interface elements and relations between them.
   3.1 Determine the type of user action of 1.4. by the meaning of the function.
   3.2 Decompose the mechanisms of action of 1.4. into the mechanism for selecting a typical interface element of 2.2 for the implementation of the interface model prototype.
4. Implementation of the user interface with the help of steps 2 and 3. Based on the implementation of steps 2 and 3, perform a software implementation of the user interface prototype (Fig. 1).

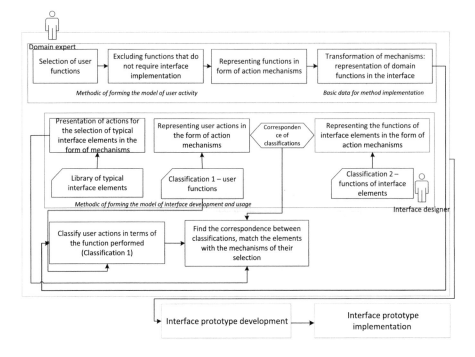

Fig. 1. Structure of method of user interface design

6 Example of Using the UI Design Method

Approbation of the user interface design method on the basis of the semantic approach [6] is currently conducted to create an interface for the information system of child psychologist workplace. Intermediate results allow to assert that the method allows to involve successfully the end user (psychologist) to create the initial data of method – the subject domain task description and their representation in the form of action mechanisms. Here is an example of developing a set of screen forms for performing analysis of the state of children group by one of the psychologist methods. Table 1 shows the initial data: this is a task, divided into subtasks, which corresponds to the first point of the presented method. Further, each subtask was represented as a mechanism of action.

Table 1. Decomposition of the task into functions on the example of the development of the psychologist's workplace interface for the task of testing a group of children by one of the methods.

User actions in the problem domain	Steps of user actions in the problem domain
1. Preparation of a set of visual materials	1.1. Find the Image Sets 1.2. Determine the indicators by which testing will be carried out 1.3. Sort images by indicators
2. Testing children by the method	2.1. Conversation with the child 2.2. Making a record of the child answers for each indicator
3. Evaluation of each child	3.1. Formation of scores in points for each child for each indicator 3.2. Scoring in points for the child for each indicator
4. Carrying out of calculations of average values by indicators on the child, on group	4.1. Get average value in points for all indicators for each child 4.2. Get an average score in points for each indicator for all children in the group
5. Report writing	5.1. Get the report in the form of a chart by indicators 5.2. Get the report as a chart for a group of children

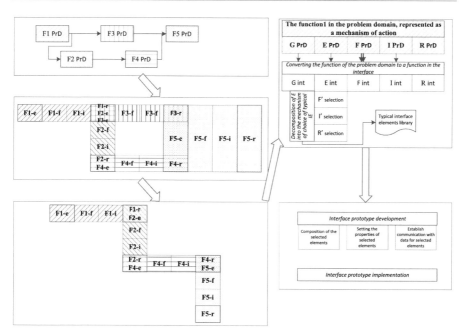

Fig. 2. Process of user interface development based on the proposed method on the example of task № 1 "Preparation of a set of visual materials"

Figure 2 graphically shows a process of performing the method for one of the tasks. Here is the list of abbreviations from the figure:

- Structure of action mechanism: g-goal, e-element, f-function, i-instrument, r-result;
- PrD – problem domain;
- Int – interface.

7   Conclusion

The paper presents a method of user interface design based on a semantic approach, and also describes the prerequisites for its development and the advantages of its using. The results of the implementation of the presented method on the presented example will make it possible to verify the effectiveness of the method and to identify possible problem areas which have to be eliminated in future.

Acknowledgements. The reported study was funded by RFBR according to the research project № 17-07-00099 A.

References

1. Standish Group 2015 Chaos Report - Q&A with Jennifer Lynch. https://www.infoq.com/articles/standish-chaos-2015. Accessed 19 Dec 2017
2. Crystal, A., Ellington, B.: Task analysis and human-computer interaction: approaches, techniques, and levels of analysis. In: Proceedings of the Tenth Americas Conference on Information Systems, New York, August 2004
3. Muller, M.J.: Participatory design: the third space in HCI. In: The human-Computer Interaction Handbook, pp. 1051–1068. L. Erlbaum Associates Inc., Hillsdale (2003). ISBN 0-8058-3838-4
4. Trætteberg, H.: Model-based User Interface Design, p. 204 (2002)
5. Rogozov, Y., Sviridov, A.: The concept of methodological information systems development. In: 8th IEEE International Conference on Application of Information and Communication Technologies, AICT 2014 - Conference Proceedings, vol. 8 (2014)
6. Belikova, S., Rogozov, Y., Borisova, E.: Semantic approach to information system user interface design. In: 17th International Multidisciplinary Scientific GeoConference SGEM 2017, Conference Proceedings, 29 June–5 July 2017, vol. 17, issue 21, pp. 673–680 (2017). www.sgem.org, https://doi.org/10.5593/sgem2017/21/s07.086, ISBN 978-619-7408-01-0, ISSN 1314-2704
7. Kaptelinin, V., Nardi, B.A.: Acting with Technology: Activity Theory and Interaction Design, p. 334. The MIT Press, Cambridge (2006). ISBN 978-0-262-11298-7
8. Katifori, A., et al.: Ontology visualization methods—a survey. ACM Comput. Surv. **39**(4), Article 10 (2007)
9. Little, A.: Storyboarding in the Software Design Process. https://uxmag.com/articles/storyboarding-in-the-software-design-process. Accessed 13 Jan 2018
10. Belousova, S., Rogozov, Y., Sviridov, A.: Technology of using properties and mechanisms of actions in user interface design. In: 15th International Multidisciplinary Scientific GeoConference Surveying Geology and Mining Ecology Management, SGEM 2015, Albena, Bulgaria, vol. 1, issue 2, pp. 339–345 (2015)

Embedded Software Monitoring Using Pulse Width Modulation as a Communication Channel for Low Pin Count Microcontroller Applications

Przemyslaw Mazurek and Dawid Bak[(✉)]

Department of Signal Processing and Multimedia Engineering,
Faculty of Electrical Engineering, West–Pomeranian University of Technology,
26. Kwietnia 10 Str, 71–126 Szczecin, Poland
{przemyslaw.mazurek,dawid.bak}@zut.edu.pl
http://www.media.zut.edu.pl

Abstract. Low pin count microcontrollers are used in numerous applications, especially in power control. Pin number reduction complicates real–time monitoring and debugging. Two modulation techniques for transmission data using PWM (Pulse Width Modulation) channel without significant influence on main application of this channel are proposed in this paper. Two modulation techniques: Additive Pulse Width Fluctuation and Differential Pulse Width Fluctuation are proposed with appropriate demodulation methods.

Keywords: Software monitoring · Embedded systems
Microcontrollers · Pulse Width Modulation

1 Introduction

Software monitoring of embedded systems is a challenging task. There are a lot of development tools that are available for developers but debugging and monitoring could be very limited for the final system. Embedded system could be completely simulated in software–based system simulator (e.g. Proteus Virtual System Modeling [16]) with limited real–world interaction. Another option could be ICE hardware (In–Circuit Emulation) for real–time simulation of particular microcontroller in the destination board [10]. Such option is very expensive but gives the ability of debugging and monitoring with complete transparency. Nowadays microcontrollers and DSPs support cost effective version of ICEs implemented in chip and JTAG port is widely used for debugging [6,10]. JTAG debugging capabilities are manufactures dependent unfortunately. Transparent real–time monitoring using a JTAG port is usually limited or not available and embedded software should be stopped during debugging. Some monitoring capabilities are not documented in chip specifications. Debug port occupies a few pins typically and serial communication is used. It is acceptable for high pin count

© Springer International Publishing AG, part of Springer Nature 2019
R. Silhavy (Ed.): CSOC 2018, AISC 763, pp. 319–330, 2019.
https://doi.org/10.1007/978-3-319-91186-1_33

microcontrollers and DSPs but not for the low pin count devices. There is no exact definition of low pin count devices but usually devices with 20 or less pins are assigned to this group. There are even a 8 or 6 pin devices available (6–pin: Atmel ATtiny family [3], and Microchip PIC10 family [15]; 8–bit are produced by most microcontroller manufactures). A low pin count devices are important for general cost reduction: cheaper device, smaller board, and smaller casing. Large microcontrollers and DSPs have 100pins or more. Low pin count devices are very specific and needs more exotic techniques for software debugging and monitoring especially for real–time systems.

A lot of simple techniques are used from many years for monitoring of microcontrollers: LEDs, output pin signaling, hardware and software serial ports, and displays. All of them could be applied for real–time monitoring and needs a small number of pins. Even a single output line could be successfully applied to delivery of monitoring data. Such techniques are cheap, but limited by microcontroller speed and developers' invention.

Most microcontrollers are grouped into families and a large pin count microcontroller could be used for development so the application of mentioned techniques is rather simple. Low pin count device is used at the end of development phase so porting from large to low pin count device is quite trivial. The monitoring is very limited after this step, what is serious problem. Real–time requirements are very often an additional limitations and device could be not stopped during monitoring.

Typical reasons for monitoring of such devices are: real–time monitoring of the control process state with temporal data; access to the log memory that consist events stored in the internal EEPROM; access to the self–test results.

2   Extending of the Input and Output Pins Capabilities

Software or hardware serial communication channel should be used but a low pin count devices use all pins typically for other purposes. Pin sharing techniques should be used for data transmission what is not trivial task because function of the appropriate pin should be not influenced. There are not available alternatives for such situation and another transmission methods should be applied. Such idea is not new and a lot old and recent professionally designed embedded systems uses such techniques for e.g. combining of input devices (push switches) with output devices (LED, displays).

Every pin is dedicated to the single purpose and it is well know constraint for beginner hardware and software developers according to most electronic design books. Breaking of this rule is possible and selected examples related to the real–time monitoring of microcontroller will be discussed.

2.1   Additional Debug/Monitoring Chip

This technique does not change a purpose of the microcontroller pins. This is extension of the input and output lines by the application of the additional chip compliant with the serial bus (Fig. 1).

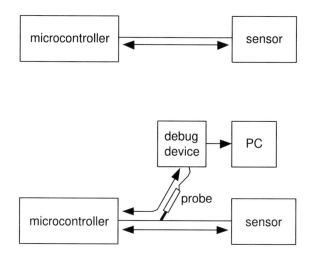

Fig. 1. Embedded system during normal work and monitoring

Serial busses are used in embedded systems especially for extending capabilities of microcontroller. Microcontrollers available nowadays uses a serial buses for the measurement purposes mainly because most serial–based peripherals used in old systems are now available inside microcontrollers (EEPROM, RTC, PWM, A/D converter, D/A converters). The connection to another microcontroller, network, or PC is not considered in this paper. The single wire (1–Wire bus) [7], two–wire (I2C bus [11]), or three wire (SPI bus) connection is used typically. Single active wire connection with additional ground line could be used for measurement purposes (e.g. Maxim/Dallas temperature sensors DS18xx [7] or connected to general purpose A/D converter DS2450 [8]). An example 1–Wire devices are addressable so connection of additional device for monitoring purposes is also possible.

There are no special software or hardware requirements for this technique, because bus lines are used according to manufacture specification. Monitoring data could be transmitted to the debug devices. This technique is very cheap but limited for embedded systems with such busses.

2.2 Hidden Cycles

This is next one bus oriented technique. Data transmission using serial, parallel or serial/parallel buses are defined by the bus cycle with strictly defined time–value (voltage) transitions. Only selected combinations are accepted by peripherals, some of them are not acceptable, but there are some ignored combinations. Ignored combinations could be used for communication purposes without disturbing effects.

This technique is more complicated and specific hardware should be used – typically a CPLD or FPGA devices should be prepared for filtering and interpretation of such bus transitions. This method is interesting because the previous one method requires transmission with complete bus cycle. Reduced bus cycles reduce bandwidth requirements also.

2.3   Overriding of the Input Logic Line

This technique could be applied to the input/output line that is configured as an input. An example is the push switch (key) or similar sensor that is non–frequently used for process control (Fig. 2).

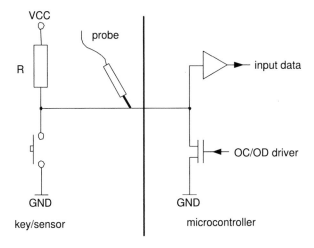

Fig. 2. Transmission using key line – schematic

Key line is typically configured as an input. High logic level is recognized by the microcontroller if the key is open and low level is recognized if the key is pushed. Configuring Key line in input/output mode with the open–collector or open–drain option a serial transmission is possible. Microcontroller should test this line and transmit data only if this line high (the key is open). Comparing input and output values gives ability of state detection of the key. Transmission should be interrupted if output value is high and input value is low what is possible in open–collector and open–drain modes. An additional delay before transmission is necessary due to switching oscillation of the real key (Fig. 3).

Transmission protocol should be application dependent and such aspects like: redundant transmission – FEC (Forward Error Correction) [5,19], CRC (Cyclic Redundancy Check) [14], short packet lengths, should be considered. Resistance of the pull–up resistor should be selected according to the specification of the chip manufacture.

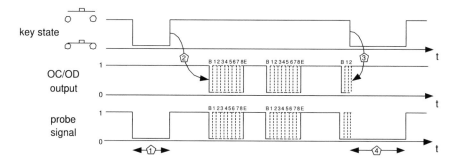

Fig. 3. Correct and broken transmission using key–input line. (1) – the key is pressed and the state is detected by microcontroller. (2) – after a key release and additional is necessary before transmission. (3) – broken transmission by the key press (4) microcontroller tests line and knows about broken transmission.

2.4   Overriding of the A/D Input Line

Some microcontroller have A/D converter input line that could be also configured as an input/output line (Fig. 4). Time domain sharing is possible if pin mode selection is accessible by software. This technique gives ability of data transmission if thermistor or similar resistance based sensor is connected.

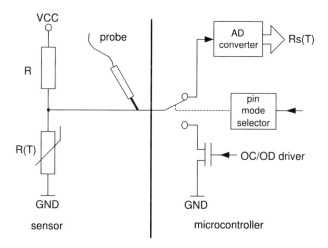

Fig. 4. Transmission using voltage input line – schematic

The resistance of the sensor should be not too low (not below $Vmin$) in relation to the pull–up resistor because input value will be very low and can not be driven by the open–collector or open–drain transistors. Output low value is defined by the ground level and the high value is variable (amplitude modulated by the sensor) what is shown in Fig. 5.

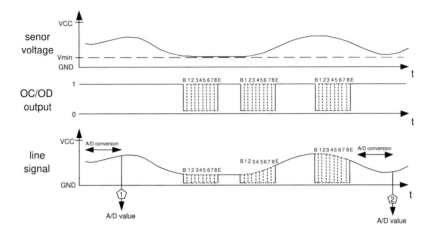

Fig. 5. Transmission using voltage input line. (1) and (2) – sampling moments known for microcontroller, transmission occurs between A/D converter cycles

Binary values could be recognized using analog comparator or Schmitt trigger in the probe input stage. Threshold value should be set to the $Vmin/2$ level. Pull–up resistor can not be too low due to input current from VCC to the ground by the turned–on transistor in the microcontroller.

Data transmission in this method could be independent on the input sensor what is important advantage.

There are also limitations of this method - for input low–pass filter. Values from the sensor could be filter by additional RC filter and transmission disturbs the state of the filter. Depending on the filter properties a settling time define minimal time period between end of transmission and a new A/D measurement cycle.

2.5   Hijacking of the Output Line

The last example of technique is based on the transmission by the output line used in parallel to the completely different purpose (not for communication).

An example is the LED driving by the output line. It is possible to use such line for low speed transmission of serial data without visible light influence. LED light will be modulated but it will be not visible for human if modulation frequency will be high.

More advanced techniques should be used for data transmission over PWM (Pulse Width Modulation) channel. PWMs are used in many microcontrollers (especially low–pin count) for power control applications (light, motor) – these very important end–market of low–pin count microcontrollers. Applications are real–time so disturbance of the power control process is not recommended. Switching mode of output line is not possible because controlled device should be driven all the while.

3   Hidden Data Transmission over the Pulse Width Modulation Channel

3.1   PWM Channels

PWM (Pulse Width Modulation) channels are quite simple devices and are implemented inside microcontrollers. They are very usefull for numerous purposes. PWM channel could be used as D/A converter with minimal number of external components [13] if e.g. conversion to voltage is necessary (single resistor and capacitor as a low–pass filter) with optional voltage follower using operational amplifier (Fig. 6). Some external devices could be controlled using PWM signal without voltage or current conversion components what is used for e.g. control of rotation speed of motors or light level. Modern microcontrollers support a few types PWM modes dedicated to specific applications. Embedded systems related to the power control are very important for today 'green' requirements (power saving), support reduction of radio interferences and power line interferences. Software algorithms give the ability of motor or light control using advanced algorithm selected for the specific application what is significantly better in comparison to the old–style on/off control.

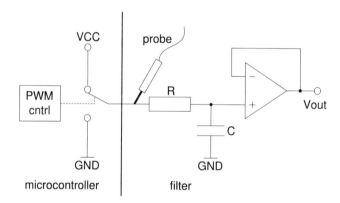

Fig. 6. Transmission using PWM Output Line – Schematic

Conventional PWM channel based on the digital comparator with the fixed width of the PWM cycle is considered for simplification in this paper. Presented techniques could be extended to another PWM modes also.

PWM channel is driven by the internal generator typically. Counter values are incremented and counting starts from beginning value if overflow occurs. Values from the counter are compared with value stored in the comparator and the output pin state $y(n)$ is modified:

$$y(n) = \begin{cases} 1 : Counter(n) \geq Comparator \\ 0 : Counter(n) < Comparator \end{cases} \tag{1}$$

Such formulation is correct for the fixed cycle width of PWM. Values stored in the comparator could be software modified for variable PWM signals. In modern microcontrollers additional register (also comparator) for limitation of the maximal possible value of the counter is used.

The PWM channel can not replace conventional D/A converter (e.g. R–2R ladder based) due to the counter clock limitations and clock width–resolution relation. Large counters (e.g. 16-bit) gives higher accuracy of output value (e.g. resolution of the D/A converter) but refresh rate is rather low, but it is acceptable depending on application.

3.2   APWF and DPWF Modulation Techniques

Hidden transmission using PWM channel is a specific kind of hijacking of the output line with twice functions at one time (PWM output and serial output transmission). The interference between both functions should be minimized using appropriate modulations of PWM signal in real–time by modification of the comparator value depending on transmitted bit. Proposed method is different from the method shown in [1] where the PWM output is used only for real–time monitoring so the PWM channel can not be used for other purposes.

The first proposed modulation technique (APWF – Additive Pulse Width Fluctuation) assumes two times higher resolution of the counter (Fig. 7).

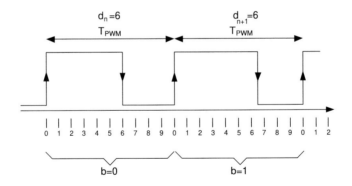

Fig. 7. Additive pulse width fluctuation technique

Pulse width PW value and the value stored in the comparator for particular time moment n is modified according to formula:

$$PW_n = 2 \cdot d_n + b_n \tag{2}$$

where b_n is the transmitted bit value (0 or 1).

Symbol rate is defined by the PWM cycle. The number of effective bits depends on the implementation of 2'nd layer ISO/OSI model. Asynchronous or synchronous transmissions are possible. Bit stuffing technique [9] for synchronous transmission is interesting option for packet forming [1,2]. Increasing the number of bits per symbol is possible by increasing of the PWM time resolution (PWM clocking).

The second proposed modulation technique (DPWF – Differential Pulse Width Fluctuation) assumes the minimization of high–frequency components (ripples) for desired constant value of PWM (Fig. 8). This technique reduces width of the particular pulse and extend width for the next one pulse if the transmitted bit has 1 value. Such operation does not occur if the transmitted bit has 0 value.

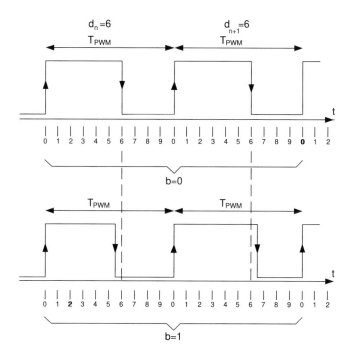

Fig. 8. Differential pulse width fluctuation technique

Transmission speed for this modulation is two times slower in comparison to the previous one.

$$\begin{cases} PW_{2 \cdot n} = 2 \cdot d_n - b_n \\ PW_{2 \cdot n+1} = 2 \cdot d_n + b_n \end{cases} \tag{3}$$

Both kinds of bits values are extended over two PWM cycles, but only one influent on signal so high frequency component occurs in signal a later are suppressed by the low–pass filter. A modification of this method is possible where bit 1 corresponds to the two PWM cycles and 0 bit corresponds to the single PWM cycle. Effective transmission speed is dependent on bit stream values. Efficiency is 25% above in comparison to the original DPWF modulation for random bit stream.

4   Demodulation Techniques

Both modulation methods are very simple for software implementations that is especially important for microcontrollers. The demodulation is more tricky because d_n value is unknown and frequency of the PWM clock is known but not reliable. A low pin count microcontrollers are driven by the external crystal or by internal RC clock. Possible calibration in some microcontrollers of the RC generator is insufficient for high stability of the core clock and PWM clock. The RC based generator is sensitive to temperature that is the additional factor that should be considered in demodulator design. RC generators are often used for low pin count applications due to pin saving (two pins are necessary for crystal, a single pin is necessary for external crystal generator input).

The demodulation for the unknown but stable frequency is also possible with assumption of the fixed width of PWM cycle. There are two edges (falling and raising) of the PWM signal and one of them is exactly related to the PWM cycle. The position of the second edge PWM is the comparator actual value dependent. Measurements of the time period between two staring PWM cycle edges give the high–quality reference clock for demodulation.

$$\hat{b}_n = \begin{cases} 1 : 0.25 < mod\left(T_P, \frac{T_{PWM}}{N}\right) < 0.75 \\ 0 : otherwise \end{cases} \tag{4}$$

where: T_P – measured time period, T_{PWM} – time cycle of PWM, mod – modulo function, N – maximal number of time quants of PWM.

Constant T_{PWM} value should be replaced by the variable time period T_{PWMn} that is measured for every PWM cycle. A following reformulation is required for non–stable frequency of the PWM clock (e.g. RC clock generator):

$$\hat{b}_n = \begin{cases} 1 : 0.25 < mod\left(T_P, \frac{T_{PWMn}}{N}\right) < 0.75 \\ 0 : otherwise \end{cases} \tag{5}$$

The demodulation of the DPWF is similar to the APWF. Obtained bit stream should be additionally processed because every bit is double transmitted. A starting edge of transmission is the first edge (bit value change) for asynchronous transmission and it is exactly related to the position of all bits. The length of the transmitted word (e.g. byte) is known a priori so only even or odd bits are important from this starting edge.

5   Conclusions

Techniques for monitoring of software presented in this paper are very important for embedded systems development. Low pin count microcontrollers require very specific methods of monitoring. The PWM channel could be used for data transmission. Proposed method of monitoring using PWM channel with APWF and DPWF modulations could be used for typical microcontrollers. The modulation of PWM channel has low cost what is very important for microcontrollers.

Transmitted data could be crypted if it is necessary for particular application. Similar technique could be used for ASIC microcontrollers and FPGA microcontrollers if it is necessary with optional hardware support of modulation.

For another PWM modes similar techniques could be proposed also. The analysis of the PWM signal is very complicated, especially a signal spectra [4,17,18] and simplified analysis is used typically [1]. Moreover, due to infinite bandwidth of PWM signal a typical numerical analysis using the FFT (Fast Fourier Transform) is also complicated (due to sampling of such signal and aliasing effect). There are two sources of errors: resolution and ripples. The resolution depends on the number of clocks count per cycle and ripples are produced by the unfiltered harmonics. Differential signal related to distortions introduced by the transmission could be obtained by subtraction of original and modulated PWM signals.

Presented demodulation techniques could be implemented using fast microcontroller or DSP with pulse width measurement. CPLD and FPGA implementations are also possible for application.

References

1. Alter, D.M.: Using PWM Output as a Digital–to–Analog Converter on a TMS320C240 DSP, SPRA490, Texas Instruments, November 1998
2. Aviran, S., Siegel, P.H., Wolf, J.K.: An improvement to the bit stuffing algorithm. IEEE Trans. Inf. Theory **51**(8), 2885–2891 (2005)
3. ATtiny4/5/9/10 8–bit AVR Microcontroller with 512/1024 Bytes In–System Programmable Flash, Atmel (2010)
4. Bech, M.M., Pedersen, J.K., Blaabjerg, F., Trzynadlowski, A.M.: A methodology for true comparison of analytical and measured frequency domain spectra in random PWM converters. IEEE Trans. Power Electron. **14**, 578–586 (1999)
5. Clark Jr., G.C., Cain, J.B.: Error-Correction Coding for Digital Communications. Plenum Press, New York (1981)
6. Dushistova, A., Rusev, A., Mehaffey, J.: Debugging with JTAG, Montavista (2009). http://tree.celinuxforum.org/CelfPubWiki/ELC2009Presentations?action=AttachFile&do=get&target=DebuggingWithJtagCelf2009.pdf
7. Maxim/Dallas DS1822 Econo 1–Wire Digital Thermometer Datasheet
8. Maxim/Dallas DS2450 1–Wire Quad ADC with Programmable Resolution
9. National Communications System Technology and Standards Division, Telecommunications: Glossary of Telecommunication Terms, General Services Administration Information Technology Services (1996)
10. Haller, C.A.: The ZEN of BDM 1996–1997. www.macraigor.com/zenofbdm.pdf
11. I2C–bus specification and user manual, UM10204, NXP 2007. http://www.standardics.nxp.com/support/documents/i2c/pdf/i2c.bus.specification.pdf
12. Mazurek, P.: Hardware supported debugging in 32–microcontrollers and digital signal processors. In: 6th International Conference on Advanced Computer Systems ACS 1999 Szczecin, pp. 169–178 (1999)
13. Palacherla, A.: Using PWM to Generate Analog Output, AN538, Microchip Technology (1997)
14. Peterson, W.W., Brown, D.T.: Cyclic codes for error detection. In: Proceedings of the IRE, vol. 49, pp. 228–235, January 1961. https://doi.org/10.1109/JRPROC.1961.287814

15. PIC10F200/202/204/206 Data Sheet 6–Pin, 8–bit Flash Microcontrollers, Microchip (2007)
16. Proteus Design Suite. Product Guide. Labcenter Electronics. http://downloads. labcenter.co.uk/proteus7brochure.pdf
17. Song, Z., Sarwate, D.V.: The frequency spectrum of pulse width modulated signals. Sig. Process. **83**(10), 2227–2258 (2003)
18. Stankovic, A.M., Lev-Ari, H.: Randomized modulation in power electronic converters. Proc. IEEE **90**(5), 782–799 (2002)
19. Wilson, Stephen G.: Digital Modulation and Coding. Prentice-Hall, Englewood Cliffs (1996)

Synthesis of the Life Cycle Stages of Information Systems Development

Alexandr N. Belikov, Yuri I. Rogozov[(✉)], and Oksana V. Shevchenko

Institute of Computer Technology and Information Security, Southern Federal University,
Taganrog, Russian Federation
{anbelikov,yrogozov,ovshevchenko}@sfedu.ru

Abstract. The problem of agreeing points of view on the target system being developed, often inseparable from the problem of mutual stakeholders under-standing in the process of information systems development, as a special case of complex technical systems, is one of the most fundamental today. The paper considers the solution of this problem and suggests a methodology for the synthesis of the stages of the information systems development life cycle. This methodology assumes two models: a model for creating knowledge (concepts), and a model for reconciling knowledge. In the paper the main aspects of these models and the process of their use are considered. The ideas proposed in the work are aimed at formalizing the process of information systems development and removing the empirical component that currently present in the information systems design.

Keywords: Information system · Mechanism of action · Life cycle
Synthesis of stages

1 Introduction

Despite the development of technology and methodological support in the field of IT, one of the key problems remains to be solved – the problem of mutual understanding between the customer and the developer (stakeholders). The essence of the problem lies in the complexity of the development process, such as conflicts between the results of project activities of development teams working together on the system design, delib-erately distorted or limited information exchange among developers, semantic barriers between development teams, protection of sensitive data of design by experts [1–3]. The process of developing complex technical systems requires interaction between several technical disciplines, the authority over which is distributed among a multitude of experts. And experts themselves can be distributed within the organization and carry out activities at various stages of the product life cycle and at different stages of its production chain. The complex nature of such projects requires the interaction of many disciplines in the context of decentralized authority to design and incomplete knowledge of development teams about the system. In fact, this problem can be extrapolated and transferred to the development team - mutual understanding between the analyst and the designer, between the designer and the programmer.

© Springer International Publishing AG, part of Springer Nature 2019
R. Silhavy (Ed.): CSOC 2018, AISC 763, pp. 331–337, 2019.
https://doi.org/10.1007/978-3-319-91186-1_34

Researchers around the world link the problem of mutual understanding of stakeholders with the lack of knowledge about the specifics of different subject areas of the system development stages. For this reason, methods, approaches and tools are created to extract and transmit meaning between subjects from two different subject areas:

- requirements extraction techniques (for the customer-analyst/project manager bundle) [4];
- model-oriented approaches (for linking analyst/project manager -designer) [5];
- interpreted models and code generators (for the bundle designer-programmer).

Also in a separate class can be selected object-oriented languages (DSL) [6], ideally aimed at a direct transition between the customer and the system. However, such languages require preliminary work of analysts, designers and programmers with a deep immersion in the subject area, and at the output of the development process they give a highly specialized tool that is not suitable for mass use.

Despite some successful attempts to solve the problem of mutual understanding between two neighboring participants of the software life cycle, all the above approaches have one essential drawback: the methods and models that are being created do not agree with each other and, most importantly, they do not allow us to single out the meaning (semantics) of system and transfer it through all stages of the life cycle.

This causes the urgency of developing new approaches to the formulation, documentation and use of a single evolutionary semantic model of the target system at all stages of the life cycle.

2   Related Work

The problem of agreeing points of view on the target system being developed, often inseparable from the problem of mutual stakeholders understanding in the process of information systems development, as a special case of complex technical systems, is one of the most fundamental today. The quality of its solution in each case is directly related to the compliance with the terms and budget of the project, and also the meeting the customer's requirements.

Improving the methods of information systems development, accelerated the implementation of each stage in a separate workplace, and significant progress was made in this. However, according to the research of the Standish Group, in 2011–2015 only 29% of projects were successful, 53% were problematic and 18% were failed projects [7]. At the same time, in the area of medium and large systems, the percentage of success is only 10–27% (Table 1), which indicates problems in the existing approaches of the information systems design and development. This is observed against the background of the availability of advanced tools for documenting, modeling, designing, frameworks for the program code development, as well as in conditions of market saturation by good specialists.

Table 1. Research of the Standish Group

	2011	2012	2013	2014	2015
Successful	29%	27%	31%	28%	29%
Required changes	49%	56%	50%	55%	52%
Failed	22%	17%	19%	17%	19%

According to the researchers, the cause of the situation is the inconsistency of the image of the target system, which arises from a misunderstanding between the customer and the developer. The overwhelming majority of modern research and practical developments are aimed at solving the problem of mutual understanding between two participants in the life cycle of information systems, located at its neighboring stages. However, this does not solve the problem globally, as evidenced by the above statistics.

The reason of this problem is at the methodological level. If we consider the life cycle of information systems development, it consists of a sequence of steps. Life cycle is characterized by a series of logically following stages: requirements development, business modeling, system design, implementation, etc.

Currently, at each stage of the life cycle it is empirically formed the so-called "finished knowledge" (Fig. 1). On the basis of "Complete Knowledge", the process of creating a form of knowledge is being formed. After that, using this process (rules of modeling), we get a form of knowledge filled with content (the same models that are known to experts of the domain, for example, IDEF, UML, etc.).

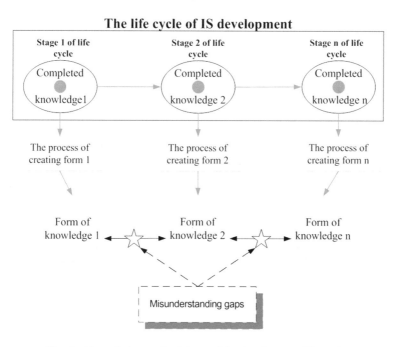

Fig. 1. The existing methodology of the development life cycle.

Such methodology of the life cycle of the information systems development leads to the fact that the obtained forms of knowledge at different stages of development can not be coordinated among themselves, because of the use of knowledge of different areas [8].

Let's consider an example: the analyst, using the most advanced methods and techniques, can get up to 100% reliable description of the customer's wishes (requirements), but he is still neither a domain specialist nor a designer (programmer). In this connection, not having at the moment tools for documenting the system meaning with a view to its further transfer to the next stages of the life cycle, he translates the knowledge received from the customer about the subject area together with a set of requirements through a new set of tools to ensure its mutual understanding with the designer. With this approach, the loss of important aspects of the subject area and the target system is important for the successful implementation of the project.

Thus, the problem of harmonizing the points of view of stakeholders (or the results of the stages of the system life cycle), eliminating contradictions and misunderstandings between participants in the information system life cycle stages is topical and leads to a significant increase in project risks.

3   Approach to the Synthesis of the Stages of the Information Systems Development Life Cycle

The paper proposes a solution to the described earlier problems, by developing a new methodology for the system development life cycle. The essence of the idea is to use abstractions to create knowledge, i.e. it is proposed to consider the process of obtaining the "Knowledge". It is proposed to use as a basic abstraction not concepts (entities, objects), but system means of means self-organization for creating means in which various forms of specific concepts are self-generated. In other words, to use as abstractions not specific concepts, but means, for example, methodical, in which specific concepts are created. And we will assume that different methodologies are used at different levels of development: methods at the higher level, methodics at the subordinate level. In this case, it is necessary to reconcile between levels not specific concepts (languages) of these levels, but methods, methodics and norms (specific actions) [9].

Let's consider the model of the knowledge creation tool in more detail (Fig. 2). So we distinguish three main components that are considered in this model:

- Knowledge is represented as a process of action constructing;
- Through these actions, knowledge acquires the form of an object;
- Any action (or mechanism of action) is a four characteristics: elements (E), functions (F), tools (I), and result (R)) [10].

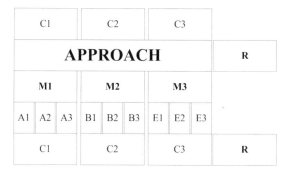

Fig. 2. Model of the means of knowledge creating.

Therefore, this model can be represented in the form of a process:

1. The approach organizes an action from the elements C1, C2, C3, which in turn form the result R. Actions C1, C2, C3 represent possible knowledge;
2. The form of knowledge is created from objects using methodology. The methodology (M1, M2, M3) in this case is a means of drawing objects into the form of knowledge;
3. Actions M1: A1, A2, A3; M2: B1, B2, B3; M3: E1, E2, E3 form the actions for creating C1, C2, C3, respectively. This stage allows you to obtain a form of knowledge, i.e. specific models with content.

Thus, in order to obtain knowledge, the subject (the domain expert) should construct his actions (he should present his knowledge in the form of action), and then shape this knowledge to the form of object, through these actions.

Using the model of the knowledge creation tool, it is possible to reconcile knowledge and forms of knowledge, which allows to synthesize the stages of the life cycle. For their coordination it is necessary to consider not even the reconciliation process, but the organization of the connection between methods and techniques used at various stages of the information systems development. The methodological model of organization of communication between methodological tools, that are used at different stages of information systems development, is proposed. This model uses as a primary abstraction not specific concepts, but concepts such as method, methodology and norm (a set of concrete actions).

It is proposed to use a single form of development stages presentation, which we will call a methodological model (means of creating concepts) (Fig. 3). This model considers not concepts and relations between them, but an instrument (means or tools) with which we create concepts.

Representation of stages in the form of a model of the "method" deprives it of its subject knowledge, which allows us to compare, find places of interpenetration and penetration of different subject domains into each other [11].

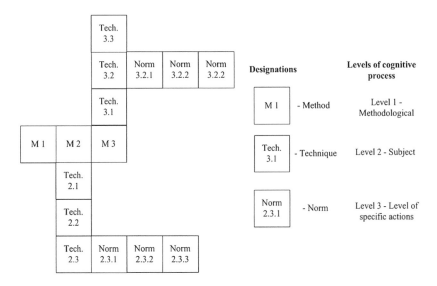

Fig. 3. Methodological model of knowledge reconciliation

The methodological model consists of a method, methodics and norms (concrete actions). The method is the main part of any development stage. Currently, there are several methods for each of the stages of system development. The method in its general form is a set of steps. Each step of the method must be represented by the methodic of its implementation. Further, each methodic is represented by norms (concrete actions). After the method, methodics and norms are allocated, it is necessary to check the reconciliation. There are two levels of reconciliation: vertical and horizontal. Vertical reconciliation allows you to track the sequence of transitions from one step to another without losing the meaning within one subject area. Horizontal reconciliation allows you to track consistency between the method, methodology and norms (specific actions).

Reconciliation of methodological models will allow to formalize the process of information systems development. The formalization of the development process will solve the problem of the emergence of "misunderstandings gaps" between different groups of actors involved in the information systems development. From the point of view of practical information systems development, it will allow to increase the percentage of successfully completed projects.

4   Conclusion

The ideas presented in the paper solve existing problems in the information systems development and present a new look at the information systems design in general. The paper proposes the solution of the problem of transition between different stages of information systems design and development. It is proposed the approach that aims at formalizing the process of information systems designing and removing the empirical component that presents during the transition from stage to stage in the information systems design and development. The models presented in the paper make it possible to synthesize the stages of the life cycle of information systems development, and to obtain information systems that meet the requirements of the customer.

Acknowledgements. The reported study was partially supported by RFBR, research project No. 17-07-00098 A.

References

1. Sunkle, S., Kulkarni, V., Roychoudhury, S.: Analyzing enterprise models using enterprise architecture-based ontology. In: Moreira, A., Schätz, B., Gray, J., Vallecillo, A., Clarke, P. (eds.) Model-Driven Engineering Languages and Systems, MODELS 2013. Lecture Notes in Computer Science, vol. 8107. Springer, Heidelberg (2013)
2. Zapata, C.M., Giraldo, G.L., Jiménez, L.D.: Ontological representation of the graphical language of Semat. In: 2014 9th Computing Colombian Conference, 9CCC 2014, pp. 137–143. Institute of Electrical and Electronics Engineers Inc. (2014)
3. Jacobson, I., Spence, I., Johnson, P., Kajko-Mattsson, M.: Re-founding software engineering – SEMAT at the age of three (keynote abstract). In: Proceedings of the 27th IEEE/ACM International Conference on Automated Software Engineering - ASE 2012, p. 15. ACM Press, New York (2012)
4. Schooley, B.L., Feldman, S.S., Alnosayan, N.S.: Development of a disability employment information system: an information systems design theory approach. In: Proceedings of the 44th Hawaii International Conference on System Sciences (HICSS-44), pp. 1–10 (2011)
5. France, R., Ghosh, S., Dinh-Trong, T., Solberg, A.: Model-driven development using UML 2.0: promises and pitfalls. Computer **39**, 59–66 (2006)
6. Gronback, R.C.: Eclipse Modeling Project: A Domain-Specific Language Toolkit. Addison-Wesley, Reading (2009). 706 p.
7. Hastie, S., Wojewoda, S.: Standish group 2015 chaos report — Q&A with Jennifer Lynch, 4 October 2015. https://www.infoq.com/articles/standish-chaos-2015. Accessed 28 Apr 2017
8. Belikov, A., Rogozov, Y., Sviridov, A.: The approach to the information systems design based on the properties of the domain. In: International Multidisciplinary Scientific GeoConference Surveying Geology and Mining Ecology Management, SGEM, pp. 361–366 (2015)
9. Kucherov, S., Rogozov, Y., Sviridov, A.: The model of subject-oriented storage of concepts sense for configurable information systems. In: Proceedings of the First International Scientific Conference "Intelligent Information Technologies for Industry" (IITI 2016), Volume 1 in Advances in Intelligent Systems and Computing, vol. 450, pp. 317–327 (2016)
10. Rogozov, Y., Sviridov, A., Belikov, A.: Approach to CASE-tool building for configurable information system development. WIT Trans. Inf. Commun. Technol. **58**(1), 173–179 (2014)
11. Belousova, S., Rogozov, Y., Sviridov, A.: Technology of using properties and mechanisms of actions in user interface design. In: 15th International Multidisciplinary Scientific GeoConference Surveying Geology and Mining Ecology Management, SGEM 2015, Albena, Bulgaria, vol. 1, issue 2, pp. 339–345 (2015)

Towards Requirements Engineering Process for Self-adaptive Embedded Systems

Zina Mecibah[(✉)] and Fateh Boutekkouk

Research Laboratory on Computer Science's Complex Systems ReLa(CS)2,
University of Oum El Bouaghi, Oum El Bouaghi, Algeria
Mecibah.zina@hotmail.fr, Fateh_boutekkouk@yahoo.fr

Abstract. At present, there are a large number of embedded systems (ES) which need to modify their behavior at run time in response to changing environmental conditions (sensor failures, noisy networks, malicious threats, unexpected input…etc.) or in the cases where the requirements themselves needs to change. i.e. ES require self-adaptive capabilities. This kind of ES has been widely used in different domain, for instance in the smart home systems, automotive systems, telecommunication systems, environmental monitoring and others. Unfortunately, up to now, there are few researchers which interest for the high level design process of the self-adaptive embedded systems (SAES) specifically in the field of requirement engineering (RE). For this reason, the objectives of this paper is to try as much as possible to recall and compare between existing works build around the RE of SAES in the sake of identifying advantages and weak points of each work.

Keywords: Embedded systems · Self adaptive · Requirements engineering
Adaptive requirement

1 Introduction

In contrast to traditional Embedded Systems (ESs), nowadays ESs are becoming more complex, more autonomous, and more self adaptable (submerged in a dynamical environment which is incompletely specified i.e. the necessary algorithm for solving the problem does not exist). Consequently, a new class of ES called Self Adaptive Embedded Systems (SAES) is occurred.

Self-adaptation is a new branch of system engineering and does not yet have standards, and there is neither common definition nor vocabulary. In this paper, we adopt the definition proposed by Cheng et al. [17]: self-adaptive systems can configure and reconfigure themselves, augment their functionality, continually optimize themselves, protect themselves, and recover themselves, while keeping most of their complexity hidden from the user and administrator.

In fact, in the ESs domain, more than 50% of the problems occur when the system is delivered [7]. Noted that, the origin of these problems related to the misconceptions in capturing requirements and not related to the correctness of implementation (conceptual requirements errors). Hence, the process of RE is decisive for the QoS of the system.

© Springer International Publishing AG, part of Springer Nature 2019
R. Silhavy (Ed.): CSOC 2018, AISC 763, pp. 338–345, 2019.
https://doi.org/10.1007/978-3-319-91186-1_35

Up to now, there are few researchers which interest for the high level design process of the SAES specifically in the field of requirement engineering (RE).

The process used for RE of ES vary widely depending on the application domain, the people involved and the organization developing the requirements. Nevertheless, there are five activities common to all processes which are: elicitation or requirements discovery, analysis, specification, V&V (Verification and Validation) and management of requirements.

According to literature, there is little research prepared around Requirements engineering of Embedded Systems. The following table presents a summary about existing works which are prepared around RE of ES (Fig. 1).

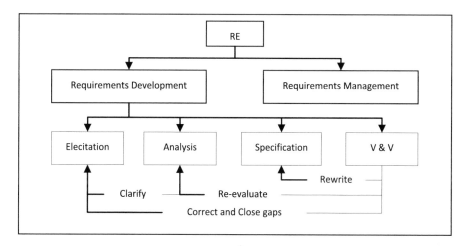

Fig. 1. RE process

1 refers to Elecitation, 2 refers to Analysis, 3 refers to Specification and 4 refers to Validation & Verification (Table 1).

Table 1. Summary about existing works which are prepared around RE of ES.

	Requirements engineering process					Supported by tool
	Requirements development				Requirements management	
	1	2	3	4		
TERASE [12]			X			Not exist
CAMA [13]			X			Not exist
GERSE [14]	X					Zaki tool
REARES [15]	X	X	X	X	X	Not exist
REPES [7]	X	X	X	X	X	Not exist

- TERASE: Portuguese acronym to "Template para Especificação de Requisitos de Ambiente em Sistemas Embarcados".
- CAMA: CAN, Martins and Almudi.

- GERSE: Portuguese acronym to "Guia de Elicitação de Requisitos para Sistemas Embarcados".
- REARES: Requirement Engineering Approach for Real-time and Embedded Systems.
- REPES: Requirements Engineering Process for Embedded Systems.

According to the table, just two researches cover all phases of the RE of ESs:

- REARES: which is proposed in 2014 by A.Abd Rahman in the sake of determines the RE approach for product based real time and embedded software projects.
- REPES: which is proposed by Pereira et al. [7] in 2016 so this proposal is still new and just now REPES not evaluated in any real industrial settings.

On the other hands, until now there are few researchers which interest for the RE of SAES. For this reason, throughout this paper we try to compare between four approaches which are: FLAGS, Adaptive RML, PERSA, and RELAX/SYSML/KAOS approach.

The rest of the paper is structured as follows: section two presents a general description of FLAGS, Adaptive RML, PERSA and RELAX/SYSML/KAOS approach. Section three presents a synthesis, and in the last section we present some conclusions and future work.

2   General Description of FLAGS, Adaptive RML, PERSA and Relax/SysML/Kaos Approach

2.1   FLAGS

In 2010, Baresi et al. propose FLAGS [1] which is an acronym of Fuzzy Live Adaptive Goals for Self-Adaptive Systems. FLAGS is a goal model that generalize the KAOS model i.e. adds the concept of adaptive goal. The main aim of FLAGS is to specifying the requirements and adaptation capabilities of self adaptive systems. FLAGS distinguish between two types of requirements which are: crisp goals and fuzzy goals. It should be noted that FLAGS divides the requirements model in different views: domain, goals, operations and adaptation [11].

In 2011, Baresi and Pasuale propose the FLAGS Designer [16] which is a graphical tool to help designers to elicit and represent the requirements of self adaptive systems. FLAGS Designer is an eclipse plug-in distributed under GNU GPL 3 license [11]. Unfortunately, this tool is still in its first release and needs further improvements.

2.2   Adaptive RML

Adaptive RML [6] is a requirements modeling language mainly used for the representation of early requirements of SAS.

In the sake of represent the runtime requirements adaptation, Adaptive RML proposes two new concepts (context and resource) and two relationships (relegation and influence).

Figure 2 presents an overview about concepts and relations used in Adaptive RML.

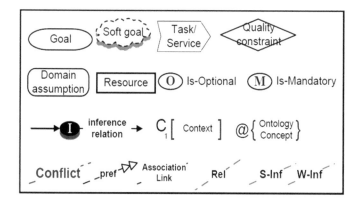

Fig. 2. Adaptive RML elements (adopted from [6])

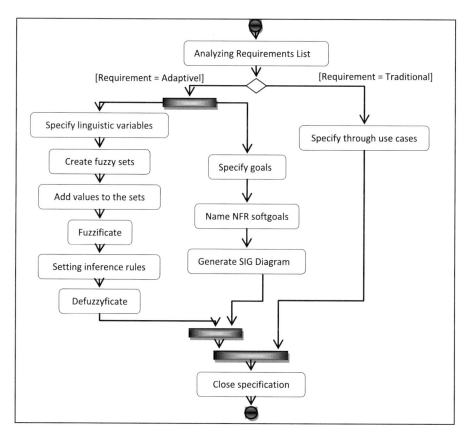

Fig. 3. Diagram of PERSA process activities [10]

2.3   PERSA

PERSA or Processo de Especificação de Requisitos Adaptativos, in Portuguese [9, 10], is an innovative approach to specify the requirements for self-adaptive systems based on the concepts of Fuzzy Logic and NFR Framework (Non-Functional Requirements Framework). So, the PERSA process being its life cycle after requirements elicitation.

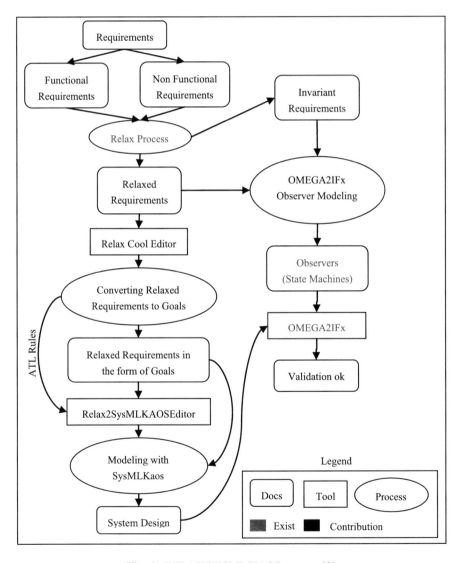

Fig. 4. RELAX/SYSML/KAOS process [2]

There are mainly three phases in PERSA subdivided into several steps.

- 1st phase: analysis of the requirements list.
- 2nd phase: Fuzzy modeling.
- 3rd phase: NFR Modeling.

2.4   Relax/SysML/Kaos

Ahmed et al. [2] propose an approach for modeling and verifying the requirements (functional and non functional requirements) of ambient SAS using Relax language [3], SysML [4] and KAOS [5].

RELAX is a requirement specification language which is proposed by Whittle et al. [3] in 2010. RELAX takes the form of a structured natural language and including operators designed in the sake of capture uncertainty. Hence, RELAX provide three types of operators to handle uncertainty: temporal, ordinal and modal operator.

The approach is supported by three tools: RELAX Cool editor, RELAX2SysML/ Kaos editor and OMEGA2IFx but these tools are not available. Figure 4 presents the overall view of this approach.

3   Synthesis

Embedded Systems has features and requirements that distinguish them particularly from conventional systems. So, the design and development of ES is different from that of software application. Indeed, ES are subject to strict technical constraints of both functional and non-functional requirements. The second type is an attribute/constraint on a system and in fact receives less attention in the design of many types of systems but in the ES they play a very important role because the design of these systems is a complex job (their resource is limited) which incurs handling a big range of Non-Functional Requirements (NFRs) such as Reliability, determinism, security, timing etc. Therefore, satisfaction of NFRs plays an important role in the correctness of the design of these systems (system design which cannot satisfy its NFRs can mean failure of the end product). So, these constraints must be taken into account at the early stages of the development cycle. In this paper we tried to compare between four approach used in the requirements engineering phase of self-adaptive systems. According to our first comparison and under the absence of some information regarding some RE approach for SAS, we can observe that:

- The use of FLAGS approach in the domain of embedded systems is very difficult because it is based on KAOS and KAOS doesn't take into account collaboration between Functional requirements and Non Functional Requirements (Functional Requirements taken into account at requirements engineering level but NFR are taken into account at the architectural level).
- Adaptive RML is a visual language for the modeling of early requirements for self-adaptive systems. Until now, this language is not supported by any tool which makes the use of this language very difficult.

- PERSA approach:
  - Used only for the specification of self-adaptive systems.
  - Not supported by any tool.
  - According to Fig. 3, PERSA not differentiate between functional and non functional requirements
  - PERSA separate between conventional requirements and adaptive requirements. Conventional requirements specified with use case??? But use case used only for functional requirements??? + PERSA process handl only the adaptive requirements???
- RELAX/SYSML/KAOS may be a good candidate for the RE of SAES, but unfortunately, it is not support the applicability of models to the runtime environment. Also, until now RELAX/SYSML/KAOS not used in the RE phase of the real world system.

4   Conclusion and Future Work

This work encompasses selected research contributions in the field of RE of self-adaptive embedded systems. Hence, this work represents a first step towards complete requirements engineering process for self adaptive embedded systems.

At present, we are trying to propose a new process (REP4SAES: Requirements Engineering Process for Self Adaptive Embedded Systems) which combine between SysML, kaos, NFR framework and fuzzy logic in the sake of provide a richer space of requirements engineering of SAES. Also, the adoption of any requirements engineering process can be facilitated if the process supported by the tool. In this sense, a tool called RE4SAES-Tool based on RE-Tool [11] is currently under development.

References

1. Baresi, L., Pasquale, L., Spoletini, P.: Fuzzy goals for requirements-driven adaptation. In: 18th IEEE International Requirements Engineering Conference, pp. 125 – 134 (2010)
2. Ahmad, M., Bruel, J.M., Belloir, N.: Modeling and verification of functional and non functional requirements of ambient, Self-Adaptive Systems. University Toulouse le Mirail - Toulouse II (2013)
3. Whittle, J., Sawyer, P., Bencomo, N., Cheng, B.H.C., Bruel, J.-M.: RELAX: a language to address uncertainty in self-adaptive systems requirement. Requir. Eng. **15**, 177–196 (2010). RE'09 Special Issue; Guest Editor: Kevin T Ryan
4. Cardenas, C.E.G.: Modeling Embedded Systems Using SysML. Universidad de Los Andes, Bogotá (2009)
5. Dias, A., Amaral, V., Araujo, J.: Towards a Domain Specific Language for a Goal-Oriented Approach based on KAOS (2009)
6. Qureshi, N.A., Jureta, I., Perini, A.: Adaptive RML: A Requirements Modeling Language for Self-Adaptive Systems. Technical report (2011)
7. Pereira, T., Albuquerque, D., Sousa, A., Alencar, F., Castro, J.: Towards a metamodel for a requirements engineering process of embedded systems. In: Computing Systems Engineering (SBESC) (2016)

8. Supakkul, S., Chung, L.: The RE-Tools: a multi-notational requirements modeling toolkit. In: Proceedings of 20th IEEE International Requirements Engineering Conference (RE), pp. 333–334 (2012)

9. Paraiba, J.D., Martins, L.E.G.: A proposal of requirements specification process for adaptive systems based on fuzzy logic and NFR-framework. In: The Eighth International Conference on Software Engineering Advances, ICSEA 2013, pp. 100–105 (2013)

10. Paraiba, J.D., Martins, L.E.G.: PERSA: a requirements specification process for self-adaptive systems based on fuzzy logic and NFR-framework. Int. J. Uncertain. Fuzziness Knowl.-Based Syst. **25**, 145–178 (2017)

11. Baresi, L., Pasquale, L.: An eclipse plug-into model system requirements and adaptation capabilities. In: 6th Italian Workshop of the Italian Eclipse Community, Milano, Italy (2011)

12. Martins, L.E.G., Souza Jr., R., Oliveira Jr., H.P., Peixoto, C.S.A.: TERASE: template para especificação de requisitos de ambiente em sistemas embarcados. In: 13th Workshop on Requirements Engineering (WER), pp. 50–61 (2010)

13. Almudi Neto, D., Martins, L.E.G.: A requirements specification template of a communication network based on CAN protocol to automotive embedded systems. J. Comput. Sci. Technol. **10**, 143–149 (2010)

14. Ossada, J.C., Martins, L.E.G., Belgamo, A., Ranieri, B.S.: GERSE: guia de elicitação de requisitos para sistemas embarcados. In: 15th Workshop on Requirements Engineering (WER), pp. 57–70 (2012)

15. Rahman, A.A.: Requirements engineering approach for real-time and embedded systems: a case study of android-based smart phone devices. In: Proceedings of the 8th International Conference on Ubiquitous Information Management and Communication, Siem Reap, Cambodia (2014)

16. Modeling the impact of Non-functional Requirements on Functional Requirements

17. Software engineering for self-adaptive systems: a research road map

An Approach to Develop Software that Uses Blockchain

Samantha Almeida[1], Adriano Albuquerque[1(✉)], and Andreia Silva[2]

[1] University of Fortaleza, Av. Washinton Soares, 1321, BL J, SL 30, Fortaleza,
Ceara 60833-155, Brazil
samantha.kelly2@gmail.com, adrianoba@unifor.br
[2] Federal Institute of Education, Science and Technology of Ceara, Limoeiro do Norte, Brazil
andreia.rodrigues@ifce.edu.br

Abstract. Nowadays Blockchain technology has a great market visibility. The popularization of this concept can be attributed to the exponential success of the Bitcoin cryptocurrency, launched in 2008 by Satoshi Nakamoto. The disruptive vision of this technology and the diverse possibilities of application in different businesses have been generating a series of changes and investments in the market, especially in the financial sector. The emergence of several Fintechs (Startups from the financial sector) that guide their software to Blockchain demonstrates the investments made to advance this technology. In this work, a bibliographical research was carried out in order to understand the state of the art of Blockchain and an experience of use was presented, where an application was developed using this technology, being source of information for the detailed definition of a software development process, based on Lean Startup, to support startups working with projects involving Blockchain.

Keywords: Software development · Process · Blockchain · Startup

1 Introduction

In recent years, much attention has been paid to the emerging concepts of Block-chain and Smart Contracts, especially after the popularization of virtual currency Bit-coin, launched in 2008 by Nakamoto [1].

With the realization of the Bitcoin virtual currency and the disruptive concepts that this action brought to the financial market, organizations such as banking institutions, financial institutions and public regulators began to discuss explicitly about the importance of Blockchain and Smart Contracts. Some observers say that it is anew era, declaring that "[…] one should think of Blockchain as another class of thing like the Internet…" [2].

This technology can also be used to create and support Smart Contracts: sets of rules defined and based on a block database that are executed only when specific actions occur. Eris Industries, a software company that has created one of the first block-based platforms for this application, describes Smart Contracts as modular components, similar to applications in a financial network, that can be combined to provide verifiability to

R. Silhavy (Ed.): CSOC 2018, AISC 763, pp. 346–355, 2019.
https://doi.org/10.1007/978-3-319-91186-1_36

any type of transaction. According to some professionals from Eris Industries, the use can be "[…] as simple as voting a post in a forum, for the more complex, and a loan guarantee and futures contracts, for the highly complex, such as the prioritization of repayment in a structured note" [3].

In this work, we defined a software development process to support startups in the development of software products oriented to Blockchain, helping to disseminate the technology and to clarify the gaps arising from the growing technological advance.

This article consists of five sections: the following presents the main concepts of Blockchain Infrastructure. Section 3 presents the Experience of Use. Section 4 describes the proposed process and finally, Sect. 5 presents the conclusions.

2 Background

As the importance of software in our society has grown in recent years, it is also necessary to evolve the duties, skills and knowledge of software engineers, as well as adjustment of software engineering itself to adapt to market evolution. [4] Nowadays there is a new type of software, the Blockchain-oriented software, that is a software that uses the implementation of Blockchain in its components [2].

2.1 Lean Startup

The Lean Startup methodology, created by Eric Ries and derived from Lean Thinking, introduced the idea of rejection of long-term planning and embraces the iterative learning and experimentation. Many studies have been taking place based on this methodology for the entrepreneurship area, mainly in the context of the software products development [5].

According to the founder, Lean startup is a set of processes used by entrepreneurs to develop products and markets, combining agile software development, customer development and existing software platforms [5]. It can be considered as lean thinking focused on entrepreneurship, as defined by the author. The practices of the Lean Startups model have become widely adopted by Startups in Brazil [6].

2.2 Blockchain Infrastructure: Blockchain, Smart Contracts and Startups

Proposed by Satoshi Nakamoto as a solution to the double-spending problem when using the digital currency Bitcoin, Blockchain is a decentralized ledger system that uses a network consensus to write and execute transactions. It is known as the platform for the Bitcoin currency, which is currently revolutionizing financial services. However, the Blockchain has recently drew attention of a growing number of business leaders, recognizing that the underlying technology of this architecture can be applied to almost any industry [7].

Blockchain technology can be understood in a variety of ways. In general, we can say that it is a distributed system of log database, maintained and managed in a distributed and shared way (through a peer-to-peer network), in which all participants are responsible for storing and maintaining the database, and they are responsible for all the information accepted as true that generates a new block chain [8].

A block database consists of two record types: transactions and blocks. Blocks have lots of valid transactions that are hashed and encoded in a Merkle Tree. Each block includes the hash of the previous block in the Blockchain, connecting the two. The linked blocks form a chain. This iterative process confirms the integrity of the previous block, all the way back to the original block. Some block chains create a new block every five seconds.

Figure 1 illustrates the standard operation of a transaction occurring in a Block-chain Network [9].

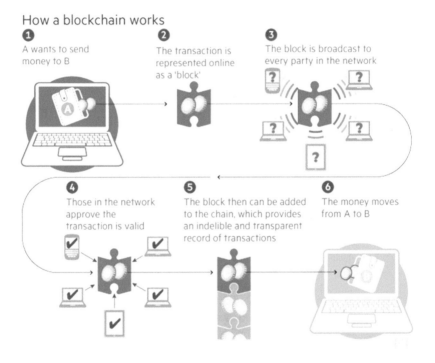

Fig. 1. Operation of a Blockchain transaction

The technology was built on four main architectural features: (i) security of operations, (ii) decentralization of storing/computing, (iii) data integrity and (iv) transaction immutability" [8].

The most distinctive feature of Blockchain is that no single agent has the ability to perform control over the system activity. The stored information is both transparent and private, which makes it a perfect architecture for IoT. All attributes of a particular person can be known without their identity being known. This is possible because the Blockchain uses avatars as a information of the processing mechanism to preserve individual privacy [7].

Any attempted to fraud the existing ledger would be easily detected by all users of the Bitcoin criptocurrency. To achieve this robustness to modifications, the elements of the ledger are linked together, forming a chain. [10] As Blockchain technology uses public-private key technology (PKI Cryptography), it is reasonably secure from fraud, theft, and corruption, but the anonymity is never a hundred percent guaranteed [11].

The Smart Contracts property and characteristics of any asset or data can be digitalized and represented by a computer code. Intangible assets, such as: rights, personal data, certificates, wills, trade balances can be stored and protected in a chain of blocks. Not only that, but relationships involving these assets can be programmed by computer, and their execution is applied through the nodes of a block without intermediaries. These operations are generally referred to as "smart contracts", that is, computer protocols that formalize the elements of a contractual relationship, being able to automatically execute the coded terms, once the predefined conditions are met. [9] The implementation of the contract should not require a direct human involvement from the moment the contract was signed [12].

Figure 2 illustrates the operation of a Smart Contract, highlighting the agreement between the parties.

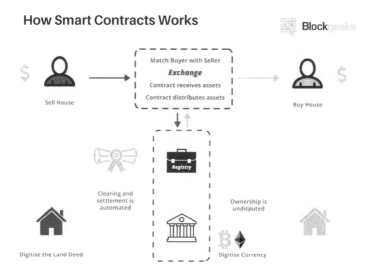

Fig. 2. Operation of a smart contract

Software startups face intense market time pressure and are exposed to "Hackatons", operating in a chaotic, rapidly evolving and uncertain context. [5] Sutton [14] presented some challenges to Software Startups: (i) Little or no operational history; (ii) Little experience in processes and organization of development; (iii) Limited resources; (iv) Different kind of influences; and (v) Dynamic technologies and markets. Accordingly to Giardino [13], the implementation of methodologies for structuring and controlling Startup development activities is a major challenge for software engineers.

3 Experience of Use: The Development of a Blockchain-Oriented Software

The main motivation for this experience was to identify the main challenges to develop applications using Blockchain Technology and to propose a process that aims to facilitate this task.

For this work, we select the context that involves the registration of the candidate and registration of the vote by the voter. With this, we aim to demonstrate the proposal of secure and transparent voting, where it is possible to check the candidate's registration and vote in a public and unchanging way in the Blockchain. It is important to highlight that it occurs without exposing the identity of the voter.

For the construction of the prototype, the following steps were followed: (i) Know the theoretical basis; (ii) Define the scope of application; (iii) Identify the personas and use scenarios; (iv) Define requirements; (v) Define the architecture; (vi) Configure the environment; (vii) Prototype coding and testing; (viii) Collect the difficulties and specificities of the process; (ix) Register the documentation of the results obtained through the construction of the process.

Nowadays, there are several active Blockchain networks in market. In addition to the Bitcoin network, there are: Ethereum, HyperLedger, Ripple etc., and all of them work in parallel. For this work we selected the Ethereum network, as a result of a bibliographical research. We found that it is a platform adequate for execution of smart contracts in the context of Blockchain. It has a decentralized virtual machine called Turing Completeness, a Ethereum Virtual Machine (EVM), which can run scripts using an international network of public nodes. In this way, it is possible to develop applications that work exactly as scheduled without any possibility of censorship, fraud or interference from third parties, because the contract is immutable [15].

Figure 3 presents the components used to construct the application.

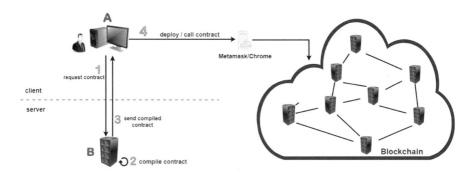

Fig. 3. Proposed architecture

In the decentralized architecture of the application, four main elements can be highlighted: the client application, the server application, the Metamask plugin on the client and the Blockchain. The client application, represented by the letter A in Fig. 3, consists of a web-application that, like any other, is executed in the user's browser. This is the application that the end client interacts with, executing requests for the application on the server and obtaining responses from it. The main request of the client application is the solicitation of the contract to be interacted, described in the figure by the arrow number 1.

The server application, represented by the letter B makes the client application available to the end user and is also responsible for compiling the requested contract code - described in the figure by the arrow number 2 - and send it to the client application, represented in the figure by arrow number 3. Receiving the compiled contract, the client application itself is able to deploy the contract in the Blockchain. After the implementation of a contract, it is possible to make calls to the public methods of this contract. The client application, in this case, interacts with the contract already available in the Blockchain without the need of intermediation of the server application, because it is a decentralized application.

However, the client application depends, in the case of this architecture, of the intermediation of the Metamask plugin (available only for the Google Chrome browser) to carry out transactions in the Blockchain, represented in the figure by the arrow number 4. Through this plugin, the user authorizes or not transactions to the Blockchain, and can even manage the account that he wants to use to authorize a particular transaction, without the need to install a Blockchain node locally. If the called method of the contract only returns some information, that is, it does not change the state of the Blockchain, it does not require user authorization, which increase the process performance.

3.1 Results Obtained with the Bibliographic Review and the Experience of Use

With the execution of this experience of use, we could understand some relevant points in the context of software development oriented to Blockchain, such as: (a) Firstly, it is necessary to analyze whether the problem to be solved is adequate to use Blockchain; (b) Blockchain oriented development requires specific and differentiated architecture, comparing with traditional web applications; (c) The configuration of the environment requires the creation of user ac-counts in the Blockchain selected for the product development; (d) The person that specifies the smart contract does not need to have a high degree of knowledge in information technology. A legal know-ledge is relevant in this context and can be a differentiating factor to the completeness and validity of the smart contract; (e) Because many Startups often fail to correctly identify their customers, we believe that the technique of identifying people and scenarios can eliminate this risk; (f) We have identified that for each Blockchain platform, there is usually a specific programming language to coding the smart contract and to interact with the Blockchain. In this study we analyze and use the language Solidity, specific to the Ethereum platform; (g) We also realize that to avoid blocking transactions addressed to the contract in a test or production environment, there must be an indicator that monitors the gas consumption throughout the development cycle, which must be collected and controlled by the project manager; (h) The contract tests need to be as accurate as possible to predict business rule failures that could lead to incorrect transactions at the smart contract execution life cycle. Preferably, these tests must be automated and, whenever there is a change in the contract code, there should be executed regression tests; (i) We noticed the relevance of using a Blockchain platform of tests to avoid wasting gas in non-real transactions. The gas to be used on test platforms can be replaced free of charge in a web address of the Ethereum platform; (j) As defined in the literature, software oriented to Blockchain is 80% business and 20% technology; therefore, we understand that the Lean Startup methodology is the most appropriate for executing projects in this context, because it provides short deliveries oriented to value and workflow redirection based on customer feedback. In this scenario, the smart contract may have a continuous evolution of its clauses aligned to business; (k) Due to the high investment in Startups that develop software products oriented to the Blockchain, we believe it to be a promising business market for these companies.

4   Development Process to Software Oriented to Blockchain

As a main result of the bibliographic review and the experience of use, a software development process oriented to Blockchain was defined, aimed to support Startup companies. It was also generated some support guides and checklists.

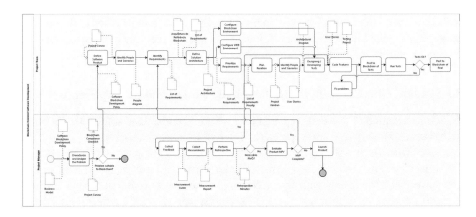

Fig. 4. Graphical representation of the proposed process

The process was mainly based on the concept of product-oriented development and on the methodology proposed by Eric Ries, Lean Startup, conceived to Startup companies. We can observe the use of these concepts in the following activities: (i) definition of the product, and (ii) collecting feedback from the client that generates return to the definition of the product, evidencing the feedback loop proposed in practice. It is also worth noting the analysis of the need for Pivot and completeness of the Minimum Viable Product - MVP of the project. Figure 4 presents the graphical representation of the proposed process.

The following Process Assets were also defined: (a) Business model; (b) Software development policy oriented to Blockchain; (c) Blockchain adequacy checklist; (d) Project Canvas; (e) List of requirements; (f) Blockchain Reference Architecture; (g)

Architecture document; (h) Kanban of the project; (i) User stories; (j) Design of tests; (k) Measurement specification; (l) Product evaluation questionnaire; (m) Retrospective Minutes.

Besides that, to control the proposed process, three metrics were suggested: (a) Work In Progress (WIP); (b) Gas Reserve Leve; and (c) Satisfaction of the Customer with the Product.

The methodology selected for the validation of the process was the peer-review, which consists of the inspection of work products by a pair (collaborator with skills similar to the author of the artifacts). [16] In this work, the peer review occurred in three stages: 1 - Presentation of the process to those involved; 2 - Completion of the questionnaire by the evaluators; and 3 - Analysis of the results by the researchers.

As results, we obtained the answers presented in Table 1.

Table 1. Results of the process evaluation.

Question	Strongly disagree	Disagree	Neither agree, nor disagree	Agree	Strongly agree
The proposed process is easy to understand (usability) for the reader	0%	0%	0%	60%	40%
The activities described in the process are adequate for the development of a software product	0%	0%	0%	50%	50%
The documentation generated in the process is adequate	0%	0%	0%	70%	30%
The process has characteristics of lightness that help the application within a Startup context	0%	0%	0%	50%	50%
The process has specific characteristics that address Blockchain technology	0%	0%	10%	60%	30%

Some suggestions of improvement and strengths were registered by the evaluators. The opportunities of improvements accepted by the researchers were: (i) The identification of the person as should be executed before the definition of the architecture; (ii) The definition of all roles involved on the project's team to improve the measurements.

And the more important strengths were: (i) The process was defined in a general way, being able to be used in different business domain that aim to be supported by Blockchain; (ii) The inclusion of an activity on the beginning of the process to analyze the viability to use Blockchain; (iii) The use of The Build–Measure–Learn loop of the Lean and (iv) The process can address a lot of problems inherent of a software development process, such as: requirements analysis and specification.

5   Final Considerations

With the bibliographic research and the experience of use, we learned that it is challenging to understand the new paradigms that involve the development of software products that use Blockchain, mainly in the context of startups that need to handle with many market constraints and uncertainties.

We realized that some concepts of traditional Software Engineering need to be evolved to meet the demand from this new technology. Professionals with an integrative vision who make the link between technology professionals and the customer need should be hired to improve the definition of the rules of the smart contract. It was also identified the need to evolve architectural proposals that better support the decision making during the software development process. It is also necessary to develop tools that allow greater automation and better support to software testing, dealing with issues related to software development oriented to Blockchain.

As future works we could identify: (i) evaluate the process in Startups with different level of maturity; (ii) construct a guideline to support the adaptation of the process to different contexts; and (iii) develop a tool to support the activities related to software testing in a software development process oriented to Blockchain.

References

1. Nakamoto, S.: Bitcoin: a peer-to-peer electronic cash system (2009). https://bitcoin.org/bitcoin.pdf. Accessed 5 May 2017
2. Porru, S., et al.: Blockchain-oriented software engineering: challenges and new directions. Departamento de Engenharia Elétrica e Eletrônica, Departamento de Informática e Matemática – Universidade de Cagliari, Itália (2017)
3. Plansky, J., et al.: A strategist guide to Blockchain (2016). https://www.strategy-business.com/article/A-Strategists-Guide-to_Blockchain?gko=0d586. Accessed 24 out 2017
4. Kazman, R., Tang, A.: On the worthiness of software engineering re-search (2017). http://shidler.hawaii.edu/sites/shidler.hawaii.edu/files/users/kazman/se_research_worthiness.pdf. Accessed 25 dez 2017
5. Ries, E.: A startup enxuta: como os empreendedores atuais utilizam a inovação contínua para criar empresas extremamente bem-sucedidas. In: RIES, Eric. [tradução Texto Editores]. – Lua de Papel, São Paulo (2012). ISBN 9788581780139
6. Ribeiro, G.: Lean Startup: análise exploratória sobre sua utilização por novas empresas brasileiras. Dissertação de Mestrado, FGV (2014)
7. Collins, R.: A new architecture for digital content (2016). http://www.econtentmag.com/Articles/Editorial/Commentary/Blockchain-A-New-Architecture-for-Digital-Content-114161.htm. Accessed 28 out 2017
8. CPQD. Centro de Pesquisa e Desenvolvimento em Telecomunicações. Tecnologia Blockchain: uma visão geral (2017). http://www.cpqd.com.br. Accessed 05 nov 2017
9. Cuccuru, P.: Beyond Bitcoin: an early overview on smart contracts. Int. J. Law Inf. Technol. 1(17) (2017)
10. Lucena, A.U.: Estudo de arquiteturas dos blockchains de Bitcoin e Ethereum. Departamento de Engenharia de Computação e Automação Industrial (DCA), IX Encontro de alunos e docentes do dca/feec/unicamp (EADCA) (2016). http://www.fee.unicamp.br/sites/default/files/departamentos/dca/eadca/eadcaix/artigos/lucena_henriques.pdf. Accessed 05 nov 2017

11. Murray, A.: All aboard the block chain express. KM World Magazine (2015). http://www.kmworld.com/Articles/Column/The-Future-of-the-Future/All-aboard-the-blockchain-express-102652.aspx. Accessed 06 dez 2017
12. Szabo, N.: Formalizing and securing relationships on public networks. First Monday **2**(9) (1997). http://firstmonday.org/ojs/index.php/fm/article/view/548/469. Accessed 05 nov 2017
13. Giardino, C., et al.: Software development in startup companies: a systematic mapping study. Inf. Softw. Technol. **56**(10), 1200–1218 (2014). http://www.sciencedirect.com/science/article/pii/S0950584914000950?via%3Dihub. Accessed 6 dez 2017
14. Sutton, G.: The 7 most common legal mistakes startups make (2017). https://www.linkedin.com/pulse/7-most-common-legal-mistakesstartups-make-garrett-sutton-1/. Accessed 15 set 2017
15. Buterin, V.: Ethereum: a next-generation cryptocurrency and decentralized application platform. Bitcoin Magazine (2014). https://bitcoinmagazine.com/articles/ethereum-next-generation-cryptocurrency-decentralized-application-platform-1390528211/. Accessed 05 dez 2017
16. Beizer, B.: Software system testing and quality assurance. Van Nostrand Reinhold Co., New York (2014)

Metrics in Software Development and Evolution with Design Patterns

Anna Derezińska$^{(\boxtimes)}$ [ID]

Institute of Computer Science, Warsaw University of Technology,
Nowowiejska 15/19, 00-665 Warsaw, Poland
`A.Derezinska@ii.pw.edu.pl`

Abstract. Software metrics are employed in software development and maintenance to assess different quality attributes, support processes of software design, testing, and reengineering. We overview software metrics used with regard to design patterns, especially these dealing with object-oriented program features. Metrics of this kind have also supported certain approaches to refactoring to design patterns. However, code refactoring to design patterns requires selection of suitable code parts and taking decisions about pattern application. In order to make this process partially or fully automated, specialized relevance metrics have been proposed. They were designed as a structure-based assessment of an adequacy of code to be transformed into a given design pattern. Relevance metrics for refactoring to selected design patterns (Replace Type Code with Class, Replace State-Altering Conditions with State) are presented in the paper. The metrics were tested in a prototype tool for automated refactoring of Java programs to design patterns, giving satisfactory results.

Keywords: Design patterns · Code refactoring · Software metrics
Object-oriented metrics

1 Introduction

Design patterns tend to encode good design practices and experience of developers. In this paper we focus on "classical" design patterns (DP in short) applied in object-oriented software [1]. DP propose solutions to common design problems. An idea is described by a structural design, in which individual components, i.e. classes, objects, relations, operations, are associated with specific roles to be taken in a pattern.

Software metrics have a wide-ranging application in software development and evaluation [2–5]. They can be used for assessment of impact of design patterns on code quality, as well as in code transformation to design patterns [6].

While design patterns, like other architectural solutions, could be simpler applied in a newly developed software than in a reengineered one, the latter case is common in practice. A need for usage of DP may emerge due to maintenance of a long living software and extension of its functionality. Therefore, DP are introduced into extended code or, in particular, a part of code is refactored to a selected design pattern [7].

Refactoring is a kind of a reengineering transformation dealing with systematic changes in object-oriented models or programs [8]. It is aimed at enhancement of

© Springer International Publishing AG, part of Springer Nature 2019
R. Silhavy (Ed.): CSOC 2018, AISC 763, pp. 356–366, 2019.
https://doi.org/10.1007/978-3-319-91186-1_37

software maintainability, while external behavior of a program remains unaffected. Refactoring can be applied during a new software development, as it is promoted in the TDD process. It can also be used during further evaluation of a program, including reengineering of legacy systems [9]. Refactoring is often inspired by software quality monitored by a developer, evaluated by software metrics, and suggested by specialized code analyzers [10–12].

Recommendation of standard refactoring [8] is primarily based on recognition of code smells, i.e. catalogued code shortcomings. General software metrics can also be used in software evaluation that supports introduction of DP into existing code. However, general metrics cannot accurately assess capability of a code to be refactored into a specified design pattern [13]. Therefore we have recommended design of relevance metrics that support developers in DP determination and code transformation.

In this paper, we have surveyed different software metrics used in code evaluation of applications with design patterns. Moreover, in order to automate code refactoring to design patterns, support for decisions on pattern introduction has been proposed. The approach is based on metrics that assess relevance of a code extract to given design patterns. Detailed metrics for selected design patterns have been presented. The solution was implemented in a prototype tool for refactoring Java programs to design patterns and tested in case studies [14].

Next section gives an overview of metrics used in evaluation of code with design patterns. A metric-based approach to code refactoring to design patterns as well as detailed metrics of selected design patterns are presented in Sect. 3. We conclude the paper in Sect. 4.

2   General Software Metrics and Related Work

Different software metrics can be employed in various phases of software lifecycle [2–5, 9, 15]. Program analysis is often based on sets of metrics dealing with object-oriented program features, such as MOOD [3], Chidamber and Kemerer [4], or QMOOD [5]. Therefore, these sets are also applied in context of design pattern utilization. The same metrics could have been components of different models, or metrics of the same names were associated with a different meaning. Therefore, primary software metrics applied in solutions concerning design patterns are summarized in Table 1.

Different metrics are used in assessment of impact of design pattern on selected quality attributes [16–19], e.g., software maintainability, flexibility to change introduction, performance, fault susceptibility, etc. One of quality models of object-oriented software is QMOOD [5]. Different object-oriented features are associated to quantitative measures, such as SIZE, NOC, DIT, DAM, CBO, CAM, MOA, MFA, NOP, RFC, WMPC (Table 1). QMOOD has been applied in research aimed at quality improvement of projects that use DP [17]. Directions of changes of object-oriented design for selected DP were determined, e.g., *Observer* lowers project coupling, *Strategy* tends to increase polymorphic features, whereas *Iterator* rises encapsulation.

QMOOD metrics have also been used for assessment of quality parameters, addressing an issue "to which extend DP help in maintaining of non-functional

Table 1. Software metrics used to assess quality of software with design patterns.

	Metric	Meaning (variant)
DIT	Depth of inheritance tree	Depth of inheritance Tree
NOC	Number of children	Number of direct descendants of a base class
NOC*	Number of classes	Number of all classes in a system/module
AC	Attribute complexity	Sum of complexity of class attributes. (Any attribute type has a specific complexity.)
COF	Coupling factor	Counts numbers of classes that a class is directly related to
CF	Coupling factor	A fraction, where the numerator represents the number of non-inheritance couplings and the denominator is the maximum possible number of couplings in a system
MPC	Message-passing couple	Call number of methods in a class
CBO	Coupling between object classes	Number of classes coupled with a class (method calls, attribute accesses, method arguments, inheritance, method arguments, exceptions)
CAM	Cohesion among method of a class	Ratio of a sum of different types in parameters of a method to a product of types in parameters of all methods of the class he method number
DAM	Data access metric	A ratio of a number of private and protected attributes to a number of all attributes of a class
MOA	Message of aggregation	Number of attributes that are instances of classes defined in a program
MFA	Measure of functional abstraction	A ratio of a number of inherited methods to a number of all methods of a class
NOP	Number of polymorphic methods	Number of polymorphic methods
RFC	Response for a class	A number of local methods and methods called by local methods of a class
LCOM	Lack of cohesion of methods	Cohesion of class is calculated for a method pair in dependence on an access to common attributes
DAC	Data abstraction coupling	A number of class attributes that have a type equal to another class
WMPC	Weighted method per class	A sum of local methods of a class that are weighted by their complexity, which can be e.q. number of method parameters, cyclomatic complexity of a method body, etc.
LOC1/SIZE	Lines of code	Number of lines in a source code file
LOC1	Lines of code without comments	Number of lines in a source file except white characters, comments, and annotations

(*continued*)

Table 1. (*continued*)

	Metric	Meaning (variant)
SIZE1	Number of statements	Number of statements calculated as a number characters ";" that end a statement
SIZE2	Number of properties	Number of fields and local methods of a class
NF	Number of files	Number of files used by a program
NM	Number of methods	Number of methods in a system/module
DS	Declarative statements	Number of statements in a source code (static measure)
ES	Executable statements	Number of executed statements (dynamic measure)

requirements". For example, *Strategy* supports a functional requirement of a kind "An object represents an algorithm that solves a specific problem", and a non-functional requirement "Easy substitution of one algorithm with another".

In [19], benefits of DP usage have been assessed with a set of metrics: LOC, NOC*, AC, WMPC, WMPC*, CF, LCOM (Table 1). It showed, that in general, usage of DP resulted in more comprehensible and easier maintainable code, although its size calculated in LOC and number of classes was higher.

General metrics are used in evaluation of software with and without design patterns in order to assess its quality. This approach is applied both in development of new software with design patterns, as well as in software refactoring to design patterns.

In some automated search-based approaches [20–23], common software metrics are used to asses a software quality, and serve as a base for a fitness function to decide about a refactoring to be performed.

In [23] software metrics were calculated on pairs of programs with and without a design pattern. An impact of a particular metric to a program quality was determined with a genetic algorithm and expressed as a weight coefficient. Comparison of weighted metric outcomes suggested whether a program should be refactored and which of design patterns should be applied.

An advantage of general metrics is an easy way to calculate and taking into account different program features. However, it is questionable weather changes in general metrics can sufficiently determine application of selected DP, and reengineering of an existing code into a selected pattern. Therefore, refactoring based on code analysis directed to specified design patterns have been promoted [10–12, 24]. Though, they still lack on an automated recommendations to code-to-pattern suitability.

3 Refactoring Towards Design Patterns

In order to automate introducing DP into existing code a refactoring process has been proposed [14]. It is based on relevance metrics that support recommendation of patterns. The process consists of three main phases:

1. code analysis and calculation of relevance metrics,
2. determination of a refactoring range,
3. proper refactoring - i.e. realization of an adequate code transformation.

A process of an automated transformation of an existing code to design patterns faces challenges not only of introducing DP but also of deciding about their applicability and a range of modifications to be completed. In order to support decisions we need to express a code suitability in a quantitative way. It can support developer decisions and establish a base of an automated refactoring.

Detailed processes with their relevance metrics have been specified for the following refactorings: Replace Type Code with Class (RTCC), Replace State-Altering Conditions with State (RSACS), Replace Conditional Logic with Strategy (RCLS) and Replace Constructors with Creation Methods (RCCM). The metrics were implemented in a prototype tool that supports refactoring to design patterns [14]. The tool extends the Eclipse environment and transforms Java programs.

3.1   Relevance Metrics to Support Recommendation of Patterns

There are various factors that influence decisions in code refactoring and selection of elements to be transformed into corresponding DP structures. Therefore, they should be taken into account in relevance metrics. An impact of selected factors can be express quantitatively in different ways.

If many factors contribute partially to a metric, we can assess their participation with a set of coefficients λ_i which sum should evaluate to 100%. A result value is calculated as a sum of components weighed by corresponding coefficients.

In another situation, many similar factors can influence an overall result only up to a certain number of the factors. Existence of more factors of the same type is important in code transformation but do not change a decision about a pattern selection. For this reason, a truncation operation can be applied. It assures that the calculation result does not exceed a certain limit, e.g. 100%.

$$\text{trunc}(x, \text{limit}) = \begin{cases} x & \text{if } x \leq \text{limit} \\ \text{limit} & \text{if } x > \text{limit} \end{cases} \tag{1}$$

A given value x usually originates from a sum of different factors. The truncation function seems to be suitable in metrics in which dependence of some factors is not linear and tends to saturation.

Different kinds of factors and their coefficients can be defined by standard values before the refactoring process, or could be adjusted by a user if demanded.

Preconditions of refactoring and appropriate metrics for selected design patterns are presented in the following subsections. Relevance metrics of two refactoring transformations (RTCC, RSACS) are discussed. The following notation is used:

X - a class currently considered as a main class of a refactoring,
f_X - a field in the class X that meets precondition of a refactoring,
$S(f_X)$ - a set of static values that can be assigned to the field f_X,
$Card(S(f_X))$ – number of elements in $S(f_X)$
k – a refactoring parameter, e.g., $k = 3$.

3.2 Relevance Metric for "Replace Type Code with Class" (RTCC)

During this refactoring, the *Type* pattern is introduced. It is useful when a variable takes selected static values. After refactoring, the variable is represented by an object of a separate class. All values that the variable can hold are defined in this class. Instances of the class are responsible for representing the values.

We can use the *Type* pattern in order to gather in one place all values characteristic for a given field (variable). Therefore, they are better manageable and a set of such values can be easily maintained. While modifying a value we only need to change a field value in a class, without looking for all occurrences of the value in code. Moreover, the refactored code appears also to be easier comprehensible. This refactoring is often performed as a preparation before other transformations, e.g. Replace State-Altering Conditions with State (RSACS).

Precondition. There exists a field f_X such as:

(a) f_X can take at least k static values from a set $S(f_X)$,
(b) values $S(f_X)$ are of a standard primitive type (e.g. in Java int, char, string, etc.),
(c) the field can take only those values, or values of corresponding fields of other instances of the class X.

In a class, there can be more than one field that satisfies the precondition, although it is a quite unusual situation.

Metric. A relevance metric is calculated for each field f_X that meets the precondition of the refactoring. Two main factors influence a possible decision about this refactoring. The first one takes into account a number of different values that could be assigned to f_X. The more such values the more reasonable is introduction of the refactoring. The second factor correlates the metric with the number of values with which field f_X is compared. Therefore, the metric for refactoring Replace Type Code with Class (RTCC) is calculated as a sum of two components (Eq. 2). Each component reflects a value of one factor multiplied by its weight coefficient. The whole sum of components is truncated using a limit of 100%.

$$RTCC\_Metric(f_X) = trunc((\theta(f_X) * \rho + \Gamma(f_X) * \gamma), 100\%) \qquad (2)$$

Where:

$\theta(f_X)$ is the number of values that could be assigned to f_X, $\theta(f_X) \leq Card(S(f_X))$,
ρ is a weight coefficient of the first factor,
$\Gamma(f_X)$ is the number of values that could be compared with f_X, $\Gamma(f_X) \leq Card(S(f_X))$,
γ is a weight coefficient of the second factor.
According to this formula, if a certain value is both assigned and compared it is counted to the RTCC metric as a sum of coefficients $(\rho + \gamma)$.

Metric Example. In case studies the weight coefficients were selected to take values equal to 9% and 6%, for ρ and γ accordingly. The approximation of the weight coefficients was based on tuning of several preliminary examples. The coefficient values could be interpreted as in the following calculation. RTCC refactoring is suggested to be 100% worthwhile if more than six values are assigned and compared (i.e. 7 * (9 + 6) = 105 > 100), or eight values are assigned and five compared (i.e. 8 * 9 + 5 * 6 = 102 > 100), etc.

Let us consider a *SystemPermission* class that includes a field *permission* of string type. The field meets the precondition of the refactoring. It can take six different static values: REQUESTED, UNIX_REQUESTED, CLAIMED, UNIX_CLAIMED, DENIED, GRANTED. It is also compared to the first four values. Therefore the RTCC_Metric of the field is equal to (6 * 9% + 4 * 6%) = 78%. There are no more such fields in the class. The relevance metric calculated for the *permission* field is high enough that the RTCC refactoring could be recommended to a developer. The refactoring would also be performed in case of an automated decision.

3.3 Relevance Metric for "Replace State Altering Conditionals with State" (RSACS)

In this refactoring the *State* pattern will be introduced. It represents different states of class objects, while transitions between states depend on a current state of a class under concern. Functionalities characteristic to particular states are encapsulated in dedicated subclasses. The pattern leads to improvement of state management, although increases the number of classes.

Precondition. The first part of the precondition corresponds to those of RTCC. The remaining constraints are as follows:

2. There exists a set $M_X(y_X)$ of methods m_X under the following constraints:
   (a) method m_X includes at least one *P-type* statement *due to a field* y_X. A statement is of *P-type* due to a field y_X (in short $PS(y_X)$) if it is a conditional instruction (e.g. *if-else*) and has a conditional predicate which is determined by the field y_X taking a value from $S(y_X)$, where y_X belongs to fields that satisfy the first precondition,
   (b) set $M_X(y_X)$ contains at least k methods m_X including *P-type* statements due to the same field y_X, where k is a refactoring parameter.

Metric. A relevance metric is calculated for each field f_X that is accepted after the precondition evaluation. It is equal to a weighted sum of two components that estimate an occurrence factor and a profitability factor (Eq. 3). The weight coefficients specify impact of both components, and their sum evaluates to 1.

$$\text{RSACS\_Metric}(f_X) = (\Psi(f_X) * \lambda_1 + \Omega(f_X) * \lambda_2), \text{where } \lambda_1 + \lambda_2 = 1 \qquad (3)$$

The first component of the metric takes into account all possible values from the set $S(f_X)$ (Eq. 4). An appropriate component is raised by α/n, if the field f_X is checked for a value in at least one *P-type* statement *due to* f_X. Assignment of the field to the value in a *P-type* statement of such kind or in a constructor causes increase of the component by α/n. The component can reach a maximum value of 100%.

$$\Psi(f_X) = \left(\sum_{s \in S(f_X)} \left(comp(f_X, s) * \frac{\alpha}{n} + assign(f_X, s) * \frac{\beta}{n}\right)\right) \qquad (4)$$

Where $n = Card(S(f_X))$
$\alpha + \beta = 100\%$

$$comp(f_X, s) = \begin{cases} 1 & \text{if } f_X \text{ is compared to s in } PS(f_X) \in m_X, m_X \in M_X(f_X) \\ 0 & \text{otherwise} \end{cases} \qquad (5)$$

$$assign(f_X, s) = \begin{cases} 1 & \text{if } f_X \text{ is assigned to s in a } constructor_X \text{ or in } PS(f_X) \in m_X, m_X \in M_X(f_X) \\ 0 & \text{otherwise} \end{cases}$$
$$(6)$$

The second component of the RSACS metric estimates profitability of the refactoring (Eq. 7). Five conditions (c_1 to c_5) are checked for each method that includes P-type statements due to the considered field f_X. If a condition is satisfied in a method, a certain value is added to the second component. The value is equal to 100% divided by the number of methods multiplied by 5. Therefore, a final value of the component ranges from 0 to 100%. The conditions of a method are devoted mostly to detailed situations when a static value of S is assigned to field f_X.

$$\Omega(f_X) = \left(\sum_{m_X \in M_X(f_X)} \sum_{i=1..5} c_i(f_X, m_X) * \frac{100\%}{q * 5}\right) \qquad (7)$$

Where $q = Card(M_X(f_X))$

$$c_1(f_X, m_X) = \begin{cases} 1 & \text{if } (f_X \text{ is assigned to } s \in S(f_X) \text{ in } PS(f_X) \in m_X) \\ 0 & \text{otherwise} \end{cases} \qquad (8)$$

$$c_2(f_X, m_X) = \begin{cases} 1 & \text{if } (\wedge_{h=PS(f_X) \in m_X.} \begin{array}{l} ((f_X \text{ is assigned to } s \in S(f_X) \text{in } h) \text{or} \\ (\text{all } PS(f_X) \text{ nested in } h \text{ have } f_X \text{ assigned to } s \in S(f_X)) \\ or\ (h \text{ are called from } PS(f_X) \text{ in which } f_X \text{ is assigned to} \\ s \in S(f_X)))) \end{array} \\ 0 & \text{otherwise} \end{cases}$$
$$(9)$$

$$c_3(f_X, m_X) = \begin{cases} 1 & \text{if } (\wedge_{h=PS(f_X) \in m_X} f_X \text{ is assigned to only one } s \in S(f_X) \text{ in any closure of } h) \\ 0 & \text{otherwise} \end{cases}$$
$$(10)$$

$$c_4(f_X, m_X) = \begin{cases} 1 & \text{if } (\exists_{h,g=PS(f_X) \in m_X} \text{ conditions of } h \text{ and } g \text{ contradicts}) \\ 0 & \text{otherwise} \end{cases} \qquad (11)$$

$$c_5(f_X, m_X) = \begin{cases} 1 & \text{if } ((\cap code \text{ of all } PS(f_X) \in m_X) \geq \mu * code \text{ of } m_X) \\ 0 & \text{otherwise} \end{cases} \qquad (12)$$

Metric Example. In case studies, the following values of parameters of the RSACS metric were assumed: $\lambda_1 = 0.25$, $\lambda_2 = 0.75$, $\alpha = 30\%$, $\beta = 70\%$, $\mu = 0.75$. They were selected after a series of experimental refactorings.

As an example, we are going to calculate the metric for the *SystemPermission* class that has a field *permission* satisfying the precondition. It can take six values from its set S (n = 6). Methods of the class and their P-type statements due to the *permission* field have been examined. Hence, we can find comparison of the field with four different values of S, and assignment to 6 different values. In result, the first component is equal to:

$$4 * 30\% / 6 + 6 * 70\% / 6 = 89.6\%$$

In calculation of the second component, i.e. profitability, three methods are considered ($q = 3$). In method *claimed()* two conditions are fulfilled ($c1$ and $c2$), in method *denied()* also two conditions are satisfied ($c1$ and $c5$), and in method *granted()* three conditions ($c1$, $c2$, and $c5$). Therefore, the second component is equal to:

$$(2 + 2 + 3) * 100\% / (3 * 5) = 46,7\%$$

Using the weight coefficients the overall RSACS metric is calculated as:

$$0.25 * 89.6\% + 0.75 * 46.7 = 57.4\%.$$

This is a borderline result, but still can be treated as a positive recommendation.

4   Conclusions

The main contribution of the paper is a novel metric-based approach to automated refactoring to design patterns. Structural code analysis is driven by the specialized metrics that assess correspondence of code to selected design patterns and its suitability to an automated code transformation. The approach can be used as a recommendation presented to a user or can provide a partially or fully automated refactoring. Preliminary experiments with the prototype on refactoring of Java programs to design patterns have confirmed the approach profitability. However, metric calculation depends on different coefficients that should be carefully adjusted.

As future research, appropriate relevance metrics for other code to DP refactoring could be designed and implemented. Evaluation of relevance metrics needs also further verification and calibration experiments. Moreover, the tool-supported refactoring process could be extended with additional recommendations and verifications based on general software metrics aimed at software quality assessment.

References

1. Gamma, E., Helm, R., Johnson, R., Vlissides, J.: Design Patterns: Elements of Reusable Object-Oriented Software. Addison-Wesley, Boston (1995)
2. Kan, S.H.: Metrics and Models in Software Quality Engineering. Addison-Wesley, Boston (1998)
3. e Abreu, B.F.: The MOOD metrics set. In: Proceedings of the Ninth European Conference Object-Oriented Programming (ECOOP 1995) Workshop Metrics, August 1995
4. Chidamber, S.R., Kemerer, C.F.: A metrics suite for object oriented design. IEEE Trans. Software Eng. **20**(6), 476–493 (1994). https://doi.org/10.1109/32.295895
5. Bansiya, J., Davis, C.G.: A hierarchical model for object-oriented design quality assessment. IEEE Trans. Softw. Eng. **28**(1), 4–17 (2002). https://doi.org/10.1109/32.979986
6. Mayvan, B.B., Rasoolzadegan, A., Yazdi, Z.G.: The state of the art on design patterns: a systematic mapping of the literature. J. Syst. Softw. **125**, 93–118 (2017). https://doi.org/10.1016/j.jss.2016.11.030
7. Kierevsky, J.: Refactoring to Patterns. Addison Wesley, Boston (2004)
8. Fowler, M., Beck, K., Brant, J., Opdyke, W., Roberts, D.: Refactoring: Improving the Design of Existing Code. Addison Wesley, Boston (1999)
9. Sommerville, I.: Software Engineering, 10th edn. Pearson Education, New York (2015)
10. Tsantalis, N., Chatzigeorgiou, A.: Identification of move method refactoring opportunities. IEEE Trans. Softw. Eng. **35**(3), 347–367 (2009). https://doi.org/10.1109/TSE.2009.1
11. Silva, D., Terra, R., Valente, M.T.: Recommending automated extract method refactorings. In: 22nd International Conference on Program Comprehension (ICPC), pp. 146–156. ACM, New York (2014). https://doi.org/10.1145/2597008.2597141
12. Bavota, G., Lucia, A.D., Marcus, A., Oliveto, R.: Recommending refactoring operations in large software systems. In: Robillard, M.P., Maalej, W., Walker, R.J., Zimmermann, T. (eds.) Recommendation Systems in Software Engineering (RSSE), chap. 15, pp 387–419. Springer, Heidelberg (2014). https://doi.org/10.1007/978-3-642-45135-5
13. Vakilian, M., Chen, N., Negara, S., Rajkumar, B.A., Bailey, B.P., Johnson, R.E.: Use, disuse, and misuse of automated refactorings. In: 34th International Conference on Software Engineering (ICSE), pp. 233–243 (2012). https://doi.org/10.1109/icse.2012.6227190
14. Derezinska, A.: A structure-driven process of automated refactoring to design patterns. In: Świątek, J., Borzemski, L., Wilamowska, Z. (eds.) Information Systems Architecture and Technology: Proceedings of 38th International Conference on Information Systems Architecture and Technology – ISAT 2017: PART II, AISC, vol. 656, pp. 39–48. Springer, Cham (2018). https://doi.org/10.1007/978-3-319-67229-8_4
15. Bluemke, I., Stępień, A.: Selection of metrics for the defect prediction. In: Zamojski, W., et al. (ed.) Dependability Engineering and Complex Systems, DepCoS-RELCOMEX 2016, AISC, vol. 470, pp. 39–50. Springer (2016). https://doi.org/10.1007/978-3-319-39639-2_4
16. Ali, M., Elish, M.O.: A comparative literature survey of design patterns impact on software quality. In: Proceedings of International Conference on Information Science and Applications (ICISA), pp. 1–7 (2013). https://doi.org/10.1109/icisa.2013.6579460
17. Pradhan, P., Dwivedi, A.K., Rath, S.K.: Impact of design patterns on quantitative assessment of quality parameters. In: Second International Conference on Advances in Computing and Communication Engineering (ICACCE), pp. 577–582 (2015). https://doi.org/10.1109/icacce.2015.102
18. Hsueh, N.-L., Chu, P.-H., Chu, W.: A quantitative approach for evaluating the quality of design patterns. J. Syst. Softw. **81**, 1430–1439 (2008). https://doi.org/10.1016/j.jss.2007.11.724

19. Ampatzoglou, A., Chatzigeorgiou, A.: Evaluation of object-oriented design patterns in game development. Inf. Softw. Technol. **49**(5), 445–454 (2007). https://doi.org/10.1016/j.infsof. 2006.07.003

20. Mariani, T., Vergilio, S.R.: A systematic review on search-based refactoring. Inf. Softw. Technol. **83**, 14–34 (2017). https://doi.org/10.1016/j.infsof.2016.11.009

21. Amoui, M., Mirarab, S., Ansari, S., Lucas, C.: A genetic algorithm approach to design evolution using design pattern transformation. Int. J. Inf. Technol. Intell. Comput., 1–10 (2006)

22. Jensen, A.C., Cheng, B.H.: On the use of genetic programming for automated refactoring and the introduction of design patterns. In: Proceedings of the 12th Annual Conference on Genetic and Evolutionary Computation (GECCO), pp. 1341–1348. ACM (2010). https://doi. org/10.1145/1830483.1830731

23. Shimomura, T., Ikeda, K., Takahashi, M.: An approach to GA-driven automatic refactoring based on design patterns. In: 5th International Conference on Software Engineering Advances (ICSEA), pp. 213–218 (2010). https://doi.org/10.1109/icsea.2010.39

24. Kim, J., Batory, D., Dig, D.: Scripting parametric refactorings in Java to retrofit design patterns. In: 31st IEEE International Conference on Software Maintenance and Evolution (ICSME), pp. 211–220. IEEE (2015). https://doi.org/10.1109/icsm.2015.7332467

Design and Validation of a Scheme of Infrastructure of Servers, Under the PPDIOO Methodology, in the University Institution - ITSA

Leonel Hernandez[1(✉)] and Genett Jimenez[2]

[1] Department of Telematic Engineering, Engineering Faculty, Institución Universitaria ITSA, Barranquilla, Colombia
lhernandezc@itsa.edu.co
[2] Department of Industrial Process Engineering, Engineering Faculty, Institución Universitaria ITSA, Barranquilla, Colombia
gjimenez@itsa.edu.co

Abstract. This document describes the process of project development named Design and validation of the Infrastructure Servers in the University Institution - ITSA, in which a modeling was developed for several important network services such as Active Directory, DNS, DHCP, Web, databases and Mail. This project establishes a series of recommendations based on the performance of several tests, which aims to help improve the performance of existing technological infrastructure and the organization of network information. In this process, the specific requirements were collected to design and develop an optimal solution for user control, the organization of active directory, centralized DHCP service on servers and validation redundancy of certain important services network server level. The current logical topology of the active directory was identified and diagrammed and an outline of the new desired structure of the same was made, including subdomains for each site following the guidelines established in the lifecycle of Cisco (PPDIOO).

Keywords: Servers · PPDIOO · Active directory · DNS · DHCP

1 Introduction

Currently, most organizations or educational entities have an established network for the connectivity of computing, network and mobile devices. All organizations have servers, which are the central computers to control the network and the access of client users to the resources offered by the network (File services, web, printing, etc.). In the ITSA university institution, there are servers where the Web service, Active Directory, Database and the Moodle service (Virtual Classroom) are hosted. Certain network services managed by the ITSA University Institution are managed and configured through virtualization, establishing virtual servers. In the corporate network of the institution, some anomalies have been detected with respect to certain services provided by the servers, which is why it is necessary to make a good design for the structure of the active directory and optimization of the network services. The important points to design this

R. Silhavy (Ed.): CSOC 2018, AISC 763, pp. 367–379, 2019.
https://doi.org/10.1007/978-3-319-91186-1_38

solution at the servers' level are Active Directory, DHCP, DNS, Web, mail service and Databases.

One of the important solutions is the implementation of the Active Directory, which is an important utility for the management of an organization and the redundancy for fault tolerance by establishing domains of an existing forest [1, 2]. The current organization of the active directory can be improved, which is one of the objectives of this proposal, establishing a logical structure of the active directory for the ITSA headquarters in Soledad and Barranquilla with their organizational units [3].

According to the DHCP service [4], everything will be centralized from a server and DHCP Failover is implemented for fault tolerance to handle the redundancy issue of this service, in case there is a failure in the headquarters due to energy reasons, Hardware problems or other types of problems.

The DNS service is another key point of this solution [5], to establish the resolution of name and delivery of domains to the computers of the university institution and register the equipment to the DNS by active directory and by means of deliveries of addresses with DHCP, with this it would help to identify the computers that are on the network.

On the side of Web services, emails and databases, the institution currently has a good solvent such as Oracle for the database, the mail service through the cloud and the Tomcat service and Apache PHP for the Web service, although not redundancy is handled in these services.

2   Designing a Server Infrastructure: A Literature Review

For the elaboration of this project it was inquired with respect to the Design and validation of other schemes of Server Infrastructure, therefore, information related to the subject was found, which is reflected in the following references.

Marañon [6], in its implementation project of an application server platform, proposes to solve a problem of the server infrastructure in which it seeks to centralize the users' applications in a single point to improve its maintenance, updating and configuration. Also, the use of server virtualization to save energy and space in the data center, to have server backup through virtualization technologies.

Wang [7], in his research about Data Center Network, exposes, among several aspects of data center design, the importance of centralized server schemes in the good performance of the network. Lim [8] in his project of development of emergent environments of warehouse-computing, shows a design of servers for this type of solutions. Kai [1], in his work on virtualization based on active directory, exposes the basic aspects to be taken into account for server virtualization solutions in conjunction with active directory. Chen [9] in his research explains the importance of virtualization for the optimization of a company's IT infrastructure.

The focus of the PPDIOO methodology is to define the minimum activities required, by technology and network complexity, that allow advising the entities in the best possible way, installing and successfully operating Cisco technologies. Likewise, it is possible to optimize the performance through the life cycle of the network [10, 11].

Hernandez [12], in his research on the design of a network infrastructure in the mining sector, exposes the disposition of the servers in a distributed infrastructure of switches based on the services and applications of the company. Also, Hernandez [13] conducted another investigation with the servers and network devices of ITSA to determine if the data center meets the standards of green data centers, including the current server of the active directory that is projected to improve with this new proposal.

Ruile in his research of campus network design [14], exposes the basic aspects of a methodology for the administration of network services, such as mail, DNS, DHCP and others.

3   Research and Experimentation Methodology Used

For the elaboration of this project, the descriptive and applied methodology was used as a research methodology [15, 16]. Descriptive since all the documentation related to server infrastructure designs have been reviewed. Applied because it is proposed to solve a practical problem in the current infrastructure. The following questions can be raised as a basis for the solution of the new design: What is the impact of organizing and standardizing the information in the active directory? What are the added value and the improvements that the users will be able to appreciate with a new infrastructure? of servers? Is there a difference related to the efficiency of the network between the current network design and the proposed one?

An interview was held that sought to interact with the ITC staff of the University Institution ITSA, in which questions were asked to obtain information about the subject under study. The objective was to get to know the situations, processes and behaviors predominant in the server infrastructure. With this information it was possible to assume a descriptive image of the real functioning of the system, thus allowing ordering, grouping, and systematizing the processes involved in the investigative work. This made it possible to continue with the investigative process, because it allowed knowing information about which technologies and devices could improve the new design proposal and physical - logical structure of the server area of the ITSA University Institution, applying them in a functional prototype in network laboratories of the ITSA University Institution.

The research design to be developed in the project corresponds to a qualitative, transactional design. Qualitative because it is based on a working hypothesis, defined as the problems in the organization of the infrastructure of servers and current network services and in the standardization of information, which is expected to be solved with the development of this proposal. Transactional since it is expected to collect data from users about the feasibility of the proposal, and in case of a start-up, measure the degree of satisfaction of the same.

4   Start-Up of the Project

4.1   Preparation of the Project

This initial phase allows us to define the technical characteristics of the network. These characteristics include users, applications and services, equipment and means of transmission [17]. This information has been obtained through the documentation of the network and the carrying out of interviews with the personnel of the organization.

For the realization of the present project, in this preparation phase several meetings were held with the clients, in this case with the professors in charge of the project, and also with the personnel of the ITSA systems department, in which it is notified that for different reasons already discussed with the directors of the Institute concluded that Soledad's headquarters would continue to be the main headquarters, and the headquarters of Barranquilla and the future new headquarters that will be on Calle 30 will be the secondary.

Likewise, the systems department of the Institute facilitated access to know the facilities of the Data Center that is in block B of the Soledad headquarters, where documentation was provided about the services provided in the system of the institute. Figure 1 shows the current topology of the servers at the ITSA University Institution:

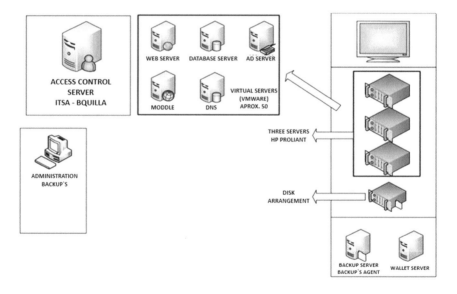

Fig. 1.   Topology of servers in the University Institution – ITSA.

4.2   Project Planning

This phase involves the analysis of the current network and the definition of the requirements of the organization. The requirements will be obtained because of the analysis of the current situation and interviews carried out with the technical staff of the organization.

Considering the information collected in the previous phase, some deficiencies have been noted in some services offered by the Institution's servers, one of which is the Active Directory. Regarding the topic of Active Directory, the Institute has a domain of itsa.edu.local, but it does not have an efficient structure and no policies are established in the organizational units, as shown in Fig. 2:

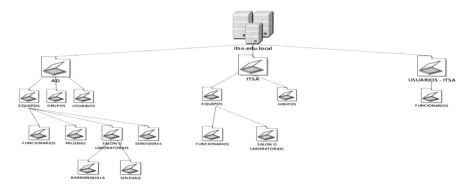

Fig. 2. Current logical topology of the Active Directory of the Institute

To assemble the new logical topology of the AD, it is necessary to know the departments and the study area (rooms, laboratories, rooms) with which the different headquarters of the university institution count through the area of Systems in the subject of the infrastructure of the LAN networks. Also, know the privileges and restrictions that have or should have the users of the Institution, as for example, the students must have their own permissions to change the IP address in the equipment of the laboratories.

The current domain of the Institute will be left, that is, itsa.edu.local, which will be the main domain and the headquarters of Barranquilla and the new headquarters will be subdomains of this and will be bquilla.itsa.edu.local and new.itsa.edu.local respectively.

With respect to the DHCP service, it was possible to establish, according to the information provided by the systems department, that currently in the network infrastructure of the university institution - ITSA, there is a DHCP service for the dynamic allocation of IP addresses, but this service is configured on the Router. That this service is provided in this type of devices would not be bad if we were talking about a small network, but as the Institution has grown in a considerable way in recent years it is not convenient that this service is provided in these devices. One of the solutions is to implement the DHCP service in a dedicated server to reduce the load of the network devices and to have centralized, and better controlled, the conceptions of IP addresses that are distributed in the network segments that are applied in the Institute.

Regarding this information, it is proposed to design the Active Directory topology using the Microsoft Visio 2016 software and to prepare virtual machines with the VMWare software and virtual box with the operating systems Windows Server 2003, Windows Server 2008 and Windows Server 2008 R2, it is also planned to simulate this scenario with the GNS3 software.

4.3   Design

According to the information gathered in the first two phases, the new ideal topology for the Active directory has been started, considering that this time it will have subdomains (new domain in an existing forest). In the design of the Active Directory, it has been established that the main domain controller would be the Soledad headquarters and from there to have the subdomains centralized by means of the Active Directory users and computers console.

According to the structure of the University institution, each headquarters will have a subdomain and its own related organizational units taking into consideration the departments (areas) and classrooms. The proposed logical topology is shown in Fig. 3:

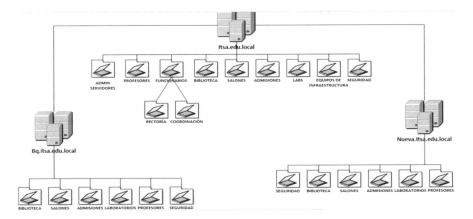

Fig. 3.   Ideal logical topology of AD in the University Institution - ITSA

Given the logical topology of the proposed AD, it has been designed the way of how the servers will be on the network if there is a router in each headquarters. In Fig. 4 the disposition of the servers in the network is shown:

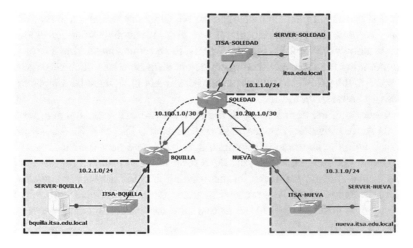

Fig. 4. Servers in the network topology

A design was also established of how the DHCP service will be established, as shown in Fig. 5, considering that the primary DHCP will be the headquarters of Soledad and the secondary ones in the other locations. The areas were established in accordance with the address provided by the systems department. This would be the DHCP Failover for fault tolerance of this service and it is suggested to do it with the Hot Standby Mode type since with this configuration one of the servers acts as primary and another as secondary. The primary is the one that carries the weight of the IPs concession and the secondary is in standby mode in case the primary fails.

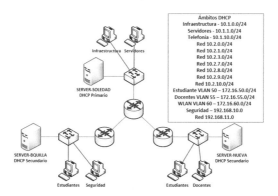

Fig. 5. Network topology, for DHCP servers of the Institute and its network areas

4.4 Implementation

In the implementation, virtual machines have been used to install the operating system and a complete network emulator and the IOS to establish the router. The Virtual Box software, the GNS3 network emulator and the IOS 3725 have been used for the routers.

The first was the implementation of an Active Directory scenario and for this, it was necessary to have these requirements: Three virtual machines with server operating system (Windows Server 2008 Standard, SP1), three routers and addresses for servers. Before mounting the Active Directory service in each Server of each site, the address of the servers, the gateways of the links and the WAN links must be assigned and the connectivity between them verified.

After checking the connectivity in the network scenario raised above, proceed to install the Active Directory service on the main server (ITSA SOLEDAD). It proceeds to install it with the dcpromo.exe command and choose a new domain controller like itsa.edu.local. Following the steps, the DNS service is also installed for the association with the Active Directory and wait for the installation. After having restarted the server with the installation of the AD, proceed to create the organizational units of Soledad, following the logical topology in the Active Directory Users and Computers console, as shown in Fig. 6:

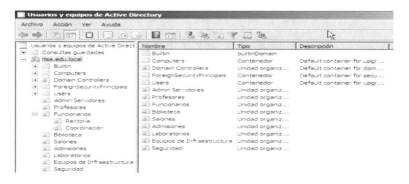

Fig. 6. Organizational units established in Soledad

Now the other Active Directory must be established to the other sites, but this time as secondary domains. For Barranquilla, it is called bquilla.itsa.edu.local and for the new venue nueva.itsa.edu.local. The installation is again applied from the dcpromo.exe.

In the same way, the domain is specified and then it is expected to restart the Server. The same procedure is applied for the secondary domain of the new headquarters. When the secondary domains are established, from the Soledad server it can be managed through the Active Directory Domains and Trusts option, as shown in Fig. 7.

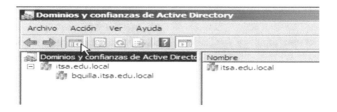

Fig. 7. Secondary domain established in Barranquilla.

In Fig. 8 it can see the active directory of the Barranquilla headquarters:

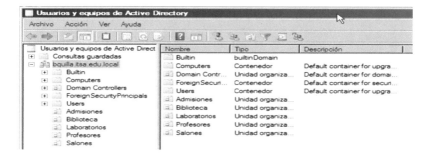

Fig. 8. AD of Barranquilla with organizational units created.

In the configuration of the DNS, according to the assigned domain, the primary zone is created where the equipment for name resolution will be registered as they enter the Active Directory, as shown in Fig. 9:

Fig. 9. Soledad's DNS primary zone.

In the DHCP service, after establishing the topology and placing the address, the DHCP service was installed on the servers. In the case of tests, two equal scopes were created on each server. Figure 10 shows scopes configured in Soledad:

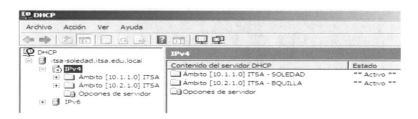

Fig. 10. DHCP service console with scopes in Soledad.

Then, on each scenario router, the ip helper-address is established to specify the DHCP server outside the network. Now, the test is done with the PC Student of ITSA Barranquilla as established in the DHCP test scenario.

5 Results

After carrying out the implementation tests, the operation was performed in an environment where there was connectivity with the rest of the departments such as security, telephony, routing and infrastructure. In the operation, the servers were established in virtual machines with VMWare and with Windows Server 2008 R2 Standard operating system, as shown in Fig. 11:

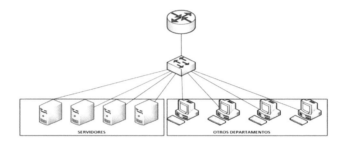

Fig. 11. The topology used in the operation with other departments.

In the Active Directory, in the same way as in the implementation, secondary domains were also created on Windows Server 2008 R2 machines, as shown in Fig. 12:

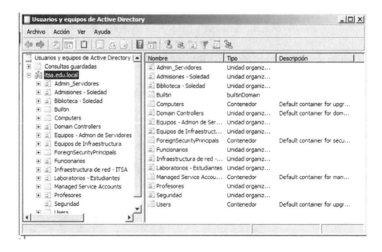

Fig. 12. Active Directory of Soledad in operation.

Equipment to the domain was included according to the department. Figure 13 shows the delivery of IP addresses to other networks:

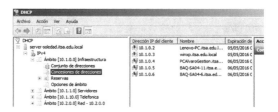

Fig. 13. Delivery of IP addresses to other networks.

Tests have also been carried out implementing mail service, databases and Web. In the mail service, Microsoft Exchange Server 2007 was installed, with its respective domain of itsa.edu.mail, as shown in Fig. 14:

Fig. 14. Exchange mail server configuration.

The database service was established with the MySQL Server service with a sample database and the Web service with Apache Tomcat with a sample page to check the services, as shown in Fig. 15:

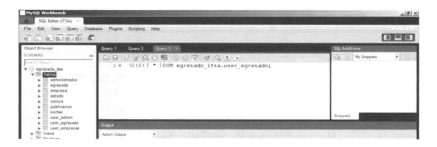

Fig. 15. MySQL database service

6 Conclusions

During the project, a network scenario was established, where the installation and operation of certain network services from a Server was worked. The servers have been worked with the family of Operating Systems for Microsoft Servers in virtualized environments with the VirtualBox and VMWARE programs. Initially, the requirements that were requested to carry out this project were checked, the network services that will be established as the Active Directory, DNS, DHCP, WEB and Database. Following the life cycle of Cisco PPDIOO, a meeting was held at the headquarters, gathering current information from the network services to proceed with the planning of the project and then carry out an ideal design of the structured topology of the Active Directory and the network where the servers will be located. Throughout the tests carried out in the GNS3 program, the connectivity between the servers and the routers that were emulated in the program was checked and there was no problem of connectivity and connection delay. The DHCP service was established from the Server with Windows Server and the operation of this service was verified. Similarly, the functionality of other services such as the mail server based on the Exchange service, MySQL databases and JSP service with Apache Tomcat was verified and all the tests were successful.

The proposed and modeled solution exposed in this project is feasible to be developed as part of a strategy to improve the technological infrastructure of the Institution, following the guidelines of the PPDIOO methodology. The internal processes would be much agiler and efficient, positively impacting the entire academic community.

References

1. Kai, H.: The design and realization of server virtualization based on active directory. In: Proceedings - 2009 International Forum on Information Technology and Applications, IFITA 2009, vol. 1, pp. 740–742 (2009)
2. Svidergol, B., Allen, R.: Active directory cookbook. Saudi Med. J. **33**, 832 (2013)
3. Lane, R., Muggli, N., Bhai, S.: Active directory domain services in the perimeter network (Windows Server 2008). Computer **April**, 52 (2009)
4. Lin, C., Su, T., Wang, Z.: Summary of high-availability DHCP service solutions. In: Proceedings - 2011 4th IEEE International Conference on Broadband Network and Multimedia Technology, IC-BNMT 2011, pp. 12–17 (2011)
5. Badach, A., Hoffmann, E.: Domain Name System (DNS). In: Technik der IP-Netze, pp. 175–238 (2015)
6. Marañon, J.: Implementación de una plataforma de servidores de aplicaciones, p. 127 (2013)
7. Wang, T., Su, Z., Xia, Y., Muppala, J., Hamdi, M.: Designing efficient high performance server-centric data center network architecture. Comput. Netw. **79**, 283–296 (2015)
8. Lim, K., Ranganathan, P., Chang, J., Patel, C., Mudge, T., Reinhardt, S.: Understanding and designing new server architectures for emerging warehouse-computing environments. In: Proceedings - International Symposium on Computer Architecture, pp. 315–326 (2008)
9. Chen, Q., Xin, R.: Optimizing enterprise IT infrastructure through virtual server consolidation. In: 2005 Informing Science and IT Education Joint Conference, vol. 19, p. 7. Accessed Jan 2005
10. Oppenheimer, P.: Top-Down Network Design, 3rd edn. Cisco Press, Indianapolis (2011)
11. Cisco Networking Academy: Cisco Networking Academy (2015). http://www.cisco.com/web/learning/netacad/index.html
12. Hernandez, L.: Distributed infrastructure for efficient management of network services. case: large company. In: Mining Sector In Colombia, 2016 2nd International Conference on Science Information Technology, Proceedings, pp. 63–68. IEEE (2016)

13. Hernandez, L., Jimenez, G.: Characterization of the current conditions of the ITSA data centers according to standards of the green data centers friendly to the environment. In: Advances in Intelligent Systems and Computing, vol. 574, pp. 329–340 (2017)
14. Ruile, L.: Design and implementation of IT service management system of college or university campus network. In: 2012 7th International Conference on Computer Science & Education (ICCSE), pp. 299–304 (2012)
15. Sampieri, H., Collado, C.F., del Lucio, P.B.: Metodología de la investigación (2010)
16. Gallardo, M.A.S.: Metodología de investigación científica. In: Conacyt, pp. 1–18 (2006)
17. Tanenbaum, A., Wetherall, D.: Redes De Computadoras, 4th edn. Pearson, México (2012)

Accelerating Massive Astronomical Cross-Match Based on Roaring Bitmap over Parallel Database System

Jianfeng Zhang[1,2], Hui Li[1,2(✉)], Mei Chen[1,2], Zhenyu Dai[1,2], and Ming Zhu[3]

[1] College of Computer Science and Technology, Guizhou University,
Guiyang, People's Republic of China
yunyouzi66@gmail.com, {cse.HuiLi,gychm,zydai}@gzu.edu.cn
[2] Guizhou Engineer Lab of ACMIS, Guizhou University, Guiyang, People's Republic of China
[3] National Astronomical Observatories, Chinese Academy of Sciences,
Beijing, People's Republic of China
mz@nao.cas.cn

Abstract. In order to reduce the large network overhead and the heavy cost of cross-match on the astronomical catalog in the database cluster, we proposed a novel method of cross-matches based on Roaring Bitmap. Firstly, we store astronomical catalog data in column-oriented storage with compression setup to reduce I/O overhead of accessing field in the parallel database system. Secondly, we create the spatial index, which maps the 2D coordinates into integer number. Then, using Roaring Bitmap convert the spatial index into a bitmap index. Finally, the received spatial range search of cross-match is translated into bitmap operations to achieve batch processing. The experiments over the real large-scale astronomical data show that the proposed method is 4 to 10 times faster than traditional method, meanwhile, only consume less than 10% of memory resource.

Keywords: Cross-match · Catalog · Roaring Bitmap · Parallel database system
Spatial index

1 Introduction

At present, astronomy has entered the era of data explosion and information-rich full-band, and the amount of data in a single astronomical catalog table has reached and exceeded one billion orders of magnitude. Scientists are heavily rely on the database-related technologies to retrieve the needed data from the data store [1, 2].

Astronomical data retrieval [2] and cross-match operation are data-intensive computation and its time complexity is $O(n^2)$. Performance bottlenecks is easily occurred in disk I/O and intensive computing. Creating the spatial index for catalog data could speed up the performance of astronomy analysis tasks e.g., cross-matches, cone searches. The commonly spatial index not only include the multidimensional index space points for the RA and DEC, such as database extensions PgSphere, PostGIS [3], but also include quad tree index like Q3C [4], Healpix [5] and HTM [6]. Aforementioned indexing techniques are have its database extensions implementation and widely used by many

© Springer International Publishing AG, part of Springer Nature 2019
R. Silhavy (Ed.): CSOC 2018, AISC 763, pp. 380–389, 2019.
https://doi.org/10.1007/978-3-319-91186-1_39

astronomical institutions. However, because of the explosive growth of astronomical data, traditional relational databases have been inefficient to process millions of astronomical analysis tasks [7]. Some researchers have begun to study the use of distributed database clusters instead of the traditional database, such as using Hadoop to store astronomy data and recoding cross-match by MapReduce program [7, 8]. However, because of the internal mechanism of MapReduce and Hadoop is not properly support for indexing extension, there are still very easy to lead to the network and disk I/O bottlenecks in astronomical application in Hadoop [8].

In this paper, we propose to solve above problems as follows: firstly, we use column-oriented compression [9–11] to store the catalog data in less disk space. It could reduce unnecessary I/O overhead for the astronomical analysis tasks. Secondly, we use Roaring Bitmap [12] to accelerate the execution of cross-match in the Greenplum parallel database system, which aims to significantly reduce the network overhead and computing resource consumed in astronomical data analysis.

The rest of this paper is organized as follows. In Sect. 2, we present the related work. Section 3 describe the detail of our work. Section 4 present the experimental results and analysis. Finally, Sect. 5 concludes the work.

2 Related Work

2.1 Cross-Match

Most of the queries processed by the Virtual Observatory [7, 13] daily require cross-matches. Multi-band astronomical data mining and statistics are heavily need the cross-matches operation, which is the core technology of multi-band data fusion. It can be said that cross-match is one of the most measures for the service performance of the virtual observatory. Cross-matches is based on location information to determine the angular distance, the main purpose is to quickly and coarsely locate homologous stars that could reducing the scope of astronomers to analyze data size.

2.2 Storage Model

The astronomy field usually uses metadata file [14] to store astronomical catalog data. The advantage of this method is that it can save the storage cost, but at the same time, it may impair the efficiency of random access. Therefore, some astronomical institutions have already chosen relational database to store massive catalog data. Due to the characteristics of the traditional relational database, astronomical catalog data is stored in the relational database in the form of row-oriented of heap table, which greatly improves the efficiency of random access. Nevertheless, it may have serious I/O waste in analytical scenarios.

Due to the limitations of the traditional relational database extension, the distributed database cluster Greenplum is proposed to store vast amounts of astronomical data. The reason for choosing Greenplum to store massive astronomical data is as following: It has the mature indexing mechanism of RDBMS, and supporting efficient column-oriented compression storage (Append Optimized table), meanwhile, has good horizontal scaling mechanism to provide parallel data processing capability. A brief storage model comparison for Greenplum [11] is as Table 1.

Table 1. Row vs. Column oriented storage

Storage model	Features
Column-oriented storage	Save disk space, low I/O consumption, non-clustered index
Row-oriented storage	Take up a lot of disk space, high I/O consumption, clustered index

2.3 Roaring Bitmap

In order to enhance the efficiency of cross-matches, bitmap technique have been integrate to achieve performance acceleration [12, 15, 17, 18]. It means that use bitmap logical operation to replaced spatial range search. Greenplum supports bitmap computation using RLE, it supports the cardinality from 100 to 1 million, which making it difficult to adapt to massive data processing.

Roaring Bitmap is a hybrid bitmap compression method and it is not based on run-length encoding. Roaring Bitmap partitions range 2^{16} where all the elements in the same chunk share the same 16 most significant bits. Roaring Bitmap encodes a chunk depending on the number of actual elements k in the chunk. In particular, if $k > 4096$, Roaring Bitmap uses 65536-bit uncompressed bitmap to encode the elements, otherwise, it used a sorted array of 16-bit short integers. It chooses 4096 as the threshold because it guarantees that each integer use no more than 16 bits to represent, because Roaring Bitmap uses either 65536 bits to represents 4096 integers or at most 16 bits per integer for array buckets. Specific examples as follows. There are three data sets.

Data_1: 1000 multiples of 62
Data_2: all integers $[2^{16}, 2^{16} + 100)$
Data_3: all even numbers in $[2 * 2^{16}, 3 * 2^{16})$

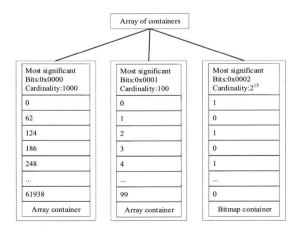

Fig. 1. Example of Roaring Bitmap containing data sets

As shown in Fig. 1. Data_1, Data_2 are sorted in the sort array containers, and Data_3 is stored in the uncompressed bitmap container in an equivalent encoding.

Since Roaring Bitmap provides two storage containers, it is important to understand how containers perform logical operations in each scenario. There are three scenarios included, Array OP Array, Array OP Bitmap and Bitmap OP bitmap. Since the upper 16 bits are stored in the array container, the operations for the upper 16 bits are Array OP Array operations [12].

3   Design and Implementation

Cross-match is a type of space range queries that can be converted to bitmap logical operations [12, 15, 20]. According to the obtained bitmap based results, the number of '1' is equal to the number of homologs obtained after cross-matches, and the position of '1' in bitmap of the result corresponds to the position of the space coordinate.

Figure 2 shows the system architecture. Astronomical tables stored in the Greenplum database cluster use column-oriented compression model, each field of the table as a separate file is stored in database. In our work, we insert the pseudo-space value and mapping the 2D coordinates to the integer number into the catalog table. Then create B-Tree index for ipix field. When the database system received request from the client, by scanning the index to access the ipix field, extract the field value using the array aggregate function (array_agg), and then translate the array to bitmap by User Define Type of Roaring Bitmap. Using CTE [19] techniques to store bitmaps with temporary tables can solve the computational skew problem caused by the imbalance of intermediate result sets, but also introduces additional memory overhead for storing temporary tables. The Roaring Bitmap interface is used Greenplum User Define Interface implement. The final search statement of cross-matches use various operations on Roaring Bitmap, including union (bitwise OR) and intersection (bitwise AND) [12]. The results returned

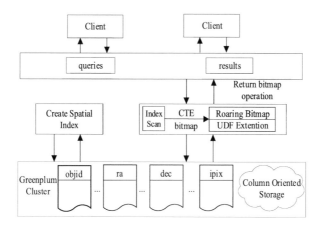

Fig. 2. System architecture

to the client. The specific calculation of how to achieve the conversion will be given in the specific examples.

Q3C is chosen for the following reasons: The data type of the Roaring Bitmap operation is integer, and the Q3C construction pseudo-space sky index is obtained by obtaining a '01' bitmap string by using the quad tree coding of the two-dimensional spatial coordinates. The bitmap string into decimal integer. Q3C in the implementation of spatial range search for cross-matches, the use of integer and integer interval comparison, thus cross-matches is easy to map bitwise operations.

Suppose we want to determine the following integer 13, 6, 8, 4 whether fall within the scope of 24 integer ranges in the range field of Table 2.

Table 2. Data sets for range query

ipix	Range	Tag
13	[1, 5], [2, 4], [8, 11], [9, 13]	B1
6	[7, 8], [3, 5], [9, 12], [8, 10]	B2
8	[1, 4], [7, 8], [3, 4], [2, 3]	B3
4	[8, 9], [7, 10], [3, 4], [2, 5]	B4
	[1, 2], [3, 4], [2, 4], [7, 9]	B5
A	[8, 10], [2, 4], [3, 5], [1, 5]	B6

The traditional way of doing space range search is to take a record from *ipix* field and compare it with the 24 integer ranges in the range column of Table 2, which returns 1 if *ipix* fall within any of these ranges, otherwise return 0. With *ipix* data volume increases, the time-consuming comparison will increase. Usually the computational time complexity is $O(n^2)$. In a distributed environment, we also need to consider the overhead of data transfer between different nodes or hosts. We use bitmap representation as shown in Fig. 3.

A	B	B1	B2	B3	B4	B5	B6	C
0	0	0	0	0	0	0	0	0
0	0	0	0	0	0	0	0	0
1	1	1	0	0	0	0	0	1
0	1	1	1	0	0	0	0	0
0	1	1	1	0	0	0	0	0
0	1	1	1	0	1	0	1	0
0	1	1	1	0	1	1	1	0
1	1	1	1	1	1	1	1	1
0	1	0	1	1	1	1	0	0
1	0	0	0	0	0	0	0	0
0	1	1	1	0	1	0	1	0
1	1	1	1	1	1	1	1	1
0	1	1	1	1	1	1	1	0
0	1	1	0	1	1	1	1	0
0	1	1	0	1	0	1	1	0
0	0	0	0	0	0	0	0	0

Fig. 3. Data sets using bitmap representation

In Fig. 3, A, B, B1, B2, B3, B4, B5, B6 represent bitmaps. Where A represents the bitmap obtained by equivalent coding of 13, 6, 8 and 4, and B1, B2, B3, B4, B5 and B6 represent the range bitmap of the integer ranges; B is the union B1, B2, B3, B4, B5 and B6. Like the formula 1; A represents the bitmap set of values in all ipix column; C represents the intersection of A and B, like the formula 2. Intersection which is used to express the AND operation between bitmaps. The '1' of C bitmap said ipix is within the range of integers, '0' means that ipix is not within the range of integers. The number of '1' in C is also used to calculate the number of ipix that satisfy the range [12, 17, 20].

$$B = (B1 \cup B2) \cup (B3 \cup B4) \cup (B5 \cup B6) \tag{1}$$

$$C = A \cap B \tag{2}$$

The above formula can help to understand the range of queries. But in the actual distributed environment, following formula also used in the distributed bitmap based operations:

$$(A \cap B) \cup D = (A \cup D) \cap (B \cup D) \tag{3}$$

$$(A \cup B) \cap D = (A \cap D) \cup (B \cap D) \tag{4}$$

$$(A - B) \cap D = (A \cap D) - (B \cap D) \tag{5}$$

The above formula can be used to avoid bitmap based segment level skew in the distributed environment.

4 Experiments

The database cluster used in this experiment consists of one master node and four slave nodes, and the configuration environment as shown in Table 3.

Table 3. Configuration of Master and Slave

Node	Parameters					
	Memory	Disk	CPU	OS	Version
Master	20 GB	150 GB	Intel Xeon	Centos7	Greenplum
Slave	16 GB	300 GB	2.60 GHZ		5.0.0

It is important to point out that the CPU supports the PopCount instruction set. Use the data set as Galaxy catalog [1]. The suffix of the table indicates the size of the catalog data. For example, the table name Galaxy_50, which indicates that the catalog contains 5 million records.

The evaluation of time-consuming during build spatial index in different storage models are depicted in Fig. 4 (a). The row-oriented indicates that the storage model of catalog data is the row-oriented of heap table, and the column-oriented shows that catalog uses the column-oriented of AO table. As the amount of data grows, the time-consuming of creating spatial index in both storage model grows linearly. When the catalog is galaxy_200, building spatial index in the *column-oriented* is 20 times faster than row-oriented.

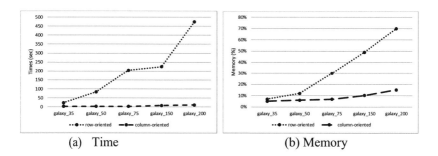

(a) Time   (b) Memory

Fig. 4. Create spatial index on row-oriented vs column-oriented storage

Due to the difference of I/O overhead in both storage model, the time-consuming of building index is different also. In Fig. 4(b), the memory usage depicts that the I/O consumption of row-oriented is significantly higher than column-oriented, which determined by different scan methods.

Row-oriented heap table supports clustered index, which can filter out unnecessary data blocks, result in the reduction of the I/O consumption of cross-matches. In relational databases, it is difficult to build clustered indexes for catalog data. From Fig. 5, with the increasing of catalog data size, the time-consuming of the aggregation operation is rising. The clustered index requires moving the entire catalog in aggregation, which re-store the catalog records according to the order of the clustered index. Furthermore, it is valid only for the access to the index field involved, when access to other fields in the catalog does not have the effect of reducing I/O cost. Therefore, it not to be a recommending approach.

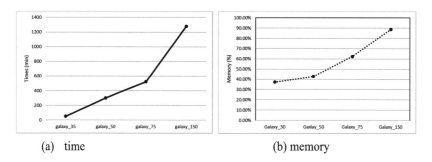

(a) time   (b) memory

Fig. 5. Cluster index time and memory consumption

As we can see from Fig. 6(a), using Roaring Bitmap to perform cross-matches is significantly faster than using a row-oriented storage to construct the spatial index. When catalog table is galaxy_35, the time-consuming of cross-matches in row-oriented is less than in column-oriented. The main reason is that clustered index in row-oriented filter out unnecessary data blocks. But its time-consuming is higher than the cross-matches using Roaring Bitmap. Overall, Roaring Bitmap do the best cross-matches.

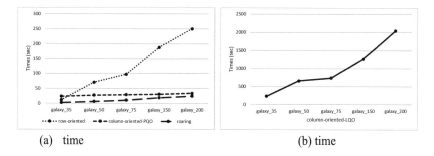

(a)   time   (b) time

Fig. 6. Cross-matches time-consuming in different method

Greenplum database provides two kinds of query optimizers [21, 22]. One is LQO (Legacy Query Optimizer), which support function index access. Another is PQO (Pivotal Query Optimizer), which does not support index expression access, but it provides more efficient column bitmap index filtering out unnecessary data blocks. In Fig. 6, when catalog data stored in column-oriented table, the time consumption of cross-matches using LQO is higher than PQO.

Figure 7 depicts that the execution plan generated by LQO takes a lot of time and memory resource for performing cross-matches. For the use of row-oriented table storage data, there is still a serious memory and network consumption problems doing cross-matches. For the use of PQO and Roaring Bitmap performing cross-matches which the memory and network consumption is less than LQO. It can be seen here that as the data grows, the Roaring bitmap method requires more memory than the PQO cross-matches, the main reason is the usage of Common Table Expressions here, which can avoid Roaring Bitmap calculations skew at segment level. Through experimental analysis, heap table row-oriented storage is no longer suitable for wide-table astronomical data storage and we should use column-oriented to store. Using Roaring Bitmap method to accelerate the effectiveness of cross-matches. As astronomical exploration continues to evolve, the data types in the catalog data will continue to change, which necessarily

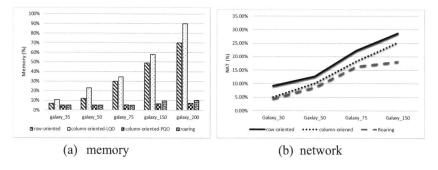

(a)   memory   (b)   network

Fig. 7. Cross-match memory and network consumption

requires further attempts to try other storage methods, such as array types, key-value pairs, column-oriented compression, and so on.

5   Conclusion

In this paper, we proposed to use Roaring Bitmap to accelerate the cross-matches operation for astronomical analysis in parallel database system. It reduces the overhead of network, CPU and memory of the database cluster. Simultaneously, we used the column-oriented compression technology to store catalog, which can reduce the I/O cost when only a few columns are accessed. In the future, we plan to work on enhance the cost estimation model for user defined types based bitmap logic operation, since current RDBMS often produce inaccurate evaluation for its intermediate result sets when the data is skewed.

Acknowledgements. This work was supported by the Fund by The National Natural Science Foundation of China (Grant No. 61462012, No. 61562010, No. U1531246), Guizhou University Graduate Innovation Fund (Grant No. 2017081) and the Innovation Team of the Data Analysis and Cloud Service of Guizhou Province (Grant No. [2015]53), Science and Technology Project of the Department of Science and Technology in Guizhou Province (Grant No. LH [2016]7427).

References

1. Metchev, S., et al.: A cross-match of 2MASS and SDSS: newly-found L and T dwarfs and an estimate of the space densitfy of T dwarfs. Astrophys. J. **676**(2), 1281–1306 (2012)
2. Detti, A., et al.: OpenGeoBase: information centric networking meets spatial database applications. In: GLOBECOM Workshops IEEE (2017)
3. Obe, R., Hsu, L.: PostGIS in Action. Geoinformatics (2015)
4. Koposov, S., Bartunov, O.: Q3C, quad tree cube – the new sky-indexing concept for huge astronomical catalogues and its realization for main astronomical queries (cone search and Xmatch) in open source database PostgreSQL. Astronom. Data Anal. Softw. Syst. **XV**, 735 (2006)
5. Calabretta, M.R., Roukema, B.F.: Mapping on the HEALPix grid. Mon. Not. Roy. Astronom. Soc. **381**(2), 865–872 (2010)
6. Gray, J., Nieto-Santisteban, M.A., Szalay, A.S.: The zones algorithm for finding points-near-a-point or cross-matching spatial datasets. Microsoft Research (2007)
7. Bonnarel, F., et al.: The ALADIN interactive sky atlas - a reference tool for identification of astronomical sources. Astron. Astrophys. Suppl. **143**(1), 33–40 (2000)
8. Zhao, Q., et al.: A paralleled large-scale astronomical cross-matching function. In: Algorithms and Architectures for Parallel Processing, International Conference, ICA3PP 2009, Taipei, Taiwan, 8–11 June 2009, Proceedings DBLP, pp. 604–614 (2009)
9. Stonebraker, M., et al.: C-store: a column-oriented DBMS. In: International Conference on Very Large Data Bases, Trondheim, Norway, 30 August–September, DBLP, pp. 553–564 (2005)
10. Abadi, D., Madden, S., Ferreira, M.: Integrating compression and execution in column-oriented database systems. In: ACM SIGMOD International Conference on Management of Data, Chicago, Illinois, USA, June, DBLP, pp. 671–682 (2006)

11. Waas, F.M.: Beyond conventional data warehousing — massively parallel data processing with greenplum database. In: Informal Proceedings of the Second International Workshop on Business Intelligence for the Real-Time Enterprise, BIRTE 2008, in Conjunction with VLDB 2008, 24 August 2008, Auckland, New Zealand, DBLP, pp. 89–96 (2008)

12. Chambi, S., et al.: Better bitmap performance with Roaring Bitmaps. Softw. Pract. Exp. **46**(5), 709–719 (2016)

13. Bayo, A., et al.: VOSA: Virtual Observatory SED Analyzer: an application to the Collinder 69 open cluster. Astron. Astrophys. **492**(1), 277–287 (2008)

14. Pence, W.D.: CFITSIO: a FITS file subroutine library. Astrophysics Source Code Library (2010)

15. Wu, K.: FastBit: an efficient indexing technology for accelerating data. Intensive Sci. **16**(1), 556–560 (2005)

16. Lemire, D., Ssi-Yan-Kai, G., Kaser, O.: Consistently faster and smaller compressed bitmaps with roaring. Softw. Pract. Exp. **46**(11), 1547–1569 (2016)

17. Wang, J., et al.: An experimental study of bitmap compression vs. inverted list compression. In: ACM International Conference ACM, pp. 993–1008 (2017)

18. Wu, K., Otoo, E., Shoshani, A.: On the performance of bitmap indices for high cardinality attributes. In: Vldb: International Conference on Very Large Data Bases, pp. 24–35 (2004)

19. Petropoulos, M., et al.: Optimization of common table expressions in MPP database systems. Proc. Vldb Endowment **8**(12), 1704–1715 (2015)

20. Nobari, S., et al.: TOUCH: in-memory spatial join by hierarchical data-oriented partitioning. In: ACM SIGMOD International Conference on Management of Data ACM, pp. 701–712 (2013)

21. Soliman, M.A., et al.: Orca: a modular query optimizer architecture for big data. ACM (2014)

22. Antova, L., El-Helw, A., Soliman, M.A., et al.: Optimizing queries over partitioned tables in MPP systems. In: SIGMOD, pp. 373–384 (2014)

Proposal of the Methodology for Identification of Repetitive Sequences in Big Data

Martin Nemeth[✉] and German Michalconok

Faculty of Materials Science and Technology in Trnava,
Institute of Applied Informatics, Automation and Mechatronics,
Slovak University of Technology in Bratislava, Bratislava, Slovakia
{martin.nemeth, german.michalconok}@stuba.sk

Abstract. The aim of this paper is to propose and describe methodology for identification of repetitive sequences in big data sets. These repetitive sequences can represent for example sequences of failures that emerge in industrial processes. Proposed methodology deals with sequences which are based on time, when the elements of particular sequence emerged. One way to approach such identification is to use so called brute-force scanning, but this approach is very demanding on computational power and computational time for big data sets cases. Our methodology approaches this issue from the side of data mining and data analysis point of view.

Keywords: Data mining · Big data · Failure · Repetitive sequences

1 Introduction

Discovering new knowledge from huge amount of data has great potential in process control and also in other areas like medicine. Many production companies, like in automotive industry, are generating large amount of process data every minute. These data has huge knowledge potential and with the use of correct techniques new knowledge about the process can be revealed. This knowledge can have impact on the control quality. Data are generated and collected at different levels of industrial process control. The process control is realized with the use of control systems with hierarchical structure. Figure 1 shows hierarchical model of process control.

Process level represents the direct interface with the process itself. This level includes devices, sensors and actuators. Communication between these elements and control systems is realized through networks and is coordinated by PLC's.

Supervisory level represents higher level of control and is often referred to as SCADA/HMI level. It integrates functions like data acquisition, command and failure alerts processing. It also integrates visualization of these data in real time and operative control interventions. This level also includes manufacturing execution systems (MES).

Business level is on the top of the hierarchy and includes components of higher hierarchy of control and management information systems. Another part of this level are systems like DDM (document data management), SCM (supply chain management) and CRM (customer relationship management) [1].

© Springer International Publishing AG, part of Springer Nature 2019
R. Silhavy (Ed.): CSOC 2018, AISC 763, pp. 390–396, 2019.
https://doi.org/10.1007/978-3-319-91186-1_40

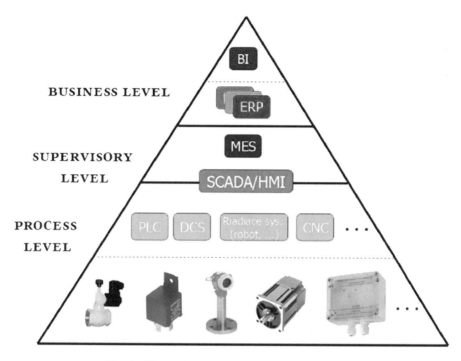

Fig. 1. Hierarchical model of distributed control [1].

This paper is devoted to processing data from the first hierarchy level. Especially in this paper we will talk about data from process level from automotive industry. The target process ready to produce large amount of data every day which will be possible

Table 1. Overview of the failure groups.

	Type A	**Type B**	**Type C**
Group	Immediate stop with out-of-order state	Immediate stop	Delayed stopping
Consequence	Causing immediate stop of one or more stations. These stations are put to out-of-order state.	Causing the stop of the cycle of the station.	Causing the stop of the station at the end of the cycle.
Example	Emergency stop. Interruption of the intangible barriers.	Failure of the operation Failure of the sensor's self-check Failure of the breaker	Faulty component Level error

with the use of sensors installed on all devices. However in current state only alarms, warnings and error messages are generated and stored locally for the period of 6 months. Next table shows the various types of these messages. The Table 1 also shows what each type could cause in the system and an example of that particular message type.

Table 1 shows three types of emerging events. These types causes the process stop and have higher priority. Table 2 shows lower priority events, which do not cause stop of the whole system, but are indicating some kind of error, or event that should be solved.

Table 2. Overview of the failure groups.

	Type D	Type E	Type F
Group	Immediate stop with out-of-order state	Warnings	Error
Consequence	They do not cause the stop, but they need to be resolved.	They do not cause the stop, and they do not need to be resolved.	They inform that the station is stopped.
Example			Waiting for loading Waiting for unloading Conveyor saturation

Table 3. Data structure

Type	Localization	Start	End	Dur.	Name
Type A	40002908 TR5MP1 ZAR.NA PLOS	19.10.2016 20:10	19.10.2016 20:10	0:00:07	AD: PORUCHA PRESUNU Z DORAZU 1 K DORAZU 2 E331.4 OP100 DB_GMM4
Type D	40002909_001 TR5MF2 NAST.F.MEDZ	23.5.2016 19:37	23.5.2016 19:37	0:00:14	AI: PORUCHA OTVORENIA DORAZU 1 E340.1 OP110 DB_GMM4
Type B	40002915 TR5CA5 RET.DOP.U.PR	3.10.2016 21:54	3.10.2016 21:54	0:00:01	AIM: Por.presunu palety z vytahu k OP70 CONVOYEUR DB_GMM5

In this paper we are aiming at proposing a methodology for finding repetitive sequences in big data. The reference data comes from automotive industry and have form of records about emerging events/failures which are described above. The event dataset consists of approximately 60000 records. These records were collected for 6 months. Each record is described by parameters like the description/ name, start date (in dd/mm/yyy hh:mm:ss format), end date (in dd/mm/yyy hh:mm:ss format), duration of the emerged event (in hh:mm:ss format), localization, where the event occurred and belongs to one of the above mentioned event groups. Table 3 shows three randomly picked records as an example of the data structure in a way, they were obtained.

2 Methods

Repetitive sequences of events and failures in the process can reveal better under-standing of the process itself and can help to understand the relationships between emerging events in the system. The reference data, describing emerging events in the process, does not contain detailed technical information, which can help to describe the cause of individual events/failures in the system. Even though the identification of event sequences can be helpful to identify relationships between individual events and can be represented as a new knowledge by a expert in given field. For finding repetitive sequences of the emerged failures in in big set of data, we have evaluated several methods like neural networks and SVM. Neither of these methods were suitable because of the data structure shown in Table 3. Both of these methods are used to solve classification and regression problems which is not this case. So, we have decided to use the cluster analysis method. This data analysis was performed using Statistica 13 software. Particularly the Data Miner module. This module offers the possibility of aggregating data using three algorithms, K-means algorithms, EM algorithm, and hierarchical clustering algorithm.

The process of finding repetitive sequence of events in the given dataset, is based on the cluster analysis at multiple hierarchical levels. At each level it is necessary to define the number of clusters in which we would like to divide our set of data. Tis parameter, number of clusters, affects what degree of similarity in aggregate records will be achieved. For this purpose, we decided to use the k-means algorithm. The Statistica software has the option of hierarchical clustering, but the workflow with this method was not suitable. The desired result of cluster analysis was to find data clusters which will contain records of a given failures occurring at similar time and which will have similar duration. The records in such cluster should however not occur within the same week. Based on such resulted clusters of data, it is then possible to monitor the surroundings of events which are in the same cluster and this way detect possible repetitive sequences of occurring events in the given system or process.

First step of the cluster analysis, was to select appropriate parameters by which the clustering algorithm would determine the similarity between the occurring events/failures. According to the given goal of the cluster analysis, we have decided to choose the parameters start, duration and the derived parameter the week number. We have derived this parameter from original dataset from the start parameter. The whole dataset was exported in a .csv format form the information system in the automotive company. We have derived the week number parameter from parameter start using an Excel function which can return week number based on the date.

For the purpose of the methodology, we have then randomly chose one of the occurring events which is called "POR. NASLEDUJUCE NASADZOVANIA ZLE OP20 DB_GMM3". This event occurs when there is a failure on a machine at OP20 (workstation 20) of GMM3 (logical zone of the entire production line).

Then it was necessary to choose the number of clusters into which we wanted to divide our data. This number was chosen according to the frequency of the chosen failure. We chose to initially divide the data about the chosen event into 10 clusters. Then we have observed the character of new clusters. The degree of similarity of records in clusters at the first level of clustering was not sufficient. The records were not similar to each other based on desired outcome. The next level of cluster analysis, was performed on one cluster from first level which had the most promising results based on similarity of the records. The records belonging to this chosen cluster were used as a input data for the next level of the analysis. The second level of cluster analysis was also initiated by choosing the right number of clusters in which the source was to be divided. This number was also determined by the number of records in the source cluster. The desired number of clusters in second level was set to 15. Table 4 shows one of the clusters obtained in the second level analysis.

Table 4. Result of the second level of cluster analysis

Week number	Start time	Start date	Duration
26	13:24:00	9.11.2016	0:00:12
27	13:22:00	24.10.2016	0:00:12
28	13:41:00	26.8.2016	0:00:12

It is clear, that the similarity between these records is significant. In some clusters the similarity was worse than in another clusters, but each cluster contained 5 records at average. This means that there was no sense in doing another level of cluster analysis. The pronounced hypothesis says that if there is a cluster of similar event records and each record is from another week, then there might be another event, which is following or previous to the records in the final cluster. If this is true, then these 2 events can be called a repetitive sequence. To evaluate this hypothesis, it was necessary to search for the records of final cluster in original dataset and then search their surroundings. Table 5 shows the found repetitive sequence. This sequence consists of 2 events occurring one after another almost every week at similar time. First of the sequence is the event mentioned above and the second is SG: PREBIEHAJUCA VYZVA ANDON OP20 DB_GMM3. This event occurs right after the first event which is a type of failure. The second event is causing the start of signalization on the same workstation, which should warn the maintenance team. From this it is clear, that these 2 events which were put together by our method are related to each other.

Table 5. Repetitive sequence example.

Failure name	Start time	Start date	Duration
AD: POR. NASLEDUJUCE NASADZOVANIA ZLE OP20 DB_GMM3	13:24:00	9.11.2016	0:00:12
SG: PREBIEHAJUCA VYZVA ANDON OP20 DB_GMM3	13:24:00	9.11.2016	0:00:14
AD: POR. NASLEDUJUCE NASADZOVANIA ZLE OP20 DB_GMM3	13:22:00	24.10.2016	0:00:12
SG: PREBIEHAJUCA VYZVA ANDON OP20 DB_GMM3	13:22:00	24.10.2016	0:00:14
AD: POR. NASLEDUJUCE NASADZOVANIA ZLE OP20 DB_GMM3	13:41:00	26.8.2016	0:00:12
SG: PREBIEHAJUCA VYZVA ANDON OP20 DB_GMM3	13:41:00	26.8.2016	0:00:12

3   Results

As it is clear from previous section, where we described the methods used for finding repetitive sequences, we were successful in finding a repetitive sequence. We assume, that this method can be used to find repetitive sequences also in other areas and with

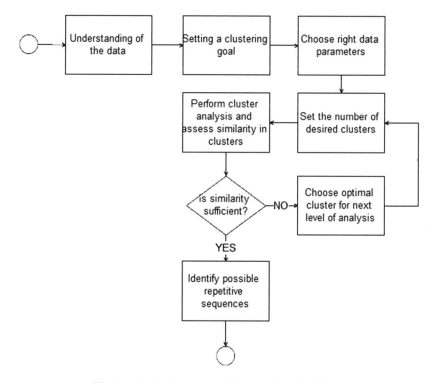

Fig. 2. Flowchart diagram of proposed methodology

other types of data. Figure 2 shows the flowchart diagram of our proposed methodology for finding repetitive sequences in big data sets.

4 Conclusion

The aim of this paper was to propose a methodology for finding repetitive sequences in big data sets. We were working with data from the automotive industry. These data were about emerging events in one production line. All events are divided into various categories and are described in Tables 1 and 2. The goal was to find relationships between these events, even though there were not detailed technical information in the dataset. Desired relationships to find were in form of repetitive sequences which we wanted to find in the original dataset. To achieve this goal, we have used clustering analysis, especially k-means algorithm. In methods section of this paper we are describing the process of finding these repetitive sequences. According to the results we have proposed a methodology for finding repetitive sequences in big datasets which is shown in the result section. The example shown in the methods section has conceptual character. To find all repetitive sequences it is necessary to repeat this process for each event type that can be found in the dataset. In our case, there were almost 200 different events spread across almost 60000 records which could be time consuming and was not necessary to asses our hypothesis and propose the methodology. The advantage of this methodology is, that it can be used also on another type of data and on data from different field.

Acknowledgments. This publication was written with financial support of the KEGA agency in the frame of the project 040STU-4/2016 "Modernization of the Automatic Control Hardware course by applying the concept Industry 4.0".

This publication is the result of implementation of the project: "UNIVERSITY SCIENTIFIC PARK: CAMPUS MTF STU - CAMBO" (ITMS: 26220220179) supported by the Research & Development Operational Program funded by the EFRR.

References

1. Tanuška P.: Tézy inauguračnej prednášky, MTF STU (2013)
2. Friedman, J.H.: Data Mining and Statistics: What's the Connection? Stanford University, Stanford, CA 94305, 10 November 2016. http://statweb.stanford.edu/~jhf/ftp/dm-stat.pdf
3. Babcock, B., Datar, M., Motwani, R., O'Callaghan, L.: Maintaining variance and k-medians over data stream windows. In: Proceedings of ACM Symposium on Principles of Database Systems (2003)
4. Kamath, C.: On the role of data mining techniques in uncertainty quantification. Int. J. Uncertain. Quantification **2**(1), 73–94 (2012)
5. Nazari, Z., et al.: A new hierarchical clustering algorithm. In: ICIIBMS 2015, Track2: Artificial Intelligence, Robotics, and Human-Computer Interaction, Okinawa, Japan
6. Alpydin, E.: Introduction to Machine Learning, pp. 143–158. The MIT Press, Cambridge (2010)

Computing Importance Value of Medical Data Parameters in Classification Tasks and Its Evaluation Using Machine Learning Methods

Andrea Peterkova[1(✉)], Martin Nemeth[1], German Michalconok[1], and Allan Bohm[2,3]

[1] Faculty of Materials Science and Technology in Trnava, Institute of Applied Informatics, Automation and Mechatronics, Slovak University of Technology in Bratislava, Bratislava, Slovakia
{andrea.peterkova,martin.nemeth,german.michalconok}@stuba.sk
[2] Faculty of Medicine, Slovak Medical University in Bratislava, Bratislava, Slovakia
allan.bohm@gmail.com
[3] Research Institute of Academy, Bratislava, Slovakia

Abstract. This paper aims to evaluate the importance values of medical data parameters for further classification tasks. One of the steps of proposed methodology for analyzing medical data is initial data analysis. One part of the initial data analysis is to determine the importance rate of parameters in given data set. The reason behind this step is to provide overview of the parameters and the idea of choosing right predictors for classification task. Statistica 13 software provides a tool for determining the importance rate of each data parameter, which can be found in feature selection module. However, it is not always clear whether is the importance rate correct or not.

Keywords: Data analysis · Classification · Predictors

1 Introduction

Data analysis and data mining are both parts of the Industry 4.0 concept. However they are used not only in the field of industry, but also in areas like medicine. Whether the target area of is industry or medicine, the methodology stays the same. First it is necessary to understand the process, which is generating the data, and to understand the data itself. Next step in data mining process is to analyze the data. Initial data analysis is necessary to be able to determine which parameters are relevant for given problem and which parameters are not. According to this step it is possible to clean data and work only with relevant portion of given data set. This is important and it has impact on the amount of time, which is needed to analyze the data. Many data sets, mainly from industrial processes, are huge and consists of hundreds of thousands data records and these data sets have huge demand on memory. Analyzing and subsequently cleaning the data can significantly reduce the demand on memory and computation power.

The original version of this chapter was revised: Acknowledgment has been included. The correction to this chapter is available at https://doi.org/10.1007/978-3-319-91186-1_50

Our research is focused on data from the area of medicine. In our research we are dealing with data related with area of cardiology. The main goal of data mining in our case is to find relationships between various medical parameters of patients with the ischemic heart disease and to be ale to predict the actual risk of myocardial infraction according to these revealed relationships. The data set is divided into 4 categories or classes of patients with the ischemic heart disease. Each category represents the final outcome of the patients after underwent treatment. First category consists of patients with chronic chest pain (angina pectoris), second category consists of patients that have encountered non-fatal myocardial infraction in past, third category consists of patients that have encountered fatal myocardial infraction because of the ischemic heart disease and fourth category consists of patients with successful treatment. Each record represents one patient. Each patient is represented by 56 parameters and there are 354 patients in data set. These parameters are mixture of laboratory blood tests, screening examinations and clinical parameters. This example shows the need of data cleaning. Dataset, which consists of 56 parameters, could have significant amount of noise, which can lower the preciseness of data mining and machine learning methods. By reducing the data noise by analyzing and cleaning data we can optimize performance of chosen data mining methods and lower the demand on computational power and memory. Table 1 shows preview of chosen parameters.

Table 1. The results of the trained and tested SVM for parameters from the first case of training

Dataset training set:	
Dependent:	Outcome
Independents:	LV EF [%], stent, diastolic, lesion 50, presence of lesion
Sample size:	
Train	339
Test	113
Overall	452
Support vector machine results:	
SVM type:	Classification type 1 (capacity = 10,000)
Kernel type:	Radial Basis Function (gamma = 0,200)
Number of support vectors	51 (0 bounded)
Support vectors per class:	3 (F-IM), 22 (N-IM), 7 (RAP), 19 (healthy)
Class. Accuracy (%):	
Train	99,705
Test	100,000
Overall	99,779

2   Background and Methods

In our research we have used chi-square test to compute importance value of each medical parameter. According to these values, we were able to specify the order of parameters based on their importance on final outcome. To evaluate the given order of important parameters we have used various methods of machine learning for a classification tasks. These methods were artificial neural networks and support vector machines (SVM). Below we discuss some theoretical background for chosen methods.

2.1   Chi-Square Test

Chi-square test is based on chi-square distribution. The chi-square distribution is a theoretical and mathematical distribution. It has wide applicability in solving statistical tasks. Graphically is chi-square distribution represented as follows. The horizontal x-axis represents the $\chi2$ values. The minimum possible value for a $\chi2$ and for x-axis is 0, but it has no maximum value. The y-axis represents the probability, or possibly the probability density, which is associated with each value of $\chi2$. The curve in this graphical representation reaches it's peak not far above 0 value, and then the curve declines as the $\chi2$ value increases. It can be said that the curve has asymmetric shape. As larger the $\chi2$ values are obtained, the curve becomes asymptotic to the x-axis. It is approaching it, but will never touch the x-axis.

The null hypothesis in chi-square test makes a statement concerning how many cases to expect in every category if this hypothesis is correct. The chi-square test is based on the difference between two values: The observed one and the expected values for each category. The chi square statistic is defined by following formula.

$$\chi^2 = \sum_i \frac{(O_i - E_i)^2}{E_i} \tag{1}$$

In previous formula Oi represents the observed number of cases in given category with the index i, and Ei represents the expected number of cases in category with the index i. Statistic for the chi-square is calculated as the difference between the observed number of cases and the expected number of cases in every category. The computed difference is then squared and divided by the expected number of cases in particular category. These values are then added for all the categories, and the sum of these values is then referred to as the chi-squared value.

The null hypothesis is a particular claim, which is concerning how the data is distributed across the data set. The null and alternative hypotheses for each chi-square test can be understood as follows:

$$H_0: O_i = E_i \tag{2}$$

$$H_1: O_i \neq E_i \tag{3}$$

The mean of a chi-square distribution is represented by its degrees of freedom. Shapes of the chi-square distributions are positively skewed. The degree of skew is decreasing with the increasing degrees of freedom. With the increasing degrees of freedom, the Chi Square distribution approaches a normal distribution. Figure 1 shows density functions for three examples of chi-square distributions. The skew decreases as the degrees of freedom increases.

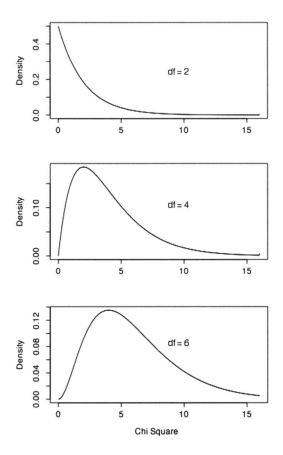

Fig. 1. The density functions for three examples of chi-square distributions

2.2 Support Vector Machines

Support Vector Machines (SVM) is a method that performs regression and classification types of tasks by constructing nonlinear decision boundaries. Taking into account the nature of the feature space where the boundaries are found, this method is able to exhibit a large degree of flexibility in solving many classification and regression tasks of varied complexities [6, 7]. However, it is mostly used in classification problems.

In Support Vector Machines algorithm, we plot each data item as a point in n-dimensional space (where n is number of features you have) with the value of each feature being the value of a particular coordinate. In our case of study, we have 56 parameters available. So each patient is represented in 56-dimensional space.

In Statistica software, there are two types how to construct the SVM. For the first, Classification SVM Type 1, training involves the minimization of the error function:

$$\frac{1}{2}w^Tw + C\sum_{i=1}^{N}\xi_i \tag{4}$$

where C is the capacity constant, w is the vector of coefficients, b is a constant, and represents parameters for handling no separable data (inputs). The index i label the N training cases. Note that represents the class labels and x_i represents the independent variables. The kernel is used to transform data from the input (independent) to the feature space. It should be noted that the larger the C, the more the error is penalized. Thus, C should be chosen with care to avoid over fitting.

Another type, hoe to construct the SVM is to use Classification SVM Type 2. In contrast to Classification SVM Type 1, the Classification SVM Type 2 model minimizes the error function:

$$\frac{1}{2}w^Tw - vp + \frac{1}{N}\sum_{i=1}^{N}\xi_i \tag{5}$$

2.3 Neural Networks

Neural networks are the sophisticated modelling techniques, which are capable of modelling extremely complex functions. Neural network learns by example so the neural network user gathers representative data. Based on these selected data is then invoked the training algorithm to automatically learn the structure of the data. Neural networks are generally applicable in virtually every situation in which a relationship between the predictor variables (independents, inputs) and predicted variables (dependents, outputs) exists. This is able to even when that relationship is very complex and not easy to articulate in the usual terms of "correlations" or "differences between groups."

In our case, the input parameters for the neural network were represented by the medical parameters of each patient.

3   Results and Validation Using SVM Method

First method to validate initial importance rate of parameters was using SVM. First step was to divide the data set to training and test stet. We have chose 70% of randomly chosen data records for training set and 30% for test set. In the next step was to choose proper parameters which to use in training and testing process. The main goal was to observe the SVM performance with different data parameters according to their importance.

In first validation case we have chose 5 most important parameters to be included in training process (DOPLNIT AKE TO BOLI). Since the goal was to predict final outcome of new patients based on known cases, the categorical dependent variable was the outcome. Next table shows the results of the trained and tested SVM for parameters from the first case.

From previous table it is clear that the SVM method performed the classification task well (99,779%). This means, that the chosen parameters do not suffer from data noise and the chi-square test results were accurate.

Table 2. The results of the trained and tested SVM for parameters from the second case of training

Dataset training set:	
Dependent:	Outcome
Independents:	LV EF [%], stent, diastolic, lesion 50, presence of lesion
Sample size:	
Train	339
Test	113
Overall	452
Support vector machine results:	
SVM type:	Classification type 1 (capacity = 10,000)
Kernel type:	Radial Basis Function (gamma = 0,200)
Number of support vectors	51 (0 bounded)
Support vectors per class:	5 (F-IM), 28 (N-IM), 9 (RAP), 20 (healthy)
Class. Accuracy (%):	
Train	96,505
Test	99,000
Overall	97,849

However to get reliable validation results it was needed to perform this test on various sets of parameters. Second validation test was performed on next 5 important parameters from the importance list. These parameters were assigned as less important than the first set of parameters.

Previous table shows the result for the second set of parameters. As it is clear, with the decreasing important rate (chi-square test) value, decreases the performance of the SVM classification method. For the last set of parameters we have chose 5 random parameters with the importancy deeply below the first 10 important parameters (Table 2).

The third test confirms that with the decreasing importance value decreases also the performance and the ability to correctly train classify new cases with particular method (Table 3).

Table 3. The results of the trained and tested SVM for parameters from the third case of training

Dataset training set:	
Dependent:	Outcome
Independents:	LV EF [%], stent, diastolic, lesion 50, presence of lesion
Sample size:	
Train	339
Test	113
Overall	452
Support vector machine results:	
SVM type:	Classification type 1 (capacity = 10,000)
Kernel type:	Radial Basis Function (gamma = 0,200)
Number of support vectors	311 (125 bounded)
Support vectors per class:	100 (F-IM), 59 (N-IM), 87 (RAP), 65 (healthy)
Class. Accuracy (%):	
Train	42,183
Test	39,823
Overall	41,593

4 Results and Validation Using Neural Networks

The validation of chi-square test was performed also with the use of artificial neural networks method. The artificial neural networks were also used for classification task exactly like in the case with the SVM method. The task for the neural network was to learn from dataset with sets of given parameters. The learned network was then used to classify new cases from test data sets.

The choice of the parameters was the same as for the SVM, which means there were 3 sets of parameters based on their importance calculated by chi-square test. A multilevel perceptron was used as the artificial neural network type. The data miner module in Statistica software was used to train the neural network. The network itself was trained according to the BFGS 43 algorithm, which stands for Broyden–Fletcher–Goldfarb–Shanno algorithm. Logistic activation function was chosen for each unit of the network (Table 4).

The first dataset with first set of parameters was divided as follows 70% training set, 15% test set, 15% validation set. Best solution for given dataset was a network with 7 hidden units and performed as follows:

Table 4. The preformation of neural network with 7 hidden units

Structure	Performance on training set [%]	Performance on validation set [%]	Performance on test set [%]	Learning algorithm	Activation function
MLP 5-7-4	95,59748	94,02985	97,01493	BFGS 43	Logistic

Second dataset consisted of 5 of less important parameters than in the first dataset. This dataset was divided same way as the first dataset. Best solution for this data was given from a network with 10 hidden units and performed as follows (Table 5).

Table 5. The preformation of neural network with 10 hidden units

Structure	Performance on training set [%]	Performance on validation set [%]	Performance on test set [%]	Learning algorithm	Activation function
MLP 5-10-4	94,65409	95,01493	95,01493	BFGS 8	Logistic

The third dataset consisted of 5 randomly chose parameters with their importance value deep below the parameters from first and second datasets. Best solution for this case was given by network with 3 hidden units. Following table shows the performance of this trained network (Table 6).

Table 6. The preformation of neural network with 3 hidden units

Structure	Performance on training set [%]	Performance on validation set [%]	Performance on test set [%]	Learning algorithm	Activation function
MLP 5-3-4	55,66038	62,68657	52,23881	BFGS 10	Logistic

5   Conclusion

This paper was aimed at one of the initial stages in data mining and gaining new knowledge from data. The step of data understanding is very important when dealing with the multidimensional data with more than 10 parameters. In such case it is very difficult to estimate all relationships between parameters that could have significant impact on the dependent categorical variable (in case of classification task). This kind of data may suffer from noise. This noise might be represented by unwanted or irrelevant parameters in data set. To be able to build a precise classification model it is important to filter out this noise, which will lead to maximization of precision and performance and minimization of computational time. One way to achieve this task is to use the methods of statistical analysis.

One of the methods, which can be used, for identifying which parameters are more important than others is the chi-square test. In this paper we were evaluating the reliability of using this method for estimating the importance of each parameter in data set. In the process of evaluation SVM method and artificial neural networks were used. We have set up classification task in which each of the method should classify new patient into the correct class of their final outcome. This scenario was repeated 3 times for each method with different data sets. Each data set consisted of 5 parameters. The first data set consisted of parameters, which were selected as the most important by the chi-square test. Second data set consisted of another 5 important parameters, which were listed right below the first set of parameters. The final data set consisted of 5 randomly chosen parameters with the importance value deep below the parameters from first two data sets. According to the results of chosen methods it is clear that with the decreasing importance computed by chi-square test the worse is the performance of selected classification method. This means that the reliability of using chi-square test for identifying the importance of parameters is sufficient.

Acknowledgments. This publication is the result of implementation of the project: "UNIVERSITY SCIENTIFIC PARK: CAMPUS MTF STU - CAMBO" (ITMS: 26220220179) supported by the Research & Development Operational Program funded by the EFRR.

This publication is the result of implementation of the project VEGA 1/0673/15: "Knowledge discovery for hierarchical control of technological and production processes" supported by the VEGA.

This publication was written with the financial support of the KEGA agency in the frame of the project 040STU-4/2016 "Modernization of the Automatic Control Hardware course by applying the concept Industry 4.0".

References

1. Geisser, S.: Predictive Inference: An Introduction. Chapman & Hall, New York (2016). ISBN 0-412-03471-9
2. Larose, D.T.: Discovering knowledge in data: an introduction to data mining. Wiley, Hoboken (2014)
3. Kotsiantis, S.B., Zaharakis, I., Pintelas, P.: Supervised machine learning: a review of classification techniques (2007)
4. Hernández, M.A., Stolfo, S.J.: Real-world data is dirty: data cleansing and the merge/purge problem. Data Mining Knowl. Disc. **2**(1), 9–37 (1998)
5. Kim, W., et al.: A taxonomy of dirty data. Data Mining Knowl. Disc. **7**(1), 81–99 (2003)
6. Meyer, D., Technikum Wien, F.H.: Support vector machines. The Interface to libsvm in package e1071 (2015)
7. Shmilovici, A.: Support vector machines. In: Data Mining and Knowledge Discovery Handbook, pp. 257–276. Springer (2005)

The Concept of Constructing the Designer for Processes of Algorithms

Yuri Rogozov[(✉)]

Institute of Computer Technology and Information Security, Southern Federal University,
Taganrog, Russian Federation
yrogozov@sfedu.ru

Abstract. The paper presents the results of theoretical research in the field of new methods of designing intelligent systems. The concept of self-projecting as a set of system and non-systemic states are introduced. Modern science tries to solve the interdisciplinary problem of eliminating the difference between the forms of representing the same meaning of finished knowledge by creating a universal abstract form of representation of knowledge. The process of obtaining knowledge is represented by the operation of creating a form of complete knowledge that are already acquired by the thinking.

Solving the problem of the semantic gap and other problems of systemic research will probably be possible only if the problem of co-building a model of the process of self-designing of knowledge that must be built on the basis of universal operations of self-knowledge, self-construction or knowledge acquisition is solved.

Keywords: Concept · Means · Algorithm · Structure · Methodology
Constructor

1 Introduction

The concept of the system is more associated with the process of designing the design processes (designing design will be called self-projecting) in the form of an organic unity of the dynamic process of obtaining knowledge and the specific structure of the stationary process of the system functioning organized in it: "The system has a systemic and non-systemic state. In the system state of its connection, the structure is also fixed. In the non-system state, the system undergoes a process of changing the structure and relationships. In the non-systemic state, the system is a symbolic form of the mechanism for changing its properties and connections" [1]. The system should represent the unity of the dynamic process, as a means of self-knowledge (the dynamic cognitive constructor of knowledge in which knowledge is constructed), and the stationary cognition process, which in the dynamic knowledge constructor constructs forms of knowledge in the form of a structure of the stationary process (means) forms of objects [2–5].

R. Silhavy (Ed.): CSOC 2018, AISC 763, pp. 406–411, 2019.
https://doi.org/10.1007/978-3-319-91186-1_42

2   Related Work

In the existing definitions and forms of the representation of the system, the instants of stability, the stationary processes of constructing forms are mostly displayed, while the ability of the system, in the process of exploitation, to change its functionality, is not reflected in the structure of the system. In existing systems, a dynamic (non-systemic) state in the form of processes for obtaining knowledge (functions) is not represented. The operations of designing or obtaining knowledge are not constructed (for example, rules are not constructed), but is represented by their result, empirically obtained forms of finished knowledge. The dynamic cognitive process of obtaining knowledge is aligned with its result: various forms of ready knowledge.

At the heart of existing abstractions is the thesis that: "… the nature and movement of cognitive thinking were already identified from the very beginning with nature and the movement of the object (and vice versa) … thinking is simply a sensory experience, dressed in verbal form, and there is no special content of thinking, except sensory images … thinking abstracted in the form of the "ability" of the individual is compared in the analysis with the individual who realizes this ability, with the world of his individual consciousness, sensuality, will, etc. As a result, the fix Only the surface of the matter is left, only the directly observed actions of an individual " [6]. The mechanisms of education of knowledge are excluded from existing abstractions of empirically obtained sign representations of forms of finished knowledge. The process of creating forms corresponds to the notion that knowledge about an object is already ready, and it is necessary to create only its form, the study of which will allow to determine knowledge about the properties of knowledge about the object. Forms of knowledge representation are studied, and not the processes of obtaining them.

3   The Process Constructor Concepts

Assuming that the design process should correspond to the process of obtaining knowledge, and the level of the implementation phase, the process of constructing specific forms of knowledge, it turns out that at the present time the cognitive process of obtaining knowledge is not represented by construction operations (self-construction) of knowledge, but empirical constructing concrete forms of ready knowledge. Roughly speaking, a paradox arises: the process of obtaining knowledge is put in correspondence with the form of the finished knowledge. Types of stages as processes self-construction of knowledge are artificially "merged" or equated with the levels of stages of designing forms of knowledge. Self-construction of stage knowledge is artificially equated with the process of constructing knowledge forms of stage level. That is why all efforts of the scientific community are aimed at finding approaches to eliminating the semantic gap between different concrete forms of representing the same sense of knowledge at different levels of one of the stages (the stage is one of the types of design operation) of design [7]. Modern science is trying to solve the interdisciplinary problem of eliminating the difference between the forms of representing the same meaning of finished knowledge by creating a universal abstract form of knowledge representation. For example, forms of the meaning of knowledge are created in the form of a more general concept (categories), which then, with the help of subject knowledge, must be transformed into its various concrete forms. The process of obtaining knowledge is represented by the operation of creating a form of ready, already acquired in the thinking knowledge.

At the stage levels, from the types of concepts of the relevant subject areas, which the subject attracts various concrete forms to impart meaning to the finished knowledge, empirical design or construction of forms of knowledge is carried out. In fact, the types of forms of representation of the meaning of the finished knowledge at different levels of stages are determined by the types of the subject knowledge involved to create the types of forms. It is still not clear what functional load these levels should carry and why exactly these or other levels and types of forms of representation of the same knowledge are used. One can only assume that each of the types of forms of knowledge, the levels of the stages of creating forms, is the result of a certain cognitive operation of creating knowledge, which is one of the operations of cognition and the process of giving the meaning of knowledge a certain physical form. But what is this operation of knowledge is not known.

Solving the problem of the semantic gap and other problems of systemic research will probably be possible only if the task of creating a model of the process of self-projecting of knowledge, which should be based on universal operations of self-knowledge, self-construction or knowledge acquisition, is solved. It is necessary to understand what the self-projecting process is as a unity of self-construction operations of knowledge and how in this operation the construction of specific forms of knowledge is organized.

Any evolutionary system, whose functionality (knowledge) can be changed in the process of its exploitation, should practically be the realized designer of the unity of evolutionary, dynamic cognitive operations of self-generation of knowledge, and the operations of self-organization of the forms of individual concepts and the stationary processes of functioning of a particular integrated system designed from them [2–4, 8]. The problem of a formalized representation of self-generation operations or self-construction of knowledge and their transition to specific forms of knowledge in the form of concrete forms of system structures has not been solved to date.

We can assume that the essence of the problem lies in the interpretation of the meaning of the basic abstraction of self-knowledge (cognition of knowledge), which underlies the explanation of existing forms of representation of systems. Existing approaches to constructing systems for interpreting the operation "cognition of knowledge" use the scheme of explanation as "knowledge about cognition," which is represented by a basic abstraction in the form of an operation for constructing forms of stationary actions [9]. The key provisions defining the existing concept of the operation "knowledge of knowledge" include the key aphorisms of Umberto Maturan [9]: "Every activity is cognition, all cognition is activity", "Every action is cognition, every cognition is an action", "Our point of departure was the realization that all cognition is action from the side of the knower "," All meditation generates the world. As such, meditation is a human act performed by a specific person in a particular place. " In fact, U. Maturana represents the thinking of the subject, in which knowledge is created, in the form of a ready-made knowledge as an operation for constructing the form of the subject's action.

Knowledge of knowledge is offered not as an operation of obtaining knowledge, but as a process of constructing the forms of the result of this operation: ready-made knowledge in the form of forms of structures of stationary actions.

As a criterion for proving the validity of this assumption, it is proposed to use a scientific explanation that will explain the process of self-generation of the cognition process: "Based on this explanatory statement, we need to understand how all the usual aspects of our cognition are generated … What are the roots and mechanisms of this process of cognition?" [9]. Practically dynamic process "knowledge of knowledge" or the process of obtaining knowledge is interpreted as a stationary process of constructing ready knowledge (experience) in the form of forms of actions that, in his opinion, are that universal operation that explains the self-generation of systems and objects.

Knowledge is understood as an empirically obtained abstract form in the form of a concrete action. It is on the notion that cognition of knowledge as a means of obtaining knowledge is the process of creating forms of structures of stationary action, modern science is based. All existing approaches to building systems empirically build the forms of activity structures, using for this abstraction the components of different subject areas.

It is the idea of the form of the structure of an action empirically organized the cognition of knowledge as from the abstractions of the components of subject domains that does not allow to solve the epistemological problem of organizing the unity of the subjective and objective, since the internal processes of the subject's thinking about obtaining knowledge are not taken into account, but only his result is ready-made knowledge, in the form of an empirically constructed form of the structure of a stationary action. Roughly speaking, in existing systems theories, the "subject" is represented as the "creator" of the form of the finished knowledge empirically obtained by him or another subject. Practically, the operation of obtaining knowledge by the subject in sign form is not represented, but only the operation of creating forms of the result of his cogitative activity in the form of a process for constructing forms of finished knowledge.

In this article, it is proposed to implement the first conceptual attempt to create an abstraction of the operation of obtaining knowledge and its unity with the operation of constructing forms of knowledge. The underlying assumption is the notion that knowledge can be represented by the form of the method of creating an operation for obtaining knowledge. It is assumed that in the process of self-construction of the operation for obtaining knowledge, specific knowledge will be created, and in it its various forms of knowledge. The knowledge presented in one form or another is proposed to be put in line with the unity of self-construction of the operation of obtaining knowledge and the operation of creating forms. Roughly: knowledge in the subject's thinking (without the form of his presentation) is suggested to be put in line with the self-construction of the operation of obtaining knowledge. Externally represented by the subject form of knowledge is proposed to put in correspondence the operation of constructing the forms of knowledge that are constructed in the operation of obtaining knowledge. For example, an explanation of the process of solving problems of a given type is a knowledge on the basis of which various forms of its solution are created. Unity in the form of the action of a "matryoshka": operations in an operation, actions from actions that are organized one in the other (light dizziness).

It is proposed to understand the "knowledge of knowledge" as a movement (the designer, the process of self-projecting), in which the dialectical unity of the operations of obtaining knowledge (by their self-construction) and the operations of constructing the forms of knowledge is organized. We will proceed with their statements that in their thinking knowledge of the subject is identified with the unity of self-construction of operations for obtaining knowledge and constructing operations for creating forms of knowledge in the form of actions through which the meaning of knowledge acquires forms of action, and through them the meaning of knowledge is transferred to the form of the object. In this unity in the form of the structure of motion, it is necessary to distinguish between the structures of operations of self-construction of knowledge and the operations of constructing the creation of forms of knowledge (forms of actions). Roughly speaking, the designer is a unity in the form of movement (action) consists of operations (constructors, movements, actions) in which knowledge and forms of knowledge are created. Without such a unity, it is impossible to describe the process of obtaining knowledge, because It is not necessary to talk about the received knowledge without obtaining the form of its presentation. Probably, knowledge not represented in one form or another (for example, in the form of language or actions) can not be considered knowledge. The operation of obtaining knowledge can not be torn away from the operation of creating its form.

This is a different scheme of explanation of the operation "knowledge of knowledge." The basis for the proposed explanation of the concept of knowledge is the provision that: "Here the works of Einstein, Schrödinger and other scientists led to the realization of the ultimate limit of physical knowledge in the form of the way we acquire this knowledge" [10]. If we develop these provisions, then we can propose that knowledge is the structure of the way of organizing the unity of the structure of the method of obtaining knowledge and the forms of the method of creating forms of knowledge obtained.

4   Conclusions

Before proceeding to the structure and forms, it is necessary to develop a way of self-building knowledge, a way of constructing knowledge forms and a way of organizing their unity in the form of a designer (actions). Based on the analysis of existing systemic notions of knowledge as ways of obtaining this knowledge, ways of organizing forms and ways of representing system properties, it is necessary to create a way of organizing the operation "knowledge of knowledge" and construct the corresponding structure for explaining the operation "knowledge of knowledge" as a dialectical unity of receiving operations knowledge (their self-construction) and operations of constructing forms of knowledge.

In subsequent works, we will consider symbolic representations of the way of organizing the unity of the structure of the method of obtaining knowledge and the forms of the method for creating forms of knowledge obtained.

Acknowledgment. The reported study was partially supported by RFBR, research project No. 18-07-00908.

References

1. Shchedrovitsky, G.P.: System movement and perspectives of the development of the system-structural methodology. In: Report at the Inter-institutional Methodological Conference of Young Scientists and Specialists, Obninsk, 31 May 1974, pp. 57–88 (1974). http://www.fondgp.ru/gp/biblio/eng/69
2. Rogozov, Y.I.: Paradigm of the semantic design of system objects. In: Proceedings of the ISA RAS 2017, T. 67, No. 3/2017, pp. 41–53 (2017)
3. Rogozov, Y.I.: General approach to the organization of definitions of systemic concepts based on the principle of generating knowledge. In: Proceedings of the 12th All-Russian Meeting on Management Problems, VSPU-2014, pp. 7822–7833. Institute for Control Sciences. V.A. Trapeznikova, Moscow, RAS (2014)
4. Rogozov, Y.I.: Methodological approach to the construction of system concepts. In: Proceedings of the ISA RAS, T. 65, No. 1 / 2015, pp. 90–110 (2015)
5. Rogozov, Y.I.: Types of methodologies. In: Informatization and Communication, No. 2, pp. 7–9 (2013)
6. Mamardashvili, M.K.: Forms and content of thinking. Middle High school, p. 6.9 (1968)
7. Rogozov, YuI: Approach to the construction of a systemic concept. Adv. Intell. Syst. Comput. **679**, 429–438 (2018)
8. Succi, G., et al.: Assessment of software developed by a third party: a case study and comparison. Inf. Sci. **328**, 237–249 (2016)
9. Rumesin, H.M., Varela, F.J.: The Tree of Human Understanding, 224 p. (2001). ISBN 5-89826-103-6. (in Russian)
10. Popkov, V.V.: G. Spencer-Brown's arithmetic of consciousness. Sci. J. Ontol. Desig, v.5, No. 1(15)/2015, pp. 85–109 (2015)

The Method of Forming Contents for a NoSQL Storage of Configurable Information System

Sergey Kucherov, Yuri Rogozov[✉], and Elena Borisova

Institute of Computer Technology and Information Security, Southern Federal University,
Taganrog, Russian Federation
{skucherov,yrogozov,eborisova}@sfedu.ru

Abstract. Configurable information systems operate under conditions of uncertainty regarding stored data and sources for filling this data. The change and expansion of functionality raises the problem of data collection and preparation, which, in turn, is connected with the integration of heterogeneous structures both in structure and in format. The technical solution to the problem is the use of ETL-systems that automate the operations of extracting, transforming and loading data into the store by rigidly defined rules. In the issues of data selection, exclusively the data specialist makes definition of rules for drive and transformation, the decision, which is a consequence of the lack of a methodological basis. This, in turn, raises such problems as the excess accuracy and inconsistency of imported data, a narrow specialization of rules (up to uniqueness) with a limited number of analytical models and known requirements for the data mart. The article presents the concept of the method for creating the content of NoSQL-storage of a configurable information system.

Keywords: Configurable information system · NoSQL · ETL
Heterogeneous data sources · Integration · Storage · Content

1   Introduction

The configurable information system is a tool for automating business processes, in which an user is able to change the functionality without programming [1]. This class of systems requires new storage technologies built on the basis of NoSQL [2]. This is due to the impossibility of implementing continuous changes in the structure of relational database management tools [3]. Considering the fact that the capabilities of configurable information systems for today are limited to the automation of individual processes [1], the problem arises of integrating such systems according to data. These data are heterogeneous, and with respect to one configurable information system, they constantly change the structure and composition (other configurable information systems that form one automated environment, can change at arbitrary times). Thus, the problem arises of integrating heterogeneous data sources. It is characterized as follows: to fill the NoSQL storage of a configurable information system with a well-known structure, it is necessary to process data from a potentially infinite set of sources with varying degrees of structuring and presentation

© Springer International Publishing AG, part of Springer Nature 2019
R. Silhavy (Ed.): CSOC 2018, AISC 763, pp. 412–419, 2019.
https://doi.org/10.1007/978-3-319-91186-1_43

formats. This, in turn, raises the task of eliminating contradictions in data, reducing the amount of useless data, filling in "gaps" in data sets, and so on.

A data specialist, who builds the process of integrating heterogeneous sources using ETL (extract-transform-load) class systems, now performs the solution of these tasks. The resulting integration process is private, suitable for a particular state of the external environment and a set of data source systems. The expert on the data, therefore, makes the formation of the meaning of the final data set, empirically, based on personal experience and assumptions. This causes an actual scientific problem of methodological support of the process of filling the NoSQL storage with content. The lack of a methodological basis for this task entails an increase in the dependence of the result of data extraction on the skills of a data specialist, which in turn causes their lack and an increase in the cost of working with data. According to some data, the ETL phase requires up to 70% of resources (financial and time) from the total volume of work with data [4–6].

Further, in the article we will consider the method of filling the NoSQL-storage of the configurable information system based on the variable ETL-procedure.

2   Related Work

World research in the field of ETL-tasks covered by this article is focused on documenting the semantics of data and the ETL procedure itself. Ontological approaches are used to solve the first problem [7–9], for the solution of the second one - BPMN and Workflow model [10, 11]. Such work, despite their effectiveness in the field of particular ETL problems, does not solve the problem in a complex manner, which does not allow them to build an entire ETL procedure model, while maintaining the data semantics and procedure, its ability to evolve. In the works of Bansal et al. [8, 9] only the question of constructing a semantic data model that affects only transformation issues (converting data to a single form of representation in a repository) is affected, this model allows to establish a correspondence between data from sources and their mappings in the data store of a configurable information system, extraction and the researchers do not affect.

Another area that is inextricably linked with filling and using NoSQL storage is the accounting of business requirements when building ETL-procedures. These requirements define sets of input data, their target representation and storage methods. The solution to this problem can be found in the works of Romero et al. [7] which, due to natural language processing methods and parsing of XML documents, formalize the requirements and on their basis perform semi-automatic generation of ETL-procedure. Despite the obvious advantages of the above methods, their main disadvantages are binding to an existing data set (necessary for building an ontology) and focusing on building one unique procedure. The issues of their evolution and development are not affected.

An interesting solution to the construction of ETL-procedures is proposed in works by sacred LDIF [12, 13] and UnifiedViews [14, 15]. The authors propose to build ETL-procedures from data processing tasks DPT (Data processing tasks), which in turn consist of data processing operations DPU (Data processing unit). Such solutions are combined into libraries of ready-made components, which is justified in the case of working with data where a finite set of analytical models and basic types of data warehouses are known.

A key feature of all known approaches, without exception, is a single-phase representation of the ETL procedure. They are aimed at modeling (at different levels of abstraction - from the conceptual to the physical) of a ready procedure designed for a specific state of the external environment and a specific set of data processing tasks. In this case, the process of building the ETL-procedure is not documented in explicit form. By creating a specification of this process in an explicit form, it is possible to ensure its development and modification in accordance with changes in business requirements and conditions of the external environment. This issue is also not considered in the works of world researchers.

3   Basic Principles of Proposed Method

Suppose that in addition to the conceptual, logical and physical representation, the ETL procedure can be in two-phase states - the go-to procedure and the construction of the procedure (its variable content). The phase state of the finished procedure can be considered the norm - a machine-processed algorithm that can be implemented without operator involvement. The phase state of the variable content of the structure of the ETL procedure describes the process of its formation. This state is currently implicitly represented in the structure of technical means (in the form of a library of "building blocks" and the basic rules for their combination), as well as reflected in the experience and qualifications of a data specialist who, using a set of technical tools, simulates the finished ETL procedure. Having the ability to explicitly specify the second phase state of the ETL procedure in which the process of its creation is declared, it is possible to ensure its development and modification in accordance with changes in business requirements and environmental conditions. It cannot be said that a single universal ETL procedure will be created in this way, since its structure (the structure in the second phase state, partially or completely devoid of content) determines the finite set of possible variants of the finished result.

The biggest interest, since this has not been considered previously in known works by authors, is the phase state of the variable content. This state of the ETL-procedure contains a description of not its components (classical decomposition), but the process of its compilation. In fact, the variable content is a representation of the experience and skills of the data scientist, which are used to solve a particular task in the face of changing situations. Therefore, the decomposition of any operation from the ETL procedure when translating it into the phase state of the variable content should reflect all possible methods, algorithms and tools that are used in this task (we call this the space of possibilities).

At the moment when such a variable content contains the concrete element that determines the situation (for example, the parameters of the data source from which the sample is taken), the matching procedure is performed. Within the framework of this procedure, the opportunity space establishes a correspondence between the input data and the target operation result, based on the established rules and constraints. Rules and restrictions, in turn, determine the permissible combinations of methods, algorithms and tools that make up the space of the capabilities of the operation that is part of the ETL procedure.

To accomplish tasks and such modeling, the authors propose to expand the methods presented in [12–15] and create the framework of the vDPU (variable data processing unit) and vDPT (variable data processing tasks) operations, which can be both in the phase state norm, and in the phase state of the variable content of the structure. That is, to make a transition from the level of variation of ETL-procedures to the variability of vDPU and vDPT, which form it.

Unlike Schultz et al. [12, 13], Oliveira et al. [16], the proposed models introduce a new phase state of "variable content", which allows you to adapt the already created operations and data processing actions to changing business requirements and environmental conditions. This allows you to model software systems through the processes of their construction, managing the content of which you can vary the capabilities of the system itself. To ensure the introduction of the second phase state - the variable content of the structure for vDPT tasks and vDPU operations, an author's approach to constructing systemic concepts is used. [17], in which the necessary basic abstraction of the cognitive procedure is proposed. This basic abstraction allows you to model the process of obtaining a concept through the organization of the unity of internal properties of the concept (which form it) and external properties (through which the subject perceives it). In the context of the problems under consideration, the concept can be considered as the vDPT task, as well as the vDPU operation for extracting, transforming or loading data, internal properties - the values of the characteristic extracted from external heterogeneous data sources, and external properties reflecting the resulting characteristics in the NoSQL storage. The main feature of such an abstraction is that the coordination of external and internal properties is ensured by including in the underlying abstraction of the method (procedure) of communication of these properties. Thus, in order to obtain from the DPU data processing operation, the vDPU operation, having a two-phase state (norms and variable content of the structure), it is necessary to include in its representation a variable content of external and internal properties and a procedure for their matching. It is worth mentioning that in implementing this basic abstraction within the framework of the set of tasks, it will require research in the field of the possibility of modern storage and data processing technologies.

The principles and rules governing the phase state of vDPU and vDPT, in contrast to MDA [18], FODA [19], CBSE [20], SOA [21], are based on a single model containing both a specific applied algorithm and its semantics, providing a change without compromising the consistency of components and the target result.

To provide the ability to model ETL-procedures for extracting, transforming and loading data from heterogeneous sources, the author's method of subject-oriented data modeling and its notation [22] is used, which allow to combine data and processes of obtaining them in a single declarative model. The concept of presenting the processes of obtaining data through a single abstraction of the subject's operation allows us, at all stages of the ETL process, to use a single and consistent way of presenting.

The above principles and methods allow one-time development of the structure of the ETL-procedure for the tasks of filling a particular storage of a configurable information system, and changing requirements and the external environment by varying the contents of vDPU and vDPT.

4   The Method of Forming Contents on the Base of Variable ETL Procedure

The process of building the procedure is as follows. At the first stage, taking into account the actual model of the configurable information system, the functional requirements and the storage structure, the data scientist builds the model of the ETL procedure using vDPU and vDPT (Fig. 1). The procedure itself is in the phase state of the variable content. This state does not require detailed data on possible sources of data.

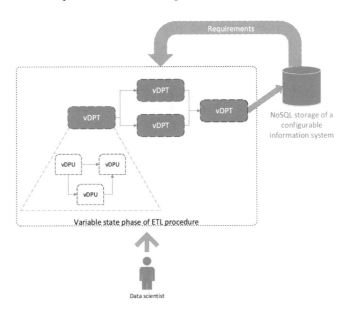

Fig. 1. Building an ETL procedure pattern

The procedure thus obtained is not a ready solution for the ETL process, but the initial data for its creation in the presence of specific data sources (Fig. 2). To form a ready-made ETL-procedure from the data sources, the domain parameters that make up the content of vDPU and vDPT.

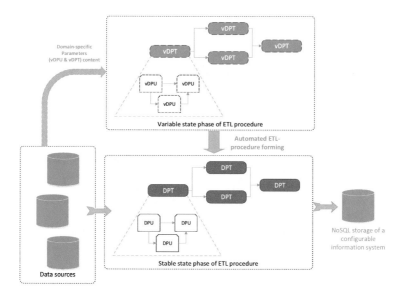

Fig. 2. Automated receipt of the finished ETL procedure

Getting ETL-procedure based on data on the domain is automated, which eliminates the need for a data scientist and leads to cost optimization.

5   Conclusions

The key difference between the solution proposed in the article is the construction of ETL-procedures possessing a two-phase state: the norms (vDPU, vDPT are received, matched and ready for machining) and the variable content of the structure (the content of vDPU and vDPT is controlled). Using the method of describing the variational actions proposed by the authors, it is possible to provide a transition between the phase states of the ETL procedure at any of the stages of its life cycle. Also, the proposed approach, unlike existing solutions, will allow:

- to perform the reconciliation and verification of the suitability of data from heterogeneous sources for solving ETL problems. The proposed structure of the ETL-procedure with a two-phase state will allow to identify unsuitable data on the content management level of vDPU and vDPT, since they will automatically disrupt the consistency of the procedure in the phase state of the norm.
- to take into account the business requirements of the configurable information system by including them in the structure of the ETL procedure with a two-phase state. vDPU and top-level vDPT will determine the business requirements of the configurable information system, and vDPU and vDPT of the lower, nested levels will determine the content of the ETL procedure that fulfills the claimed business requirements.
- to maintain the semantics of the procedure for extracting, transforming and loading data by including in the vDPU and vDPT model the phase state of the "variative

content" of objects, concepts, functions and rules that form the operation or action for data processing.

Acknowledgment. The reported study was partially supported by RFBR, research project No. 17-07-00105.

References

1. Garcia, J., Goldszmidt, G.: Building SOA composite business services. http://www.ibm.com/developerworks/webservices/library/ws-soa-composite/
2. Kucherov, S., Rogozov, Y., Sviridov, A.: NoSQL approach to data storing in configurable information systems. In: Communications in Computer and Information Science, vol. 584, pp. 120–134 (2016)
3. Kucherov, S., Rogozov, Y., Sviridov, A.: The model of subject-oriented storage of concepts sense for configurable information systems. In: Proceedings of the First International Scientific Conference "Intelligent Information Technologies for Industry" (IITI 2016). Advances in Intelligent Systems and Computing, vol. 450, pp. 317–327 (2016)
4. Simitisis, A., et al.: Data warehouse refreshment. In: Data Warehouses OlAP Concepts, Architectures, and Solutions, pp. 111–134 (2006)
5. Kimball, R., et al.: The Data Warehouse ETL Toolkit: Practical Techniques for Extracting, Cleaning Conforming, and Delivering Data, 526 p. Wiley, NewYork (2004)
6. Inmon, W.H., Strauss, D., Neushloss, G.: DW 2.0: The Architecture for the Next Generation of Data Warehousing, 400 p., New York (2008)
7. Romero, O., Simitsis, A., Abelló, A.: GEM: requirement-driven generation of ETL and multidimensional conceptual designs. Lecture Notes in Computer Science (including subseries Lecture Notes in Artificial Intelligence and Lecture Notes in Bioinformatics) LNCS, vol. 6862, pp. 80–95 (2011)
8. Bansal, S.K.: Towards a semantic extract-transform-load (ETL) framework for big data integration. In: Proceedings - 2014 IEEE International Congress on Big Data, BigData Congress 2014, pp. 522–529 (2014)
9. Bansal, S.K., Kagemann, S.: Integrating big data: a semantic extract-transform-load framework. Computer **48**(3), 42–50 (2015)
10. El Akkaoui, Z., et al.: BPMN-based conceptual modeling of ETL processes. Lecture Notes in Computer Science (including subseries Lecture Notes in Artificial Intelligence and Lecture Notes in Bioinformatics) LNCS, vol. 7448, pp. 1–14 (2012)
11. Kabiri, A., Wadjinny, F., Chiadmi, D.: Towards a framework for conceptual modeling of ETL processes. Communications in Computer and Information Science CCIS, vol. 241 (2011)
12. Schultz, A., et al.: LDIF - linked data integration framework. In: Proceedings of the 11th International Semantic Web Conference, ISWC 2011, vol. 782, pp. 1–6 (2011)
13. Schultz, A., et al.: LDIF - A framework for large-scale linked data integration. In: 21st International World Wide Web Conference (WWW 2012), Dev. Track, Lyon, France, pp. 1–3 (2012)
14. Knap, T., et al.: UnifiedViews: an ETL framework for sustainable RDF data processing. Lecture Notes in Computer Science (including subseries Lecture Notes in Artificial Intelligence and Lecture Notes in Bioinformatics), vol. 8798, pp. 379–383 (2014)
15. Knap, T., et al.: UnifiedViews: towards ETL tool for simple yet powerfull RDF data management. In: CEUR Workshop Proceedings, vol. 1343, pp. 111–120 (2015)

16. Oliveira, B., Belo, O.: A domain-specific language for ETL patterns specification in data warehousing systems. Lecture Notes in Computer Science (including subseries Lecture Notes in Artificial Intelligence and Lecture Notes in Bioinformatics), vol. 9273 (2015)
17. Rogozov, Y.: Approach to the construction of a systemic concept. In: Advances in Intelligent Systems and Computing, vol. 679, pp. 429–438 (2018)
18. Elleuch, N., Khalfallah, A., Ben, A.S.: Software architecture in model driven architecture. In: 2007 International Symposium on Computational Intelligence and Intelligent Informatics, pp. 219–223 (2007)
19. Kang, K.C., Lee, J., Donohoe, P.: Feature-oriented product line engineering. IEEE Softw. **19**(4), 58–65 (2002)
20. Jifeng, H., Li, X., Liu, Z.: Component-Based Software Engineering. Theor. Asp. Comput. **2005**, 70–95 (2005)
21. Legner, C., Heutschi, R.: SOA adoption in practice - findings from early SOA implementations. In: ECIS, № 2007, pp. 1643–1654 (2007)
22. Kucherov, S., Rogozov, Y., Sviridov, A., Rasol, M.: Approach to data warehousing in configurable information systems based on action abstraction. In: Communications in Computer and Information Science, vol. 535, pp. 139–148 (2015)

Computation of Nonlinear Free-Surface Flows Using the Method of Fundamental Solutions

Mohamed Loukili$^{(\boxtimes)}$, Laila El Aarabi, and Soumia Mordane

Polymer Physics and Critical Phenomena Laboratory,
Faculty of Sciences Ben M'sik, University Hassan II,
P.O. BOX 7955, Casablanca, Morocco
md.loukili@gmail.com

Abstract. A meshless numerical model for nonlinear free surface water waves is presented in this paper, to demonstrate that the localized method of fundamental solutions (MFS) is a stable, accurate tool for simulating and modeling the nonlinear propagation of gravity waves in the approximation of irrotational, incompressible and the fluid is assumed to be inviscid. Using the fundamental solution of the Laplace equation as the radial basis function, the problem is solved by collocation of boundary points. The present model is a first applied to simulate the generation of monochromatic periodic gravity waves by applying a semi-analytical or semi-numerical method to resolve the nonlinear gravity waves propagation, have verified by different orders of linear problems. As an application we are interested in the mechanisms of the interaction of a rectangular obstacle fixed on the bottom of the numerical wave tank (NWT) in the presence of the waves in order to provide information on attenuation process, and validate the numerical tool that we have developed for the treatment of this problem.

Keywords: Nonlinear water waves · Method of fundamental solutions
Radial basis functions · Gravity waves · Reflection

1 Introduction

The propagation of the nonlinear free surface water waves is an important phenomena in coastal and ocean engineering, the rapid development in offshore activities, along with increasingly extreme environments, demands more accurate prediction of the hydrodynamic performance of offshore structures [1, 2]. The study of the fully nonlinear water wave problem in general uses either the fully nonlinear potential approach, and they are usually handicapped or prohibited by the nonlinear features of the kinematic and dynamic free surface boundary conditions. The different calculation methods are mostly based on a potential flow approach and only the resolution methods are different. Most mathematical models used in several previous works are developed based on conventional methods such as finite element method (FEM), finite differences method (FDM) [3], and finite volume method (FVM) remained still today the most used methods for solving partial derivatives equations (PDE) modeling the physical phenomena. These methods have a very solid theoretical foundation and many techniques have been improving them over the years. However, their implementation is

© Springer International Publishing AG, part of Springer Nature 2019
R. Silhavy (Ed.): CSOC 2018, AISC 763, pp. 420–430, 2019.
https://doi.org/10.1007/978-3-319-91186-1_44

difficult and expensive in some cases, particularly in the field of modeling large deformation, fracture mechanics, free surfaces, swell, etc. Indeed, these methods are based on an area of the mesh [4] in which the problem must be discrete. In the case of some problems mentioned above, the mesh is necessarily very deformed, which might lead to the distortion of the elements and volumes purposes. For the spirit to overcome this types of problems the meshes method have been developed. These methods especially based on discretizations in points and not in parts or volumes.

The popular meshless numerical schemes to simulate the nonlinear water waves are the boundary element method (BEM) or boundary integral equation method (BIEM), are most efficient to solve the Laplace equation in the presence of a free surface flow. The main advantage of BEM/BIEM lies in having only to discretize the boundary of fluid domain (and body) [5]. It is also the RBF methods [6, 7] that uses the fundamental solution of Laplace's equation has been proposed for simulating the nonlinear free surface water waves and wave–structure interactions [7, 8]. The method is further employed by Xiao et al. [9] to simulate nonlinear irregular waves in shallow water by introducing a new form of the numerical wave tank (NWT).

In this work, we propose a semi-analytical or a semi-numerical method to resolve the nonlinear water waves problem in a numerical wave tank (NWT). This technique allows us to have verified linear problems for the different orders. The numerical resolution of the linear problems is performed by the method of fundamental solution (MFS), using the fundamental solution of the Laplace equation as the radial basis functions, in order to present solution for higher orders (order 1, order 2 …). A comparison with the exact solution of stokes is presented.

As an application we are interested in the mechanisms of the interaction of an obstacle fixed on the bottom of a numerical wave tank (NWT) in the presence of the linear gravity waves. This topic is still in its infancy as far as meshless numerical modeling of linear and nonlinear water waves, this numerical results obtained show that the present model is successful in the simulation of the fully nonlinear water waves.

2   Position of the Problem

We consider a monochromatic incident wave of small amplitude propagating in the presence of the flat bottom of a numerical wave tank (Fig. 1).

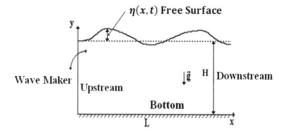

Fig. 1. Sketch of numerical wave tank (NWT).

As a part of the nonlinear wave theory, the movement is supposed to be plane and periodic in time, irrotational, incompressible and the fluid is assumed to be inviscid. The formulation can be made in terms of the velocity potential φ and the elevation of the free surface η, these two variables will be assumed complex, with harmonic dependence on time. The velocity potential $\varphi(x, y, t)$ can be written as [10]:

$$\varphi(x, y, t) = \phi(x, y)e^{i\omega t} \tag{1}$$

Under these assumptions, the problem of propagation of the nonlinear water waves is reduced to resolve the system of equations:

$$\Delta\varphi = 0 \qquad\qquad \text{in the fluid domain} \tag{2}$$

$$\frac{\partial\varphi}{\partial y} = 0 \qquad\qquad \text{at the bottom} \tag{3}$$

$$\frac{\partial\varphi}{\partial t} + \frac{1}{2}||\vec{\nabla}\varphi||^2 + g\eta = 0 \qquad\qquad \text{at the free surface} \tag{4}$$

$$\frac{\partial\varphi}{\partial y} = \frac{\partial\eta}{\partial t} + \frac{\partial\eta}{\partial x}\frac{\partial\varphi}{\partial x} \qquad\qquad \text{at the free surface} \tag{5}$$

Where: (x,y) the Cartesian coordinates, H the depth of the water; g the gravity and $\vec{\nabla}$ gradient operator. To these equations are added the upstream and downstream conditions as described in the references [11].

3   Analytical Formulation

To overcome the difficulty of the nonlinearities of this problem, we will look for a solution in trigonometric series of the fundamental frequency $f_0 = \frac{\omega_0}{2\pi}$, ω_0 as the fundamental harmonic. More precisely, the potential φ (x, y, t) and the free surface η (x, t) are sought in the form:

$$\varphi(x, y, t) = \sum_i^n \phi_n(x, y)\, e^{in\omega_0 t} \tag{6}$$

$$\eta(x, t) = \sum_i^n \eta_n(x)\, e^{in\omega_0 t} \tag{7}$$

For this procedure, the free surface itself presented as a boundary unknown to the domain of integration of the equations of motion. To overcome this difficulty and to facilitate the writing of the boundary conditions, the free surface is projected on the mean line by expressing the potential φ and its derivatives by a Taylor expansion in y = H, as in the Stokes theory [10]. The different operators involved in equations of motion can be written as follows:

$$\left.\frac{\partial\varphi}{\partial t}\right|_{y=H+\eta} = \left.\frac{\partial\varphi}{\partial t}\right|_{y=H} + \eta\left.\frac{\partial}{\partial y}\frac{\partial\varphi}{\partial t}\right|_{y=H} + \frac{1}{2}\eta^2\left.\frac{\partial^2}{\partial y^2}\frac{\partial\varphi}{\partial t}\right|_{y=H} + \frac{1}{6}\eta^3\left.\frac{\partial^3}{\partial y^3}\frac{\partial\varphi}{\partial t}\right|_{y=H} + \cdots \tag{8}$$

By injecting (6 and 7) into the expression (8), we obtain:

$$\frac{\partial \varphi}{\partial t}\bigg|_{y=H+\eta} = i\omega_0 \phi_1|_{y=H}\, e^{i\omega_0 t} + i\omega_0 \left(2\phi_1 + \eta_1 \frac{\partial \phi_1}{\partial y}\right)\bigg|_{y=H} e^{i\omega_0 t} + \cdots \tag{9}$$

$$\frac{\partial \varphi}{\partial x}\bigg|_{y=H+\eta} = \frac{\partial \varphi}{\partial x}\bigg|_{y=H} + \eta \frac{\partial}{\partial y}\frac{\partial \varphi}{\partial x}\bigg|_{y=H} + \frac{1}{2}\eta^2 \frac{\partial^2}{\partial y^2}\frac{\partial \varphi}{\partial x}\bigg|_{y=H} + \frac{1}{6}\eta^3 \frac{\partial^3}{\partial y^3}\frac{\partial \varphi}{\partial x}\bigg|_{y=H} + \cdots \tag{10}$$

By injecting (6 and 7) into the expression (10), we obtain:

$$\frac{\partial \varphi}{\partial x}\bigg|_{y=H+\eta} = \frac{\partial \varphi}{\partial x}\bigg|_{y=H} e^{i\omega_0 t} + \left(\frac{\partial \varphi_2}{\partial x} + \eta_1 \frac{\partial}{\partial y}\frac{\partial \varphi_1}{\partial x}\right)\bigg|_{y=H} e^{i2\omega_0 t} + \cdots \tag{11}$$

$$\frac{\partial \varphi}{\partial y}\bigg|_{y=H+\eta} = \frac{\partial \varphi}{\partial y}\bigg|_{y=H} + \eta \frac{\partial}{\partial y}\frac{\partial \varphi}{\partial y}\bigg|_{y=H} + \frac{1}{2}\eta^2 \frac{\partial^2}{\partial y^2}\frac{\partial \varphi}{\partial y}\bigg|_{y=H} + \frac{1}{6}\eta^3 \frac{\partial^3}{\partial y^3}\frac{\partial \varphi}{\partial y}\bigg|_{y=H} + \cdots \tag{12}$$

By injecting (6 and 7) into the expression (12), we obtain:

$$\frac{\partial \varphi}{\partial y}\bigg|_{y=H+\eta} = \frac{\partial \varphi}{\partial y}\bigg|_{y=H} e^{i\omega_0 t} + \left(\frac{\partial \varphi_2}{\partial y} + \eta_1 \frac{\partial}{\partial y}\frac{\partial \varphi_1}{\partial y}\right)\bigg|_{y=H} e^{i2\omega_0 t} + \cdots \tag{13}$$

By replacing these expressions in Eqs. (4) and (5), after combining them And by injecting the series (6 and 7) into the Eqs. (2) and (3), we obtain, after term-by-term identification according to the increasing power of $e^{in\omega_0 t}$, a succession of a linear problem Well posed verified by different orders written as:

Problem at the order 1 (Linear problem):

$$\Delta \phi_1 = 0 \qquad \text{in the fluid domain} \tag{14}$$

$$\frac{\partial \phi_1}{\partial n} = 0 \qquad \text{at the bottom} \tag{15}$$

$$\frac{\partial \phi_1}{\partial y} - \frac{\omega_0^2}{g}\phi_1 = 0 \qquad \text{at the free surface} \tag{16}$$

$$\eta_1 = \frac{i\omega_0}{g}\phi_1 \tag{17}$$

Problem at the order 2:

$$\Delta \phi_2 = 0 \qquad \text{in the fluid domain} \tag{18}$$

$$\frac{\partial \phi_2}{\partial n} = 0 \qquad \text{at the bottom} \tag{19}$$

$$\frac{\partial \phi_2}{\partial y} - \frac{4\omega_0^2}{g} \phi_2 = 2i \frac{\omega_0}{g} S_2^2 - S_1^2 \qquad \text{at the free surface} \qquad (20)$$

$$\eta_2 = \frac{1}{g} S_2^2 - 2i \frac{\omega_0}{g} \phi_2$$

Where:

$$S_1^2 = \eta_1 \frac{\partial^2 \phi_1}{\partial y^2} - \frac{\partial \eta_1}{\partial x} \frac{\partial \phi_1}{\partial x} \qquad (21)$$

$$S_2^2 = \frac{1}{2} \left[\left(\frac{\partial \phi_1}{\partial x} \right)^2 + \left(\frac{\partial \phi_1}{\partial y} \right)^2 \right] \qquad (22)$$

Problem at the order 3:

$$\Delta \phi_3 = 0 \qquad \text{in the fluid domain} \qquad (23)$$

$$\frac{\partial \phi_3}{\partial n} = 0 \qquad \text{at the bottom} \qquad (24)$$

$$\frac{\partial \phi_3}{\partial y} - \frac{9\omega_0^2}{g} \phi_3 = 3i \frac{\omega_0}{g} S_2^3 - S_1^3 \qquad \text{at the free surface} \qquad (25)$$

$$\eta_3 = \frac{1}{g} S_2^3 - 3i \frac{\omega_0}{g} \phi_3 \qquad (26)$$

Where:

$$S_1^3 = -\frac{\partial \eta_1}{\partial x} \frac{\partial \phi_2}{\partial x} - \frac{\partial \eta_2}{\partial x} \frac{\partial \phi_1}{\partial x} - \eta_1 \frac{\partial \eta_1}{\partial x} \frac{\partial^2 \phi_1}{\partial x \partial y} + \frac{1}{2} \eta_1^2 \frac{\partial^3 \phi_1}{\partial y^3} + \eta_1 \frac{\partial^2 \phi_1}{\partial y^2} + \eta_2 \frac{\partial^2 \phi_1}{\partial y^2}. \qquad (27)$$

$$S_2^3 = -\frac{\partial \phi_1}{\partial x} \frac{\partial \phi_2}{\partial x} - \eta_1 \frac{\partial \phi_1}{\partial x} \frac{\partial^2 \phi_1}{\partial x \partial y} - \frac{\partial \phi_1}{\partial y} \frac{\partial \phi_2}{\partial y} - \eta_1 \frac{\partial \phi_1}{\partial y} \frac{\partial^2 \phi_1}{\partial y^2}. \qquad (28)$$

4 Numerical Formulation

The fully nonlinear water waves problem is reduced to solve the system of equations which is written in the general form [8]:

$$\begin{cases} \Delta \phi(x, y) = 0 & \text{In } \Omega \quad (29) \\ \frac{\partial \phi(x,y)}{\partial n} = g(x, y) & \text{On } \partial\Omega \quad (30) \end{cases}$$

Ω: represents the computational domain
$\partial\Omega$: represents Neumann boundaries.

The fundamental solution for the Laplace 2D is:

$$\phi_i(x,y) = \sum_{j=1}^{N} \alpha_j \, G(r_{ij}) \tag{31}$$

$$G(r_{ij}) = \tfrac{-1}{2\pi} \ln(r_{ij}) : \qquad\qquad \text{is the Green's function.} \tag{32}$$

$r_{ij} = \sqrt{\left((x_i - \beta_j)^2 + (y_i - \delta_j)^2 + c^2\right)}$: is the distance between a field point and a boundary point.

Where N is the number of points in the boundaries and α_j are the undetermined coefficients that represent the strengths of singularities, respectively; $\overrightarrow{x_i} = (x_i, y_i)$ is the position of the field points, $\overrightarrow{s_j} = (\beta_j, \delta_j)$ is the location of the boundary points, and c is the spatial parameter introduced to be free from the ill-conditioning effect, is considered to be small. To determine the unknowns α_j of the problem, we will use a collocation method to the boundary conditions. Once these coefficients are determined, the potential $\varphi(x,y)$ and the elevation of the free surface $\eta(x,t)$ can be obtained by the linear combination of the fundamental solutions. The problem solving procedure at the order 1 are generalized to the different orders, by constructing a recurring formulation in the general form: $L_n = L(\varphi_n(x,y,z)) = S(\varphi_{n-1}, \varphi_{n-2}, \ldots)$, which is solved in a sequential manner. In this work, we typically select the location of boundary points, and centre points as illustrated in Fig. 2.

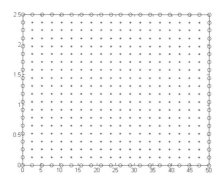

Fig. 2. Distribution of field points (red symbols) and boundary points (blue symbols).

5 Numerical Simulations and Discussions

The present numerical model is applied to simulate the wave propagation for different-orders Stokes waves (order 1, order 2, and order 3…), as a verification model and first application. The second application is the numerical modeling of wave-structure interaction using the method of fundamental solutions.

5.1   Numerical Simulation of the Linear Water Waves

The first step of this work is to simulate the linear water waves propagation as a first validation test of the present model, taking as amplitude $a = 0.01$ m, length L = 50 m, height H = 2.5 m and T = 4.42 s, we present the distributions of the elevation of the free surface along the length of the numerical wave tank (NWT) as shown in Fig. 3.

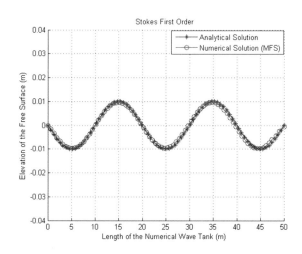

Fig. 3. Numerical and analytical elevation of the free surface (First order).

The results illustrated in the Fig. 3, demonstrated that the present numerical results affirm that the adopted numerical approach is valid tool for the treatment of the linear problem. This affirmation based on the good agreement with the analytical solutions for progressive waves in a water domain of homogeneous depth.

5.2   Numerical Simulation of Gravity Waves for Higher Orders

In this section, we present a comparison of the numerical model with the analytical solution of the Stokes gravity waves for higher orders (order 2, order 3…), along the numerical wave tank (NWT), taking as amplitude $a = 0.01$ m, length L = 50 m, height H = 2.5 m, and $T = 3.07$ s, in order to analyze the capacity of the present model to resolve the problem of the nonlinear propagation of the waves in the presence of a flat bottom.

The Figs. 4 and 5 demonstrated that the MFS is a simple and powerful meshless numerical approach to simulate the weakly nonlinear water wave problems. The present numerical model is validated against in several cases of wave propagations. Our interests in future work that the present model will be extended to the development of a 3D NWT for the prediction of hydrodynamic interaction between water waves and offshore structures in extreme seaways.

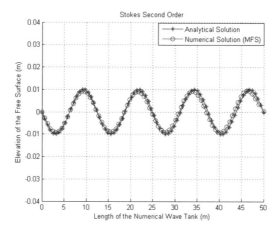

Fig. 4. Numerical and analytical elevation of the free surface (Second order).

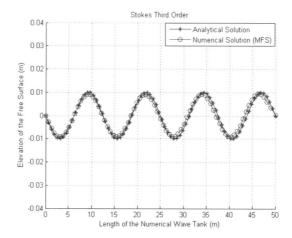

Fig. 5. Numerical and analytical elevation of the free surface (Third order).

6   Application: Wave Structure-Interaction

We consider a monochromatic incident wave of small amplitude propagating in the presence of an obstacle fixed on the bottom of a numerical wave tank (NWT) (Fig. 6).

In order to better analyze the effect of the presence of the obstacle on the waves, we have chosen to plot the reflection coefficient as a function of the dimensionless product kh. We will thus express the characteristics of the reflection coefficient as a function of the dispersion parameter associated with the immersion of the obstacle. We present a comparison between our numerical results and the experimental measurements [12]. For all of our results, we chose: H = 2.5 m, length of the obstacle l = 25 cm.

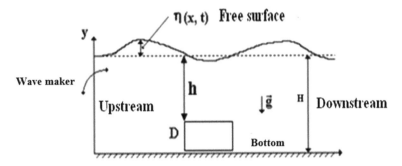

Fig. 6. Sketch of numerical wave tank (NWT) with obstacle on the bottom.

We present in Fig. 7 (h/H = 50%) and Fig. 8 (h/H = 32%) a comparison of the numerical and experimental results by plotting the variation of the reflection coefficient versus kh. These curves show that the numerical results are in acceptable agreement with the experimental measurements.

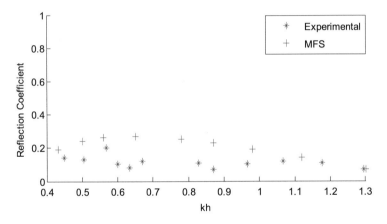

Fig. 7. Reflection coefficient R as a function of kh (h/H = 50%).

The numerical calculation code developed thus proves to be efficient for the calculation of the reflection coefficient during the passage of a monochromatic swell on an obstacle constituted by rectangular steps fixed on the bottom of a channel within the linear theory framework. This assertion is based on good agreement with the experimental results.

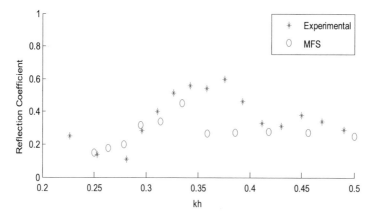

Fig. 8. Reflection coefficient R as a function of kh (h/H = 32%).

7 Conclusion

In this work a numerical scheme was proposed to resolve the nonlinear wave propagation equations using the method of fundamental solutions (MFS). Two examples in 2D situations are discussed:

– Propagation of progressive monochromatic waves in a NWT with a flat bottom where the MFS approach coincides with the analytical solution when the values of the parameter c is considered to be in the range 0.03 m and 0.1 m. outside this range the matrices happen to be ill-conditioned.
– Propagation of the wave in a NWT with a rectangular obstacle lying on the bottom, including the analysis of the reflection coefficient R as a function of kh in order to provide indications on the wave attenuation process after the obstacle.

References

1. Craik, A.D.D.: The origins of water wave theory. Annu. Rev. Fluid Mech. **36**, 1–28 (2004)
2. Xiao, L.F., Yang, J.M., Peng, T., Tao, T.: A free surface interpolation approach for rapid simulation of short waves in meshless numerical wave tank based on the radial basis function. J. Comput. Phys. **307**, 203–224 (2016)
3. Loukili, M., Mordane, S., Chagdali, M.: Formulation semi analytique de la propagation non linéaire de la houle. XIVèmes Journées Nationales Génie Côtier – Génie Civil (2016)
4. Zhang, X., Song, K.Z., Lu, M.W., Liu, X.: Meshless methods based on collocation with radial basis functions. Comput. Mech. **26**, 333–343 (2000)
5. Hongmei, Y., Liu, Y.: An efficient high-order boundary element method for nonlinear wave-wave and wave-body interactions. J. Comput. Phys. **230**, 402–424 (2011)
6. Wu, N.J., Tsay, T.K., Young, D.L.: Meshless simulation for fully nonlinear water waves. Int. J. Numer. Methods Fluids **50**, 219–234 (2006)
7. Wu, N.J., Tsay, T.K., Young, D.L.: Computation of nonlinear free-surface flows by a meshless numerical method. J. Waterways Port Coast Ocean Eng. **134**, 97–103 (2008)

8. Loukili, M., Mordane, S.: Modélisation de l'interaction houle-marche rectangulaire par la méthode des solutions fondamentales. 13ème Congrès de Mécanique Meknès, MAROC (2017)

9. Xiao, L.F., Yang, J.M., Peng, T., Li, J.: A meshless numerical wave tank for simulation of nonlinear irregular waves in shallow water. Int. J. Numer. Methods Fluids 61, 165–184 (2009)

10. Mordane, S.: Contribution numérique à la résolution du problème d'interaction houle-obstacles. Thèse de Doctorat d'Etat, Université Hassan II- Mohammedia, Casablanca, Maroc (2001)

11. Orlansky, I.: A simple boundary condition for unbonded hyperbolic flows. J. Comput. Phys. 21(3), 251–269 (1976). https://doi.org/10.1016/0021-9991(76)90023-1

12. Chahine, C.: Contribution à l'étude expérimentale de la houle avec des obstacles immergés. Thèse de Doctorat d'état, Université Hassan II-Mohammedia, Casablanca-Maroc (2002)

Quality of Service for Aggregated VoIP Streams

Przemysław Włodarski[✉]

Department of Signal Processing and Multimedia Engineering,
Faculty of Electrical Engineering,
West Pomeranian University of Technology in Szczecin,
Sikorskiego 37, 70-313 Szczecin, Poland
przemyslaw.wlodarski@zut.edu.pl

Abstract. A new method for assuring the quality of service for aggregated streams of VoIP in ICT packet networks is presented. This approach takes into consideration the presence of the self-similarity phenomenon in network traffic and introduces a new, approximated relation between the level of self-similarity and its influence on the performance of queueing systems with finite buffer size. A new algorithm for determining of reservation parameters for aggregated VoIP streams is proposed and discussed.

Keywords: VoIP · Quality of Service · Self-similarity
Queueing performance

1 Introduction

Quality of Service (QoS) in VoIP transmission is the most important factor for the end user who pays for the service. QoS-aware applications based on network mechanisms and decision processes improve the user perception of the human speech. In order to provide assumed QoS level, one has to deal with the network problems such as congestion, bottlenecks, failures, bit error rates, etc. Since the network traffic is usually self-similar [1,2], the complexity of this issue increases because of the non-trivial relationship between the level of self-similarity and queueing systems measures like latency and packet loss rate. The advantage of using packet IP networks instead of circuit switching (PSTN) for VoIP transmission is that packet switching allows to utilize the same transmission medium to transport multiple VoIP calls as well as other data. The main benefits are cost reduction (communication infrastructure), portability (the same phone number everywhere, especially when traveling) and mobility of the service (voicemail, call logs, security). VoIP technology is deployed using either centralized or a distributed architecture, but the most commonly user architecture is based on client-server centralized model, where the server plays a role of PBX (Private Branch eXchange). The example tasks for PBX are phone device registering, establishing session, call redirection, IVR (Interactive Voice Response), etc. To provide a suitable quality of voice traffic that is sensitive to

© Springer International Publishing AG, part of Springer Nature 2019
R. Silhavy (Ed.): CSOC 2018, AISC 763, pp. 431–437, 2019.
https://doi.org/10.1007/978-3-319-91186-1_45

delayed or lost packets, especially when the level of self-similarity in the background traffic is high, one should apply additional mechanisms that satisfies QoS requirements [4].

Another issue concerned with the assuring of desired level of QoS is the resource reservation mechanism [5,6]. RSVP (ReSerVation Protocol) delivers PATH and RESV parameters to each router through the network path. In our case, messages sent by the VoIP gateways include parameters of resource reservation in the routers for many simultaneously transmitted VoIP streams, saving bandwidth for packets of VoIP signaling protocols (SIP, H.323) that would be necessary for each RSVP session. In this article, a simple algorithm of determining resource reservation parameters for providing certain QoS level of aggregated VoIP streams, that takes into consideration the self-similar nature of network traffic, is presented and discussed.

2   Proposed Algorithm

The aim of the proposed algorithm is to provide a simple method for determining the required bandwidth as well as buffer size for aggregated VoIP streams shown in Fig. 1. The problem is further complicated by the fact, that self-similar traffic has an adverse effect on queueing performance. In order to find the simple relationship between the level of self-similarity and the mean delay or mean packet loss of the queueing system, a number of experiments was carried out. All random sequences used in simulation process consisted of 10000 samples that was utilized to drive the input process of the queueing system with one server and different buffer sizes. Because the all packets in VoIP streams have the same size, the service times was deterministic.

The results of the simulations are presented in Fig. 2. For the example traffic load (0.893) one can see that for different buffer sizes there is a big difference in delay. The highest delay is for infinite buffer sizes, because all packets find a space in the queue, the lowest delay is when there is only a place for one packet in the queue. A bigger difference can be observed for packet loss values (Fig. 3). The service times are deterministic due to the constant length of VoIP packets. As far as for the impact the self-similarity on these measures goes, it was determined by the shift of traffic load for the same, $M/D/1/K$ queueing system:

$$S_D(H, K) = d_0(K) \cdot H + d_1(K)$$
$$S_P(H, K) = p_0(K) \cdot H + p_1(K), \tag{1}$$

where S_D and S_P are shift functions of delay and packet loss, respectively, H - Hurst exponent responsible for the level of self-similarity, K - buffer size. The following functions: $d_0(K)$, $d_1(K)$, $p_0(K)$ and $p_1(K)$ are the binomial functions that were obtained by minimizing the mean square error for delay and packet loss with respect to the shift in traffic load:

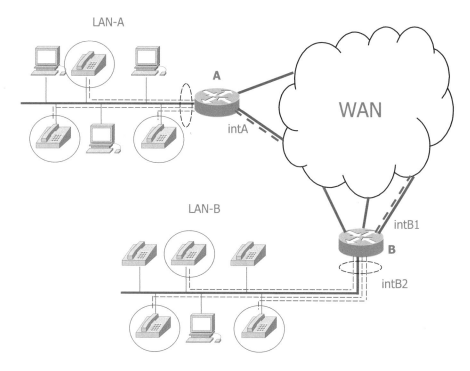

Fig. 1. Diagram of the VoIP network

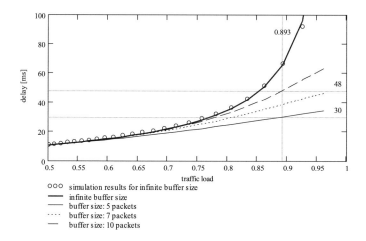

Fig. 2. Simulation results for delay in M/D/1 system and analytical curves for M/D/1/K systems

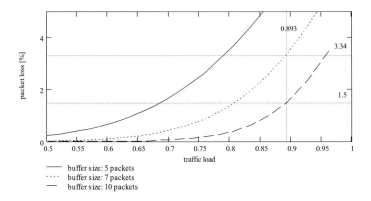

Fig. 3. Packet loss in M/D/1/K systems

$$
\begin{aligned}
d_0(K) &= 10^{-3}\left(-0.9K^2 + 35.7K + 18.7\right) \\
d_1(K) &= 10^{-3}\left(0.5K^2 - 19.6K - 4.7\right) \\
p_0(K) &= 10^{-3}\left(-1.84K^2 + 61.1K + 14.2\right) \\
p_1(K) &= 10^{-3}\left(0.96K^2 - 30.9K - 23.1\right).
\end{aligned} \tag{2}
$$

The above equations make it possible to determine delay and packet loss values for the self-similar traffic based on the analytical dependencies. The theoretical delay and packet loss of the M/D/1/K can be calculated using the following formulas [7]:

$$
D(\rho, K) = \frac{t_s}{\lambda} \sum_{k=0}^{K} kP(\rho, k, K) \tag{3}
$$

$$
P_B(\rho, K) = 1 - \frac{\psi(\rho, K - 1)}{1 + \rho\psi(\rho, K - 1)}, \tag{4}
$$

where:

$$
\begin{aligned}
P(\rho, 0, K) &= \frac{1}{1 + \rho\psi(\rho, K - 1)} &&\text{for}\quad k = 0 \\
P(\rho, k, K) &= \frac{\psi(\rho, k) - \psi(\rho, k - 1)}{1 + \rho\psi(\rho, K - 1)} &&\text{for}\quad k = 1, \ldots, K - 1 \\
P(\rho, K, K) &= P_B(\rho, K) &&\text{for}\quad k = K
\end{aligned} \tag{5}
$$

$$
\psi(\rho, k) = \sum_{i=0}^{k} \frac{(-1)^i}{i!}(k - i)^i e^{(k-1)\rho}\rho^i \tag{6}
$$

The full algorithm of determining the bandwidth and buffer size parameters is presented in Fig. 4 and will be explained by means of an example. The input parameters for the algorithm are:

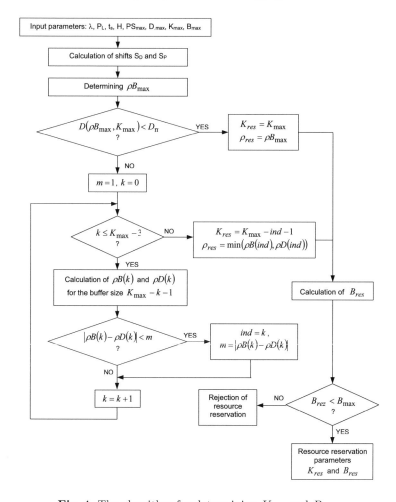

Fig. 4. The algorithm for determining K_{res} and B_{res}

- λ: input traffic intensity [pkts/time unit]
- P_L: packet length [bytes]
- t_s: time unit - interval when packets are counted [s]
- PS_{max}: maximum allowed packet loss, probability $[0\ldots1]$
- D_{max}: maximum delay [s]
- K_{max}: maximum buffer size [packets]
- B_{max}: maximum allowed bandwidth [kbits/s]

Regarding a certain level of QoS defined by the maximum delay and packet loss rate, one can determine reservation bandwidth and buffer size for the outgoing aggregated VoIP packet stream from $LAN\text{-}A$ network towards router B ($A \rightarrow B$). The same should be done in the opposite direction. Because the algorithm is the same for both routers, only the case for the router B will be investigated. Network traffic parameters are measured at the $intB1$ interface. Then, the QoS parameters are determined: PS_{max} and D_{max}, so that after taking into account the average level of packet loss and delay between routers A and B, do not exceed these maximum values for VoIP devices in LAN-A and LAN-B networks. After determining the input parameters, the shift values of the traffic load are calculated with respect to the theoretical delay and packet loss of the M/D/1/K queueing system (Eqs. 3 and 4). These shift values corresponds to the adverse impact of self-similarity in network traffic, where the Hurst exponent value belongs to the range $H \in (0.5; 1)$. After many simulations, a simple relationship between H and shift was developed (Eq. 1). The next step is to calculate the maximum value of traffic load $\rho_{B_{max}}$ resulting from the following condition:

$$P_B\left(\rho_{B_{max}} + S_P(H, K_{max}), K_{max}\right) = PS_{max}, \tag{7}$$

where $P_B(\rho, K)$ is the blocking probability in M/D/1/K system. Additionally, if the analytic delay value for $\rho_{B_{max}}$ and K_{max} is less than the assumed maximum delay D_{max} then the reservation bandwidth B_{res} is determined and the reserved buffer size $K_{res} = K_{max}$. The reservation is possible only when the B_{res} is less than B_{max}, otherwise the reservation is rejected. The next condition is related to the delay, if the following inequality holds: $D(\rho_{B_{max}}, K_{max}) > D_{max}$, where $D(\rho, K)$ denotes theoretical delay, then the maximum value of traffic load is determined by decreasing the buffer size, having regard the Hurst exponent based shift. The $D(\rho, K)$ delay is calculated based on the probability mass function of the queue length in M/D/1/K system.

The maximum value of traffic load, which corresponds to the minimum value of reserved bandwidth, can be determined for the smallest absolute difference of traffic loads calculated for delay and packet loss, i.e.:

$$|\rho_B(k) - \rho_D(k)| \to min., \quad k = 0, \ldots, K_{max} - 2 \tag{8}$$

where $K = K_{max} - k - 1$ is the buffer size. The bandwidth B_{res} is related to the less value of traffic load:

$$B_{res} = \left\lceil \frac{\lambda \cdot P_L \cdot 8}{\rho_{res} \cdot t_s} \right\rceil, \tag{9}$$

$$\rho_{res} = \min\left(\rho_B(ind), \rho_D(ind)\right), \tag{10}$$

where $\lceil x \rceil$ denotes the lowest integer greater or equal x and ρ_{res} is the reservation load corresponding to the smaller value obtained for delay and packet loss for the reservation buffer size $K_{res} = K_{max} - ind - 1$. The index variable ind can be found applying Eq. 8. Example values for the aforementioned procedure of obtaining reservation load value is shown in Table 1. All results in this table were performed for the following input parameters: $B_{in} = 266.54$ kbps (aggregated VoIP traffic rate), $H = 0.8$. For the assumed QoS parameters: $P_{max} = 1.5\%$, $D_{max} = 30$ms, using the proposed algorithm one can obtain the reservation buffer size and bandwidth: $K_{res} = 8$ packets and $B_{res} = 366$ kbps.

Table 1. Example values of traffic load and bandwidth for different buffer sizes

Buffer size	2 pkts	3 pkts	7 pkts	8 pkts	9 pkts
Load for packet loss	0.158	0.374	0.72	0.745	0.763
Load for delay	0.972	0.964	0.75	**0.729**	0.716
Difference in loads	0.813	0.59	0.03	**0.016**	0.047
Bandwidth for packet loss [kbps]	1684	714	371	358	350
Bandwidth for delay [kbps]	275	277	356	**366**	373
Difference in bandwidth [kbps]	1409	437	15	**8**	23

3   Conclusions

The most important issue for the user that uses VoIP technology is the quality of service, which depends mostly on delay and packet loss rate. To provide the good quality, the voice packets should not be dropped at all, or the drop rate should be very low (1–2%). Another quality factor is delay (latency), which should not exceed certain value. Unfortunately, the performance of queueing systems at each of the router interface is degraded due to the self-similarity phenomenon. Proposed algorithm takes into account this important factor and makes it possible to determine buffer size as well as bandwidth for aggregated VoIP stream and subsequently, to make a resource reservation in routers that are involved in the transmission. Furthermore, the proposed method is relatively simple, which makes it easy to implement by network designers as well as software developers of VoIP applications. There is one more factor that has not yet been incorporated into proposed algorithm - the varying delay called jitter, which could be the subject of the future research.

References

1. Polaganga, R.K., Liang, Q.: Self-Similarity and modeling of LTE/LTE-A data traffic. Measurement **75**, 218–229 (2015)
2. Zhang, Y., Huang, N., Xing, L.: A novel flux-fluctuation law for network with self-similar traffic. Physica A: Stat. Mech. Appl. **452**, 299–310 (2016)
3. Sousa-Vieira, M.E., Suárez-González, A., López-García, C., Fernández-Veiga, M., López-Ardao, J.C., Rodríguez-Rubio, R.F.: Fast simulation of self-similar and correlated processes. Math. Comput. Simul. **80**(10), 2040–2061 (2010)
4. Vuletić, P.V., Protić, J.Ž.: Self-similar cross-traffic analysis as a foundation for choosing among active available bandwidth measurement strategies. Comput. Commun. **34**(10), 1145–1158 (2011)
5. Pana, F., Put, F.: Performance evaluation of RSVP using OPNET Modeler. Simul. Model. Pract. Theory **49**, 85–97 (2014)
6. Belhoul, A., Şekercioğlu, Y.A., Mani, N.: Mobility-Aware RSVP: a framework for improving the performance of multimedia services over wireless IP-based mobile networks. Comput. Commun. **32**(4), 569–582 (2009)
7. Seo, D.W.: Explicit formulae for characteristics of finite-capacity M/D/1 queues. Electron. Telecommun. Res. Inst. J. **36**(4), 609–616 (2014)

Improving Case Based Software Effort Estimation Using a Multi-criteria Decision Technique

Fadoua Fellir[1(✉)], Khalid Nafil[2], Rajaa Touahni[1], and Lawrence Chung[3]

[1] Lastid Laboratory, Faculty of Sciences, Ibn Tofail University, Kenitra, Morocco
fadoua_fellir@live.fr
[2] Mohamed V University ENSIAS, Rabat, Morocco
[3] Erik Johnson School of Engineering and Computer Science,
The University of Texas at Dallas,
P.O. Box 830688, Richardson, TX 75083-0688, USA

Abstract. Producing an accurate effort estimate is essential for effective software project management, and yet remains highly challenging and difficult to achieve, especially at the early stage of software development, because very little detail about the project are known at its beginning. To cope with this challenge, we present a novel framework for software effort estimation, which takes an incremental approach on one hand, using a case-based reasoning (CBR) model, while considering a comprehensive set of different types of requirements models on the other hand, including functional requirements (FRs), non-functional requirements (NFRs), and domain properties (DPs). Concerning the use of CBR, this framework offers a multi-criteria technique for enhancing the accuracy of similarity measures among cases of multiple past projects that are similar to the current software project, towards determining and selecting the most similar one. We have tested our proposed framework on 36 (students') projects and the results are very encouraging, in the sense that the difference between the estimated effort and the actual effort was lower than 10% in most cases.

Keywords: FRs (Functional requirements)
NFRs (Non-functional requirements) · Software effort estimation
Case based reasoning (CBR) · Multi-criteria decision analysis (MCDA)

1 Introduction

Software effort estimation has been noted as among the most critical factors that influence the success of software development projects, and yet equally challenging. Although software effort estimation is not a rocket science, a good technique for software effort estimation can be a basis for helping avoid overruns in cost, time and manpower, while helping enhance the level of confidence in the overall process that may be defined and used for the particular software development project. To date, several software effort estimation models have been developed,

© Springer International Publishing AG, part of Springer Nature 2019
R. Silhavy (Ed.): CSOC 2018, AISC 763, pp. 438–451, 2019.
https://doi.org/10.1007/978-3-319-91186-1_46

applying different techniques, such as expert judgments [1,2], analogy based estimation [3], algorithmic effort estimation [4–6], and AI (artificial intelligence) based models that have been extensively used in software effort estimation [7,8], one promising method among them is the Case-Based Reasoning (CBR) approach [9,10], in which effort estimates are made by retrieving a similar project from the past and adapting its actual effort for estimating the effort of a new project. CBR techniques have been found to have certain advantages over other machine learning techniques, thanks to their capability of handling large and multi-modal datasets [3] better than other machine learning techniques, while also offering the ability for the data to be easily updated continuously. Moreover, they are more reliable than, and preferable to, other methods because they can simulate human problem solving behaviors [11,12]. Unfortunately, all estimation methods are subject to uncertainty [13], and CBR approaches are no exception to this. A CBR model consists of a set of steps, including presentation, retrieval, adaptation, validation and System learning -that are executed one after another. Each step has some uncertainty associated with it. In this article, we try to cope with the uncertainty in the first and second phases of the CBR cycle. Before starting the measurement process, it is important to analyze the requirements, and also to check the relationships between the different components to be estimated. Therefore, presentation is an important step. The most reliable estimates come from understanding and taking into consideration most, if not all, of the essential software features: functional requirements (FRs), non-functional requirements (NFRs) and domain properties (DPs). FRs describe what the software system should do – e.g., "The system shall provide an interface to help people obtain their positions and a route to their homes". NFRs describe how FRs are met – e.g., "The system shall provide an interface to help people easily obtain their positions and routes to their homes". DPs describe the characteristics of the intended application domain and what should hold in it – e.g., "An old person using the smartphone app should be able to use the app". Relative to FRs, NFRs are mostly ambiguous in their meaning and cannot absolutely be verified for their satisfaction (See, for example, [14] for more on this). NFRs are sometimes also called quality attributes or constraints [15,16]. The term "quality attributes" typically is used to refer to attributes in relation to the quality of the services and the functions offered by the system (e.g., safety and security), and "constraints" to refer to restrictions on the features offered by the system (e.g., team capability and experience, methodology to be followed in building the system, such as RUP/UP and agile process). In addition to FRs and NFRs, domain properties (DPs) - characteristics of the context or environment in which the projected system is to function - should also be taken into account during software development [17], hence accordingly in the measurement process. Provision of a comprehensive description of, and about, the software under development (SUD), would help reduce risks and uncertainty in the estimate. Another important phase is the retrieval phase. In fact, the success of a case-based reasoning system highly depends on the performance of the retrieval step used [18]. In case-based effort estimation, several techniques are used to calculate the similarity between the new and past cases. However, existing literature [19,20] on the similarity measurement phase of CBR for software effort estimation has not yet provided a consensus on what precisely should be the method for selecting the most similar case in the presence of alternatives. To cope with this problem, in this paper, we propose the use of a Multi-Criteria Decision Analysis (MCDA) methodology [21,22]. To the best of our knowledge, multi-criteria decision techniques have not yet been used in the CBR retrieval phase for effort estimation. In this paper, we show how our proposal can help with selecting the most similar project, towards providing more accurate effort estimates. The objective of this paper is to present a novel software effort estimation framework, which (i) uses CBR, (ii) takes into consideration most, if not all, of the different classes of requirements and characteristics of the software under development that have been recognized as being essential, and (iii) deals with the issue of selecting the most similar case in the presence of a set of similar projects, during the measurement process. Our approach employs an incremental methodology of estimation in the different requirements gathering and development cycles, hence providing not a one-shot final estimate but instead an estimate to each cycle. The rest of this paper is organized as follows. Section 2 gives an overview of casebased effort estimation methods. Section 3 presents our methodology. Section 4 presents our experimental results, together with analysis. Finally, Sect. 5 summarizes our work and discusses future work.

2   Related Work

Recently, many different methods have been proposed to estimate software effort based on machine learning techniques such as Neural Networks [23] and fuzzy logic [24]. Some methods have also promoted the use of models based on case-based reasoning (CBR) that have been extensively used in software effort estimation [25]. Accordingly, many researchers have tried to provide methods, which combine CBR and other techniques, in order to improve the accuracy of effort estimation, for example, combinations of CBR with Genetic Algorithms [26], with decision trees [27], with Neural Networks or Fuzzy Logic [28–30]. Our work adopts the spirit of these previous studies, by the use of CBR technique, but unlike other previous work cited above, in this paper, we describe how to benefit from the strengths of CBR to measure requirements effort taking into account essential elements of requirements, i.e., functional requirements (FRs), non-functional requirements (NFRs), and domain properties (DPs). Our work goes beyond by using multi-criteria decision making for measuring similarities between the requirements of the system under development, and past projects. In CBR models, estimates rely on previous cases that are similar, or close to the new project to be estimated, as such, similarity is a critical phase. To measure the similarity between projects, semantic similarity finding, nearest neighbor techniques (K-Nearest Neighbor) and distance metrics (Euclidean, Manhattan and Minkowski distances), are the widely used techniques in analogy-based estimations [20].

In software effort estimation, selecting the most appropriate case is very important, in order to adjust the reused effort, many previous works are proposed to deal with uncertainty in similarity measures, like approach adopting the genetic algorithm method [31], and studies proposed to help selecting the most similar case [32,33]. In the most researches, the similarity is calculated, in particular, by considering similarity measures between software projects' features in a global and an independent way. But we believe this is not enough, because the selection of similar case is not only a FR dependent but also NFRs and DPs dependent. The method we investigate here is different from the ones presented in the literature so far. The basic idea is that FRs, NFRs, DPs, need to be all included in the similarity measurement process. In this paper, we want to extend the CBR model by presenting new model that uses a multi-criteria decision analysis (MCDA) technique in order to solve the issue of selecting the most similar case, especially, in the presence of a set of alternatives. This proposed solution depends not only on the requirements' similarity values in global, but also mostly on the degree of similarity between different component (FR, NFRs, DP). After we apply a semantic similarity between the different component, using the semantic similarity STS [34], we assign weights to each criterion, that will be used for measuring the final similarity value of the entire requirement to be estimated. Multicriterion decision making (MCDM) comprises a set of alternatives, among which we have to select, according to the weights of a finite set of criterion or attributes [35]. Since usually not all criteria are equally similar, so criteria weights are used to express the degree of similarity of each criterion. In this study we propose making use of the general spirit of the weighted product model (WPM) [22] to address this issue. The illustration of the proposed approach is described in the next section

3   The Methodology

In this paper, a new hybrid intelligent framework for software effort estimation is proposed to deal with uncertainty of development effort estimation based on the combination of CBR and multi-criteria analysis method.

3.1   CBR Model

Recently CBR has attracted more attention in the field of software effort estimation. In general, CBR solution is based on solving a new problem by recalling a previous similar situation(s) and reusing information and knowledge from that situation(s). In our paper, the problem consists on measuring the effort of new software; this problem is solved by seeking a similar previous case, known as the source case, and reusing its solution to solve the target case which is the new project to be estimated. A case is represented by the description of a problem and its solution. The CBR cycle used in this paper comprises five phases: Presentation. Consist of elaborate and formalize the description of the current project to be measured. In this stage, requirements are represented according to requirement structure as shown in Fig. 1. Each NFR (quality attribute or constraint) is related to a FR(s) that depend heavily on the context properties: the DPs. This representation is common to requirements of completed projects stored in the case base. This step helps understand exactly what you will be measuring and helps visualize the software requirements. Retrieval. It is about measuring the similarity between the requirement to be measured and requirements candidates. In this paper similar cases are selected and ranked based on their semantics similarity value, by applying the STS similarity technique [34] to each requirement. Adaptation, Validation and Learning [36]. In the first stages of project development, requirements may not be completely defined and other changes in the project's requirements may still take place. So, features are incomplete or inaccurate, the resulting, the prediction is also inaccurate. As requirements keep changing and improving during cycles, a technique to face this problem is to measure the software effort increasingly (which means estimates should also be done in cycles), thus our estimation process is performed in incremental phases, in each phase we provide an estimate. Nowadays, software systems are becoming more and more complicated and software requirements are widely increasing. To estimate the effort, it is important to understand requirements to be measured and also the context in which the system will operate [38]. Our first step is to produce a description of the requirements to be estimated according to Fig. 1. In a form that we relate each FR to the quality attributes and constraints: NFRs and constraints, and to the context in which it will be developed: DPs; which would facilitate the analysis of requirements and that could be updated easily; as the requirements continue to change. To present an adequate model for effort and cost estimation, that propose more reliable and more accurate estimates, it is important to include all the parameters in the estimation process, and take into consideration the interdependency relationships between them.

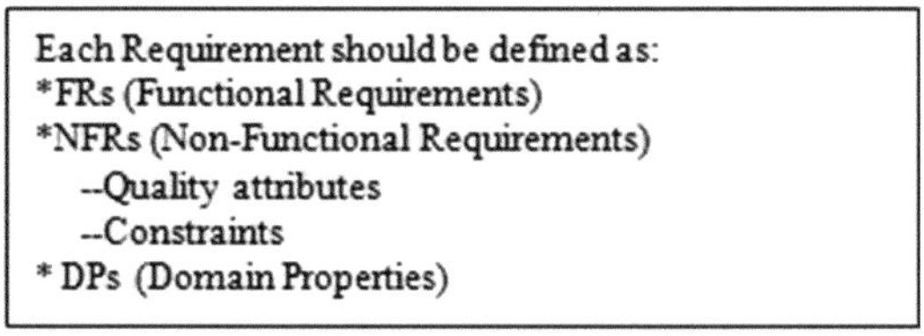

```
Each Requirement should be defined as:
*FRs (Functional Requirements)
*NFRs (Non-Functional Requirements)
   --Quality attributes
   --Constraints
* DPs (Domain Properties)
```

Fig. 1. Requirements structure representation.

The success of a CBR system depends critically on the performance of the retrieval process used and, more specifically, on similarity measure used to retrieve requirements that are similar. In this paper, we address the basic question of software estimation practitioners (basically who perform their estimations based on historical projects), which is: What case should I use? Our effort estimation methodology is based on comparing current project under development with past projects in the database. See Fig. 2.

According to the domain of the new project, we start by ranking it among the existing projects domains, in a way that we choose the respective domain.

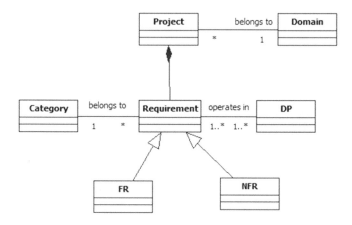

Fig. 2. General structure of projects in the database.

Then we measure the similarity between the current project and all the projects within that domain. The similarity is measured, in a bottom-up process, by measuring the similarity between requirements belonging to the same category. For each requirement, we start by comparing the category of the current requirement, with the categories of the other past requirements (as we cannot compare requirements that are incomparable: requirements belonging to different categories). Concerning NFR quality attribute, the classification is done according to the ISO 25010:2011 [37]; the category and sub-category. For The NFR constraint, the classification is done according to the list of categories of constraints (Programming language, Operating system or platforms supported, Use of a specific framework, Budget, Team composition, Methodology) [38]. If the category is not found, in this case it is a new one (a new category is created), and the effort to develop requirements will not be included in the process of effort measurement, as our approach do not deal with completely new requirements. After finding requirements in the same category, we measure the similarity between FRs, using STS. We get a list S of requirements that are functionally similar. Then from this list we measure the similarity between the NFRs and the DPs related to each FR in S and those related to the FR of the requirement to be estimated. So, We get as a result, for each FR, the values of similarity between NFRs related to the current FR, and the NFRs related to each FR in the case base, and also the value of similarities between DPs.

Let A and B are two requirements. To get the total value of similarity of NFRs, we apply the formula (1).

$$NFR\_similarity(A, B) = \frac{\sum_{i=1}^{n} max(STS(NFR_{A,i}, NFR_{B,i'=1}^{z}))}{n} \quad (1)$$

Let n the number of NFRs related to the current requirement to be estimated A, and z the number of NFRs related to B To get the total similarity value of domain properties related to the two requirements to be compared, we apply the formula (2).

$$DP\_similarity(A, B) = \frac{\sum_{i=1}^{m} max(STS(DP_{A,i}, DP_{B,i'=1}^{w}))}{m} \quad (2)$$

m is the number of DPs related to A, and w is number of DPs related to B. Thus, when comparing requirements, we get for each comparison a vector of similarity composed of the value of similarity of FR, the value of similarity of NFRs and the value of similarity of DPs. The vector is given as (FR_similarity, NFRs_similarity, DPs_similarity). In some cases, we can face various requirements that have the same value of similarity, but with different vectors. And in the presence of multiple alternatives, we may face a problem of decision, and the same question arises; which one to choose? To deal with this problem, a new process of similarity measurement is proposed, in which the final similarity values are considered with respect to weights assigned to the similarity value of each criterion. Accordingly the final similarity value provided takes into consideration not only the value but also the degree of similarity of each criterion, by integrating a multi-criteria decision technique.

3.2   Multi-criteria Decision Optimization

In this section, we propose a multi-criteria decision analysis (MCDA) technique to aid analyst in selecting the most similar case. Weights of criteria define the degree of similarity. Each criterion is graded separately using a scale (from 5 to 1) according to its similarity value. In order to determine the weights, the following process is followed: In each case there is an FR related to a list of NFRs and DPs. To measure the similarity, a pairwise comparison is performed between a new requirement and requirements from historical dataset, and the comparisons are made based on similarity vectors. To each criterion, in this step, we associate a weight, in a way that the more is the value of similarity of each criterion, the lower will be the degree or the weight associated with it. We use a scale of weights w: from 0 to 5 : 5 means very low (the case where the value of similarity varies from 0 to 0,2), 4 means low (]0.2, 0.4]), 3 means Medium (the value between]0.4, 0.6]), 2 means High (]0.6, 0.8], and 1 means Very high (]0.8 to 1], as the similarity value is between 0 and 1. In order to determine the final value of similarity of the requirements, we follow this process: Having in mind that to get the final similarity value, we apply the spirit of WPM; we divide the value of similarity of each criterion by the weight; we assign a weight to each criterion (W1, W2, and W3) as described above. Then we multiply these values results to calculate total value of similarity. The optimization formula of requirement similarity measurement is described in (3):

$$Requirements\_similarity(A, B) = \frac{FR\_similarity(A, B)}{w_1} * \frac{NFR\_similarity(A, B)}{w_2} *$$
$$\frac{DP\_similarity(A, B)}{w_3} \quad (3)$$

In comparing alternatives A1 and A2, the degree of similarity of the component of A1 against the degree of similarity of A2 is the criteria's weighting in favor of A1, and A2. Essentially, this requirement similarity measurement optimization involves finding the weight or degree of similarity of the entire requirements components, and using them to provide the final value of similarity using (4). To measure the similarity we use STS that respects the axioms of similarity measures in the literature as in [39]. (Reflexivity), (symmetry), and (transitivity). Noting that the similarity value is between 0–1 (positive). We apply the formula (3) for all the requirements' similarity vectors, and then we choose the requirement that have the maximum value (4).

$$similar\_requirement = max(Requirements\_similarity) \qquad (4)$$

Multi-criteria similarity measurement allows a better, easier, and more efficient framework for the identification of selection criteria, calculating their weights and analysis. This will help us to choose one requirement from requirements alternatives. Then based on the requirement chosen we estimate the effort, by applying directly the formula (5).

$$Effort(A, B) = priority_A * requirements\_similarity(A, B) * \frac{effort(B)}{priority_B} \quad (5)$$

In the process of effort measurement, requirement priority has been taken into account, as a variable that affect the effort, since requirement with high priority may take an effort higher than the effort required; usually superior or equal to the effort required. The approach adopted to measure the priority of requirements is based on AHP [40].

4   Results

The performance of the proposed estimation model is evaluated through estimating development efforts for 36 projects, which have been collected by the RE laboratory in the computer science department, at the University of Texas at Dallas. All these are students' projects, which are teamwork-based and developed in two phases, one semester long. At the first step, projects are ranked and compared with respect to their project domain. First, the projects are classified into two different project domains as shown in Table 1. In the second step, we analyzed these 36 projects with respect to: project domain, effort, team size, and process model. The details of the data sets are presented in Table 2.

In this case study, the data sets are used to generate the estimation by using our methodology. The dataset of the first lines represent projects from the same project domain: DMS refers to a web-based Dynamic Meeting scheduler system, while HOPE refers to a smartphone app for Helping Old/Other People. The results of this case study are shown in Table 2, Figs. 3 and 4. Two different evaluation methods have been used. These methods include the Mean of Magnitude of Error Relative to the Estimate (MMER) and the Prediction Level (PRED).

Table 1. Dataset domains

Domain	Attributes	Number of cases	Units	Min	Median	Mean	Max
Hope (Helping other people easily)	3	11	Man-hours	135	351.675	430.939	1004.9
DMS (Distributed Meeting Scheduler)	3	25	Man-hours	153	300	331.8	624

MMER is one of the criteria used for cost estimation models evaluation that can sometimes be more accurate than the Mean of the Magnitude of Relative Error (MMRE), which does not select the best prediction model [41]. MMER is the mean of MER [42], defined in (6) and (7).

$$MER_i = \frac{Actual effort_i - Estimated effort_i}{EstimatedEffort_i} \tag{6}$$

$$MMER = \frac{1}{n} \sum_{i=1}^{n} MER_i \tag{7}$$

phase1 MMER = 10%
phase2 MMER = 9%
Pred(X) is defined as follows:

$$Pred(X) = \frac{L}{n} \tag{8}$$

where n is the total number of projects, and L is number of projects with an error (MER) less than or equal to l. Pred(x) is used as a complementary criterion. The estimation accuracy is directly proportional to Pred(X). The common values used for X are 25%, 20% and 10%. In this study we used Pred(10). The results are shown in Table 2.

Phase1 Pred(10) = 55%
Phase2 Pred(10) = 97%

Although the evaluation criteria cannot confirm whether the results are significant or not, they are nonetheless useful for indicating the accuracy of the model being evaluated. Overall, all results for the use of the proposed framework were good, and corroborate the hypotheses: if we measure all the requirements and handle the uncertainty, which is associated with the similarity measurement, we can obtain estimates with less uncertainty. In addition, we can see from Table 2 that, as the project progresses and requirements become clearer, eventually estimates become more accurate. In Phase1, the maximum MER value with the proposed framework is 0,14 (Project 10) and the minimum value is 0,06 (Project 1, Phase2) see Figs. 3 and 4.

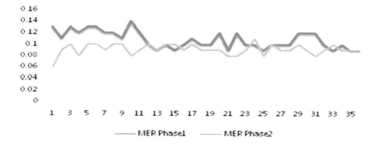

Fig. 3. MER Phase1 versus MER Phase2.

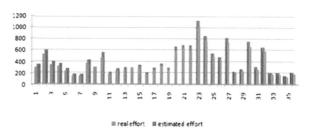

Fig. 4. The estimated effort against the real effort.

5   Limitations and Threats to Validity

In our opinion, the greatest threats to the validity of the results of our case study are related to the use of students' projects. We are aware of the potential problems of solely relying on students' projects for the validation. Firstly, the constraints given for the students' projects, e.g., the deadlines for project deliverables and the number of student participants are more or less fixed, according to the length of the particular semesters and the size of the class. These factors would have some effect on the time spent, as well as the amount of the effort. In an industrial setting, in contrast, the amount of time and effort to be spent is not necessarily fixed but vary and should be estimated, in consideration of the possible project scope and other characteristics, such as the nature of the application. Secondly, a more serious threat has to do with the degree of stakeholder realism (or lack thereof), especially concerning elicitation which typically requires interactions with real stakeholders – be they customers and/or users with unclear and changing needs, through a range of activities (and notably with a high influx of oral face-to-face activities, such as interviews and workshops). In many students' projects, requirements elicitation is instead mainly focused on documents (e.g., some task description provided by the teaching staff), where students play different roles, e.g., customer, user and developer. Additionally, the knowledge and skill levels of the project participants differ between students and industrial practitioners, where the latter often times may have years of experience in similar system development projects that the former lack. Also, concerning reuse,

Table 2. Results of experiments

P. No.	Project domain	Team size	Process model methodology	Effort phase1 (man-hours)	Effort phase2	Calculated effort phase1	Calculated effort phase2	MER phase1	MER phase2
1	DMS	7	Spiral	301	28	349.30	30	0,13	0,06
2	DMS	8	Agile/UP	523	16	592.99	17.66	0,11	0,09
3	DMS	5	Agile/UP	350	20	405.06	22.30	0,13	0,10
4	DMS	5	Agile/UP	325	16.75	369.36	18.29	0,12	0,08
5	DMS	4	Agile	240	8	278.11	8,96	0,13	0,09
6	DMS	3	Agile	153	6	176.9	6,71	0,13	0,10
7	DMS	3	Agile	153	9	174.26	9.9	0,12	0,09
8	DMS	8	Agile	378.4	10.4	432.16	11.62	0,12	0,10
9	DMS	7	Agile	308	7	285,185	7,85	0,11	0,10
10	DMS	9	Agile	477	36	556.30	39.30	0,14	0,08
11	DMS	3	Spiral	186	12	213.22	13,25	0,12	0,09
12	DMS	4	Incremental	248	8	277.24	7,25	0,10	0,10
13	DMS	5	Incremental	260	15	297.54	16.54	0,09	0,09
14	DMS	5	Incremental	300	10	272,72	11.22	0,10	0,10
15	DMS	5	Incremental	300	11.5	343.32	12.79	0,09	0,10
16	DMS	3	Incremental	180	12	201.53	13.20	0,10	0,09
17	DMS	4	Incremental	260	16	292.24	17.84	0,11	0,10
18	DMS	5	Incremental	325	15	364.61	16.53	0,10	0,09
19	DMS	4	Incremental	260	16	290.55	17.65	0,10	0,09
20	DMS	10	Waterfall	580	40	661.22	44.05	0,12	0,09
21	DMS	12	Waterfall	624	48	687.09	52.22	0,09	0,08
22	DMS	10	Spiral	600	30	682.69 32.77	0,12	0,08	
23	HOPE	13	Spiral	1004.9	42.9	1124.05	47.28	0,10	0,09
24	HOPE	10	Incremental	760	60	850.11	67,46	0,10	0,11
25	HOPE	9	Spiral	486	27	537.9	29.5	0,09	0,08
26	HOPE	9	Spiral	433.35	18	484.73	20,25	0,10	0,10
27	HOPE	11	Spiral	730.4	33	816.45 37.05	0,10	0,09	
28	HOPE	5	Spiral	201.5	20	223.6	22.08	0,10	0,09
29	HOPE	5	Spiral	235	10	267.77	11,22	0,12	0,10
30	HOPE	9	Spiral	657	36	751.87 39.86	0,12	0,09	
31	HOPE	5	Iterative	270	30	310.25	32.63	0,12	0,08
32	HOPE	10	Spiral	580	30	645.04	33.07	0,10	0,09
33	HOPE	4	Spiral	176	16	195.2	17,92	0,09	0,10
34	HOPE	4	Spiral	176	16	195.55	17.73	0,10	0,09
35	HOPE	3	Spiral	135	18	149.86	19.85	0,09	0,09
36	HOPE	4	Spiral	188	12	208.5	13.24	0,09	0,09

industrial practitioners may reuse previous or similar prototypes more seriously and with more time and efforts, especially if reuse is mandatory (e.g., for a military or governmental application), than students in an academic setting, where reuse is considered plagiarism and prohibited. When reuse is intentionally encouraged, it may influence the amount of time and effort spent. The lack of feedback from industrial practitioners is another limitation of our study, for example, in discovering issues with our model of effort estimation and with its calibration accordingly. Another possible threat to validity has to do with the effort needed to analyze relationships and dependencies between requirements, and to get the content of Fig. 1, which will require an additional time and effort. However, this effort will not be included in the estimates provided by the proposed approach, as it is considered as input and not involved in the estimation process.

6   Conclusion and Future Work

In this paper, we proposed a novel framework for software effort estimation, using a combination of CBR and a multi-criteria decision analysis technique. The first stage in the framework, presentation, represents different cases – i.e. Different requirements of the system to be estimated, in terms of functional requirements (FRs), non-functional requirements (NFRs) and domain properties (DPs), as well as relationships between them. The second stage, retrieve, is for comparing and selecting cases from previous projects that are similar to the current project. As has been recognized, a key difficulty with CBR lies in obtaining similarity measures for retrieving the most appropriate case. Knowing which case is the most relevant and useful is not always obvious but rather challenging, especially in the presence of various alternatives to choose from. Unfortunately, making a selection based on ordinal similarity measures is a difficult and hard task, and would not provide accurate estimates. So, in order to select the most suitable case among various alternatives, our proposal is to consider meaningful criteria in a manner of multi-criteria analysis (MCDA). The use of MCDA in CBR potentially leads to a more accurate and efficient retrieval of cases that are most similar to the current case. By providing weights to each criterion, and by including as many essential types of requirements and features of the software to be estimated, the framework provides additional reliability. We are encouraged by the results obtained from the present study, albeit its limitations, as among the first of its kind – if not the first - in evaluating the utility and validity of the Case-based effort estimation framework, when coupled with MCDA. One line of future work concerns carrying out case studies involving industrial projects and practitioners, in particular, to find out the strengths and limitations of the use of CBR and MCDA together. Exploration of deploying other techniques, such as swarm optimization and other machine learning techniques, is another area of research, which could be used for comparative purposes. A tool implementation will help maintain a repository of cases and retrieve most cases (semi)automatically, especially for complicated projects with many requirements, where calculations of similarity measures and retrieval of cases can be time-consuming and error-prone.

References

1. Hughes, R.T.: Expert judgement as an estimating method. Inf. Softw. Technol. **38**(2), 67–75 (1996)
2. Lopez-Martin, C., Abran, A.: Applying expert judgment to improve an individual's ability to predict software development effort. Int. J. Software Eng. Knowl. Eng. **22**(04), 467–483 (2012)

3. Shepperd, M., Schofield, C.: Estimating software project effort using analogies. IEEE Trans. Softw. Eng. **23**(11), 736–743 (1997)
4. Briand, L.C., Wieczorek, I.: Resource estimation in software engineering. Encyclopedia of software engineering (2002)
5. Silhavy, R., Silhavy, P., Prokopova, Z.: Evaluating Subset Selection Methods for Use Case Points Estimation. Information and Software Technology (2017)
6. Silhavy, R., Silhavy, P., Prokopova, Z.: Analysis and selection of a regression model for the Use Case Points method using a stepwise approach. J. Syst. Softw. **125**, 1–14 (2017)
7. Wen, J., Li, S., Lin, Z., Hu, Y., Huang, C.: Systematic literature review of machine learning based software development effort estimation models. Inf. Softw. Technol. **54**(1), 41–59 (2012)
8. Singh, Y., Kaur, A., Bhatia, P.K., et al.: Predicting software development effort using artificial neural network. Int. J. Software Eng. Knowl. Eng. **20**(03), 367–375 (2010)
9. Wu, D., Li, J., Liang, Y.: Linear combination of multiple case-based reasoning with optimized weight for software effort estimation. J. Super Comput. **64**(3), 898–918 (2013)
10. Aamodt, A., Plaza, E.: Case-based reasoning: foundational issues, methodological variations, and system approaches. AI Commun. **7**(1), 39–59 (1994)
11. Finnie, G.R., Wittig, G.E., Desharnais, J.M.: A comparison of software effort estimation techniques: using function points with neural networks, case-based reasoning and regression models. J. Syst. Softw. **39**(3), 281–289 (1997)
12. Heemstra, F.J.: Software cost estimation. Inf. Softw. Technol. **34**(10), 627–639 (1992)
13. Fenton, N., Bieman, J.: Software Metrics: A Rigorous and Practical Approach. CRC Press, Boca Raton (2014)
14. Chung, L., do Prado Leite, J.C.S.: On non-functional requirements in software engineering. In: Conceptual Modeling: Foundations and Applications, pp. 363–379. Springer, Heidelberg (2009)
15. Jacobson, I., Booch, G., Rumbaugh, J., Rumbaugh, J., Booch, G.: The Unified Software Development Process, vol. 1. Addison-wesley, Reading (1999)
16. Kotonya, G., Sommerville, I.: Requirements Engineering: Processes and Techniques. Wiley, New York (1998)
17. Gunter, C.A., Gunter, E.L., Jackson, M., Zave, P.: A reference model for requirements and specifications. IEEE Softw. **17**(3), 37–43 (2000)
18. Smiti, A., Elouedi, Z.: Overview of Maintenance for Case based Reasoning Systems (2011)
19. Burkhard, H.-D., Richter, M.M.: On the notion of similarity in case based reasoning and fuzzy theory. In: Soft Computing in Case Based Reasoning, pp. 29–45. Springer, London (2001)
20. Walkerden, F., Jeffery, R.: An empirical study of analogy-based software effort estimation. Empirical Softw. Eng. **4**(2), 135–158 (1999)
21. Greco, S., Figueira, J., Ehrgott, M.: Multiple Criteria Decision Analysis. Springer's International Series (2005)
22. Triantaphyllou, E.: Multi-criteria Decision Making Methods: A Comparative Study, vol. 44. Springer Science & Business Media, New York (2013)
23. López-Martín, C.: Predictive accuracy comparison between neural networks and statistical regression for development effort of software projects. Appl. Soft Comput. **27**, 434–449 (2015)

24. Idri, A., Amazal, F.A.: Software cost estimation by fuzzy analogy for ISBSG repository. In: Uncertainty Modeling in Knowledge Engineering and Decision Making, pp. 863–868 (2012)
25. Azzeh, M., Elsheikh, Y.: Learning best K analogies from data distribution for case-based software effort estimation. In: The Seventh International Conference on Software Engineering Advances, pp. 341–347 (2012)
26. Huang, S.J., Chiu, N.H., Chen, L.W.: Integration of the grey relational analysis with genetic algorithm for software effort estimation. Eur. J. Oper. Res. **188**(3), 898–909 (2008)
27. Huang, S.J., Lin, C.Y., Chiu, N.H.: Fuzzy decision tree approach for embedding risk assessment information into software cost estimation model. J. Inf. Sci. Eng. **22**(2), 297–313 (2006)
28. Mendes, E., Mosley, N., Counsell, S.: A replicated assessment of the use of adaptation rules to improve Web cost estimation. In: International Symposium on Empirical Software Engineering, ISESE 2003, Proceedings, pp. 100–109. IEEE, September 2003
29. Li, Y.F., Xie, M., Goh, T.N.: A study of the non-linear adjustment for analogy based software cost estimation. Empirical Softw. Eng. **14**(6), 603–643 (2009)
30. Azzeh, M.: Adjusted case-based software effort estimation using bees optimization algorithm. In: International Conference on Knowledge-Based and Intelligent Information and Engineering Systems, pp. 315–324. Springer, Heidelberg, September 2011
31. Li, Y.F., Xie, M., Goh, T.N.: A study of project selection and feature weighting for analogy based software cost estimation. J. Syst. Softw. **82**(2), 241–252 (2009)
32. Chiu, N.H., Huang, S.J.: The adjusted analogy-based software effort estimation based on similarity distances. J. Syst. Softw. **80**(4), 628–640 (2007)
33. Azzeh, M., Neagu, D., Cowling, P.: Software project similarity measurement based on fuzzy C-means. In: International Conference on Software Process, pp. 123–134. Springer, Heidelberg, May 2008
34. Han, L., Kashyap, A., Finin, T., Mayfield, J., Weese, J.: UMBC EBIQUITY-CORE: semantic textual similarity systems. In: Proceedings of the Second Joint Conference on Lexical and Computational Semantics, vol. 1, pp. 44–52, June 2013
35. Chang, T.H., Wang, T.C.: Using the fuzzy multi-criteria decision making approach for measuring the possibility of successful knowledge management. Inf. Sci. **179**(4), 355–370 (2009)
36. Finnie, G., Sun, Z.: R 5 model for case-based reasoning. Knowl.-Based Syst. **16**(1), 59–65 (2003)
37. ISO/IEC, ISO/IEC 25010:2010 SOFTWARE ENGINEERING-Software Product Quality Requirements and Evaluation (Square)System and Software Quality Models, ISO/IEC JTC 1/SC 7 (2010)
38. Boehm, B.W.: Software Engineering Economics, vol. 197. Prentice-Hall, Englewood Cliffs (1981)
39. Zadeh, L.A.: Similarity relations and fuzzy orderings. Inf. Sci. **3**(2), 177–200 (1971)
40. Fellir, F., Nafil, K., Touahni, R.: System requirements prioritization based on AHP. In: 2014 Third IEEE International Colloquium in Information Science and Technology (CIST), pp. 163–167. IEEE (2014)
41. Foss, T., Stensrud, E., Kitchenham, B., Myrtveit, I.: A simulation study of the model evaluation criterion MMRE. IEEE Trans. Softw. Eng. **29**(11), 985–995 (2003)
42. Kitchenham, B.A., Pickard, L.M., MacDonell, S.G., Shepperd, M.J.: What accuracy statistics really measure [software estimation]. IEE Proc. Softw. **148**(3), 81–85 (2001)

Automated Logical-Probabilistic Methodology and Software Tool as Component of the Complex of Methodologies and Software Tools for Evaluation of Reliability and Survivability of Onboard Equipment of Small Satellites

Vadim Skobtsov[1(✉)], Natalia Lapitskaja[2], Roman Saksonov[3], and Semyon Potryasaev[4]

[1] United Institute of Informatics Problems,
National Academy of Sciences of Belarus, Minsk, Belarus
vasko_vasko@mail.ru
[2] Belarusian State University of Informatics and Radioelectronics,
Minsk, Belarus
lapan@mail.ru
[3] Geoinformation Systems, Minsk, Belarus
roman_saksonov@tut.by
[4] St. Petersburg Institute of Informatics and Automation,
Russian Academy of Sciences (SPIIRAS), St. Petersburg, Russia
spotryasaev@gmail.com

Abstract. The paper presents solutions for current problems with estimation and analysis of indicators of reliability and survivability in onboard equipment (OE) of small satellites (SS) based on the logical-probabilistic approach to the reliability and survivability estimation of complex systems. There were developed modified logical-probabilistic method and software tool for evaluating the reliability and survivability of OE SS systems. The correctness of suggested method and software tool was shown by computational experiments on some systems of OE SS similar to Belarusian SS, later compared with "Arbitr" software complex results. The software tool was integrated into the complex of methodologies and software tools for evaluation, analysis and prediction of the values of reliability and survivability indicators of OE SS in local desktop and distributed web versions.

Keywords: Logical-probabilistic method · Reliability · Survivability
Onboard equipment for small satellites · Diagram of functional integrity
System operability function · Probability of failure-free operation
Software tool

1 Introduction

The evaluation of reliability and survivability of the complex technical systems, like OE for SS, is an important task of their safe and reliable design and operation.

© Springer International Publishing AG, part of Springer Nature 2019
R. Silhavy (Ed.): CSOC 2018, AISC 763, pp. 452–463, 2019.
https://doi.org/10.1007/978-3-319-91186-1_47

SS design features limit the possibilities of using traditional approaches to providing and evaluating fault tolerance, reliability and survivability of OE SS, since SS mass restriction significantly reduces the possibility of equipment elements redundancy. Logic-probabilistic modeling is a method for analyzing the sensitivity of a complex system operating under conditions of uncertainty. The method boasts clarity and a wide range of possibilities to detect the influence of any argument on the reliability and survivability of the entire system. The logical-probabilistic theory of reliability and survivability allows objective identification of the "thinnest" and most dangerous places, causes and preferred combinations of initiating conditions, the protection against which prevents the system from getting into a dangerous state [1, 2].

In this paper the automated logical-probabilistic methodology and software tool for evaluating the reliability and survivability of the OE SS systems is suggested. The developed software tool was implemented in two versions – a desktop-version for local workstations and a web-version for distributed application through web-services and web-interface. The logical-probabilistic software tool was combined with the OE SS telemetry data analysis software tool, represented in CSOC'2017 proceedings [3], and some other methodologies and tools in the complex of methodologies and software tools for evaluating the reliability and survivability of the OE SS. The complex was integrated into the software complex for multi-objective assessment, analysis and prediction of values of reliability and survivability for OE SS developed in SPIIRAS.

2   The Logical-Probabilistic Methodology

The usage of logical-probabilistic approach in evaluating the reliability and survivability of a structurally complex system provides sequential construction of two computational models types [1, 2]:

– Boolean function of the system operability

$$Y_F = Y_F(\{\tilde{x}_i\}, i = 1, 2, \ldots, H) \tag{1}$$

– polynomial of the estimated probability function

$$P_F = P_F(\{p_i, q_i\}, i = 1, 2, \ldots, H) \tag{2}$$

2.1   Logical Function of the System Operability

In logic-probabilistic methodology a Boolean function as a logical model of system reliability and survivability is applied. It is called a logical function of the system operability (FSO).

A logical function of the system operability represents a set of states in which the system implements an appropriate criterion of its functioning (system health state). A logical criterion can be determined by different properties of the system.

The initial data for determining a logical FSO are:

- diagram of functional integrity (DFI),
- logical criteria of functioning (LCF) of system.

The features and restrictions of OE SS design allow to apply one of the simplest methods of direct analytic substitution for FSO construction.

It provides a consistent replacement in the logical FSO of all integrative functions by their equations selected from the system. Such substitution is performed until there are no undeclared functions y_i in the resulting expression. In other words, all integrative functions y_i will be replaced by simple logical variables x_i.

Let's demonstrate the operation of this method using the example of constructing the logical function of the bridge system. Its DFI is shown in Fig. 1. Complete system of logical equations corresponding to the DFI (for the output functions of each vertex) is given below (3).

$$
\begin{cases}
y_1 = x_1 \\
y_2 = x_2 \\
y_3 = x_3 \cdot (y_1 \vee y_5) \\
y_4 = x_4 \cdot (y_2 \vee y_5) \\
y_5 = x_5 \cdot (y_1 \vee y_2)
\end{cases}
\tag{3}
$$

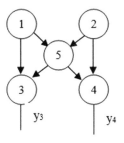

Fig. 1. DFI of the bridge system

In a given bridge system a logical FSO is defined. It should correspond to the criterion when the operability of the entire system is determined by the implementation of the output functions by two elements 3 and 4 simultaneously (4).

$$
Y_p = y_3 \cdot y_4
\tag{4}
$$

Having performed all substitutions in (4), we obtain

$$Y_p = y_3 \cdot y_4 = x_3 \cdot (y_1 \vee y_5) \cdot x_4 \cdot (y_2 \vee y_5) = x_3 \cdot (x_1 \vee x_5 \cdot (y_1 \vee y_2)) \cdot$$
$$\cdot x_4 \cdot (x_2 \vee x_5 \cdot (y_1 \vee y_2)) = x_3 \cdot (x_1 \vee x_5 \cdot (x_1 \vee x_2)) \cdot x_4 \cdot (x_2 \vee x_5 \cdot (x_1 \vee x_2)) \quad (5)$$
$$= x_3 \cdot x_4 \cdot x_1 \cdot x_2 \vee x_3 \cdot x_4 \cdot x_1 \cdot x_5 \vee x_3 \cdot x_4 \cdot x_2 \cdot x_5.$$

In the resulting expression there are no unsolved integrative functions. The substitution process is complete and the FSO of the bridge system is obtained.

2.2 Computational Probability Model of System Reliability

Polynomial of the estimated probability function (PF) P_F (2) is used as a computational probabilistic model of system reliability. The polynomial should clearly define (within limitations and assumptions) the probability of failure-free system operation, as well as all health states represented using the logical FSO Y_F (1).

In terms of physical meaning, P_F determines the probability of implementing the given logical criterion Y_F of system functioning. A polynomial P_F is a rule for aggregation of particular parameters of system elements, i.e. compositions of elementary probabilities p_i in a system probability characteristic. P_F defines an algorithm for calculating the probability of a complex event Y_F consisting of products, sums and inversions of its simple random events, whose own probabilistic parameters p_i are known [1, 2].

The parameters p_i, $q_i = 1 - p_i$ of the P_F polynomial are intrinsic probabilities of failure-free operation or availability coefficients of system elements.

As for P_F determining, the combined method is applied, one of the most effective and simplest approaches [1]. In general, two successive transformations of the original FSO are performed for accurate determination of P_F using a combined method:

- quasiorthogonalization of the FSO by a single logical variable;
- a symbolic transition to the P_F polynomial.

Conjunctions of a logical function are called orthogonal if their logical product is zero. Therefore, the events representing these conjunctions are inconsistent, and the probability of their sum equals a simple sum of the own probabilities of each event.

Assuming independence in the aggregate of all elementary binary events $\tilde{x}_i$, $i = 1, 2, \ldots, H$, the algorithm for performing these two steps is as follows.

Quasiorthogonalization by a single logical variable. All pairs of non-orthogonal conjunctions of the original FSO are checked for possibility of their orthogonalization by the following rule

$$\gamma \cdot \tilde{x}_i \vee \gamma \cdot \phi = \gamma \cdot x_i \vee \gamma \cdot \phi \cdot \overline{\tilde{x}_i}, \quad (6)$$

where γ and ϕ are parts of the tested conjunctions, in which there is no variable $\tilde{x}_i$. Orthogonalization does not increase the total number of conjunctions in the original FSO. Having performed these transformations with the example illustrated over the function (5), we obtain

$$Y_p = y_3 \cdot y_4 = x_3 \cdot x_4 \cdot x_1 \cdot x_2 \vee x_3 \cdot x_4 \cdot x_1 \cdot x_5 \vee x_3 \cdot x_4 \cdot x_2 \cdot x_5$$
$$= x_3 \cdot x_4 \cdot x_1 \cdot x_2 \vee x_3 \cdot x_4 \cdot x_1 \cdot x_5 \cdot \overline{x}_2 \vee x_3 \cdot x_4 \cdot x_2 \cdot x_5 \cdot \overline{x}_1. \tag{7}$$

The resulting function is completely orthogonal, so all its conjunctions represent incompatible events.

A symbolic transition to P_F polynomial. A logical FSO is a rigorous analytical form of describing a complex random event in which:

- a conjunction is a product of random events;
- a disjunction is their sum;
- an inversion is the opposite outcome of an event whose probability is to be determined with the help of the desired P_F.

In order to obtain the P_F polynomial from the FSO, it is enough to perform some transformations to labeling of logical variables and operations in the labeling of probability variables and arithmetic operations. These symbolic transformations must be carried out in a strict accordance with the laws of probability theory for calculating the probabilities of products, sums and additions of random events.

Having applied these rules to the orthogonalized FSO (7), we obtain the polynomial of the required P_F:

$$P_p = p_F(y_3 \cdot y_4) = p_3 p_4 p_1 p_2 + p_3 p_4 p_1 p_5 q_2 + p_3 p_4 p_2 p_5 q_1. \tag{8}$$

2.3   Survivability Model

Survivability [2] is an ability of a system to keep operating capacity at random damages of its elements owing to random emergence of the striking factors (blows, explosions, the fires, etc.). In order to account for the random damaging factor in the structural model of the system survivability, another new event x_5 is added to the developed DFI: the emergence of the damaging factor (Fig. 2). The occurrence of this event is a prerequisite for a subsequent impact and accidental destruction of the system elements 3 and/or 4. If the damaging factor does not arise, elements 3 and 4 are not significantly affected.

There is an operability persistence equation at the output of the fictitious vertex 7:

$$y_7 = \overline{y}_3 \vee \overline{y}_4. \tag{9}$$

Operability persistence is sustained by at least one element, which is the criterion Y - the survivability of the system for accidental occurrence and impact of the damaging factor. If the onset of the damaging factor occurs with a probability of $p_5 = 0.7$, $p_3 = 0.3$, $p_4 = 0.4$, a polynomial of survivability probability function is as follows:

$$P_s = q_4 + p_4 q_5 + p_4 p_5 q_3 = 0,916. \tag{10}$$

It should be noted that the DFI of the total survivability of the considered two-element system (see Fig. 2) is nonmonotonic in construction. It means, in particular, that an analogous structural model of total survivability cannot be constructed using standard block patterns or typical failure trees. At the same time, the obtained logical and probabilistic survivability functions are still monotonous, within the limits of the accepted substantive definitions for the used elementary events.

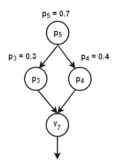

Fig. 2. The damaging factor for a two-element system

3   Method and Software Tool for Estimating Reliability and Survivability

3.1   Method for Calculating Reliability and Survivability

The software implementation of the considered methodology used for determining the logical FSO and P_F polynomials for the analysis of real structurally complex systems of OE SS with a large number of elements is difficult because of the complexity of automatic analytical simulation processes.

Therefore, we made a modification the DFI graph comparing with classical representation of DFI [1, 2]: in addition to the linking arcs and functional vertices, the logical vertices "AND" and "OR" were introduced instead of according arcs (see Fig. 3).

Firstly, it helps to see more clearly which systems are redundant and which are not.

Secondly, logical vertices can be used more effectively in algorithms for automatic calculating the probability of failure-free operation and survivability of systems. In this case, the FSO Y_F and probability function P_F polynomials are computed implicitly for computing complexity reducing.

Thirdly, there are systems consisting of X vertices that remain operative when any of the Y vertices work. For example, the flywheel group of satellite orientation system very often consists of 4 flywheels and in order to operate correctly, presence of any three operable flywheels out of 4 is required. It is possible to create and analyze such systems using logical vertices.

The graph will be designed in such a way that the functional elements commute with each other only through logical elements.

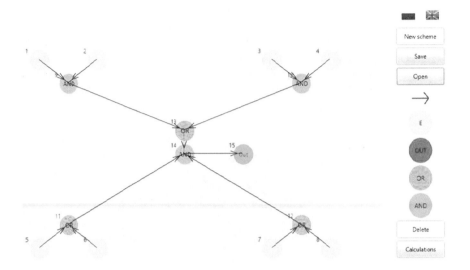

Fig. 3. Example of oriented graph as DFI of the multispectral camera subsystem (MCS) of the Belarussian SS, DFI graphical input window, desktop-version

Introduction of these vertices does not change the probability of failure-free operation and the reliability and survivability of the systems at all. As a rule, the scheme is created on the basis of four logical constructions (see Fig. 4).

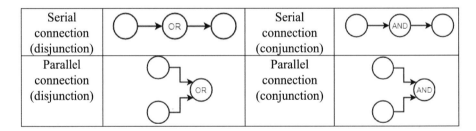

Fig. 4. Basic types of logical constructions

In order to create more complex structures, logical elements can be commuted with each other in any quantity. The width and depth search algorithms are used for graph navigation.

At each stage for each structural node, the probabilities of failure-free operation and survivability are recalculated and transferred to the next vertex. Thus, each vertex contains information about calculations on all previous vertices. And whenever we get to any vertex, we can always get the probabilities obtained for the previous vertices. Algorithm operation is finished when all final vertices are reached [4] (see Fig. 5).

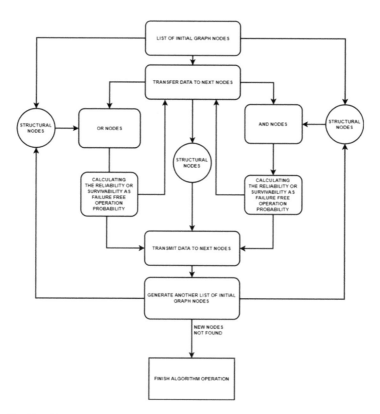

Fig. 5. Algorithm for calculating probability of failure-free operation and survivability

3.2 Software Tool of Logical-Probabilistic Estimating the Reliability and Survivability of System Operating

Correspondent software tool, based on the logical-probabilistic approach and the proposed algorithm, was developed in two versions: a desktop version and web one. The software tool was designed to automate the process of evaluating the reliability and survivability of OE SS systems. The tool implements the following functions:

- Graphical input and editing DFI representing a device under analysis (Figs. 3 and 7);
- Input and editing the system elements parameters including failure-free operation probabilities p_i and redundancy level of the system elements;
- Calculating the single value of the system reliability for the given single arguments of failure free operation probability p_i for structural nodes;
- Reliability analysis – calculating the values of the OE SS systems reliability for the series of elements probability arguments and drawing graphics (Figs. 6 and 7), in this case elements probability parameters p_i are equally changed in the range ($p_{ini} > 0$; $p_{fin} \leq 1$) by step Δp. During reliability analysis failure free operation probabilities p_i of some elements could be fixed to value in their properties by setting on according checkboxes in elements list (Figs. 6 and 7).

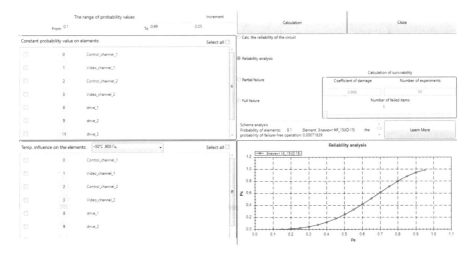

Fig. 6. Window for calculating, reliability analysis, desktop-version

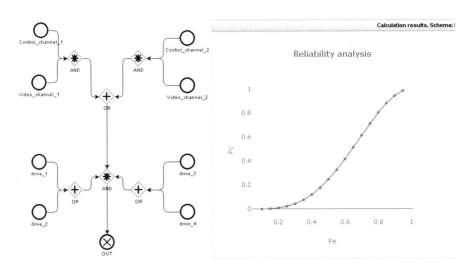

Fig. 7. Web-interface, example of graphical input and editing the DFI of the multispectral camera subsystem (MCS) of the Belarussian SS, reliability analysis results

- Survivability analysis – calculating the survivability of OE SS systems with full and partial failure and drawing graphic. The damaging factor is applied to randomly selected k elements which reduces the probability of failure-free operation p_i of these elements to 0 in case of full failure and in case of partial failure to $\tilde{p}_i < p_i$ that can OE set throw interface (Figs. 6 and 7). The experiment is repeated N (50 by default) times (Figs. 8 and 9). During survivability analysis some elements can be excluded from the impact of damaging factors by setting corresponding checkboxes in the elements list.

Moreover, during the analysis of reliability and survivability of OE SS systems, the impact of thermal and mechanical factors on some system elements could be taken into account. It can be achieved through setting corresponding checkboxes in the elements list.

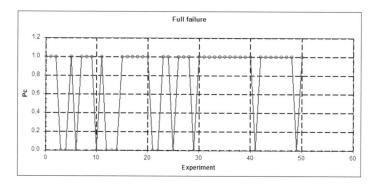

Fig. 8. Determining the survivability of the MCS under full failure

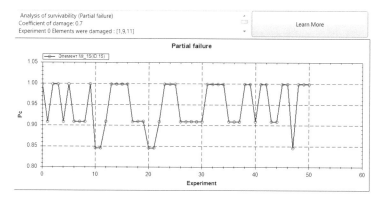

Fig. 9. Determining the survivability of the MCS under partial failure

4   Computational Experiments

The correctness of the developed methodology and software tool was tested on some test systems of OE for Belarusian SS, later compared with "Arbitr" software complex results [1, 2], certified in 2007 by Rostechnadzor (RF) for industrial application. In Tables 1, 2 and 3 the comparison results for MCS – the component of OE for Belarusian SS (Fig. 3) are represented. In Table 4 the reliability analysis comparison results for other three systems of OE SS similar to Belarusian SS are represented.

Represented computational experiments show equality of the results for the developed methodology and software tool with the results of the software complex "Arbitr" and hence the correctness of suggested method and developed tool.

Table 1. Comparison results for MCS – component of OE of Belarusian SS. Single reliability index value

Reliability calculating	Developed tool: Pc = 0,99940407919804	Software complex "Arbitr": Pc = 0,999404079

Table 2. Comparison results for MCS – component of OE of Belarusian SS. Reliability analysis

Elements probability p_i	Pc, Developed tool	Pc, Software complex "Arbitr"	Term and mechanical influence on the elements:	Constant reliability on the elements:	Pc, Developed tool	Pc, Software complex "Arbitr"
0,1	0,0007183	0,000718	–	–	–	–
0,3	0,0447111	0,044711	–	–	–	–
0,6	0,4165862	0,416586	–	–	–	–
0,8	–	–	3,6,7	1,8	0,88408152961	0,88408153
0,9	–	–	1	1	0,958474759968	0,95847476
1	–	–	1,2,3,4,5,6,7,8	2	0,999461280139	0,99946128

Table 3. Comparison results for MCS – component of OE for Belarusian SS. Survivability analysis. The damaging factor 0,998, failure of 3 elements

The following elements are failure ed:	Pc, Developed tool	Pc, Software complex "Arbitr"	Temperature influence on the elements:	Constant reliability on the elements:	Pc, Developed tool	Pc, Software complex "Arbitr"
7,2,4	0	1,9404E-50	–	–	–	–
8,2,1	0,97020197	0,97020197	–	–	–	–
8,6,1	0,96059601	0,96059601	–	–	–	–
8,5,6	–	–	1,3	2,4	0	1,95869E-50
6,7,3	–	–	1,3	2,4	0,95291124192	0,952911242

Table 4. Comparison results for systems of OE SS similar to Belarusian SS. Reliability analysis

Elements probability p_i	Navigation system equipment		Multispectral camera system		Orientation and stabilization system	
	Pc, Developed tool	Pc, Software complex "Arbitr"	Pc, Developed tool	Pc, Software complex "Arbitr"	Pc, Developed tool	Pc, Software complex "Arbitr"
0,60	0,027546	0,027546	0,064524	0,064524	0,090334	0,090334
0,70	0,087048	0,087048	0,164648	0,164648	0,214043	0,214043
0,80	0,228170	0,228170	0,347892	0,347892	0,417471	0,417471
0,90	0,512256	0,512256	0,630247	0,630247	0,693272	0,693272

5 Conclusion

In the paper an effective logic-probabilistic methodology and approach to estimating reliability of complex systems was considered. The methodology was applied to the problem of estimating the reliability and survivability of on-board equipment for small satellites systems. A modified logical-probabilistic method and a software tool for evaluating the reliability and survivability of OE SS systems were developed. The developed software tool automatizes the reliability and survivability estimating process. It also enables graphical input of DFI data and survivability analysis through drawing the graphics of obtained results. The correctness of the suggested method and software tool was shown by computational experiments on some systems of OE SS similar to Belarusian SS, later compared with "Arbitr" software complex results. The logical-probabilistic software tool was combined with the OE SS telemetry data analysis software tool, represented in CSOC'2017 proceedings [3], and some other methodologies and tools in the complex of methodologies and software tools for evaluating the reliability and survivability of the OE SS.

Acknowledgments.   The research described in Sect. 2 of paper is supported by project No. 17-11-01254 of Russian Science Foundation, the research described in Sect. 3 of paper is supported by the state research 0073–2018–0003. All represented in paper research results were supported by Program STC of Union State "Monitoring-SG" (project 6МСГ/13-224-2, the Belarusian side).

References

1. Mozhaev, A.S.: The technology of automated structural and logical modeling of reliability, survivability, safety, efficiency and risk of functioning the systems. Instrum. Syst. Monit. Control Diagn. **N9**, 1–14 (2008). (in Russian)
2. Mozhaev, A.S., Grommov, V.N.: Theoretical foundations of the general logical-probabilistic method of automated systems modeling. SPb, VITU (2000)
3. Skobtsov, V., Novoselova, N., Arhipov, V., Potryasaev, S.: Intelligent telemetry data analysis of small satellites. In: Kacprzyk, J., et al. (eds.) Proceedings of the 6th Computer Science On-line Conference 2017 (CSOC2017), vol. 2 in Advances in Intelligent Systems and Computing, vol. 574, pp. 351–361. Springer International Publishing Switzerland (2017)
4. Cormen, T.H., Leiserson, C.E., Rivest, R.L., Stein, C.: Introduction to Algorithms, 3rd edn. The MIT Press, Cambridge (2009)

A Multicriteria Structured Model to Assist the Course Offering Assertiveness

Rodrigues Forte de Lima Silva Junior[1,2] and Plácido Rogério Pinheiro[1,2(✉)]

[1] Graduate Program in Applied Informatics, University of Fortaleza, Fortaleza, Brazil
placido@unifor.br
[2] National Commercial Apprenticeship Services - NCAS/CE, Fortaleza, Brazil
rodrigues@ce.senac.br

Abstract. Decision-making processes within companies are increasingly complex. This is due to the fact that the amount of data available and the various variables that should be considered in the decision-making process. In the field of professional education, it is no different, the decision to choose which course should be offered becomes an activity of high complexity because of the various criteria that must be considered for the assertiveness of the offer, such as profitability, client profile, macroeconomic characteristics, program content, characteristics of the locality of the course, among others. Once the scenario is presented, the objective of this work is to propose a model of selection and prioritization of variables for the choice of courses offered in the professional education institution National Commercial Apprenticeship Services through the Multiple Criteria Decision Analysis. We present the context of the decision of the problem and, applying the cognitive mapping, we identify the most relevant criteria to detect which characteristics that most impact in the decision making.

Keywords: Assistance in decision making · Multicriteria decision analysis
Assertiveness of the course offering · Modeling of the criteria
Cognitive mapping

1 Introduction

The National Commercial Apprenticeship Services - NCAS was created on January 10, 1946, by the National Confederation of Commerce of Goods, Services, and Tourism (CNC), through Decree-Law 8,621. From the following year, NCAS began to develop a work that had been innovative in the country until then: to offer, on a large scale, professional education for the training and preparation of workers for trade, acting without any imputation to the public safes or to the taxpayers, since all the expenses would be exclusive to the account of the commercial entrepreneurs, calculated on the wage sheets of its employees. The National Commercial Apprenticeship Services - NCAS has regional departments in all the federative units having its headquarters in the city of Rio de Janeiro.

It operates in nine Technological Axes (Environment and Health, Educational and Social Development, Management and Business, Information and Communication, Infrastructure, Cultural Production and Design, Hospitality and Leisure, Food Production,

Security) and its segments (Beauty, Health, Environment, Education, Social, Languages, Management, Commerce, Computing, Telecommunications, Conservation and Janitorial, Arts, Fashion, Communication, Design, Tourism, Leisure, Hospitality, Gastronomy/PAS, Food Production, Security), with more than 800 courses in its different typologies.

The elaboration of its offer plan requires a complex planning, given the variety of its portfolio and its performance in several municipalities of the State of Ceará. The analysis and decision of which courses will be offered, in which units, on what dates and at what times are a joint responsibility between the managers of the operative units, technical consultants (collaborators of NCAS specialists in certain areas) and the planning team of the institution. Factors such as the development of the region, history of courses, demands by area, investments, business strategy, are criteria for creating the business map. In addition to logistical issues such as environmental capacity, available resources, various periods and schedules, which are factors considered for the decision process of the course offereding in the operation phase.

The objective of this work is to create a proposal to assist the assertiveness of the choice of the offer, aiming to launch for the market the courses with the greatest degree of accuracy in its sale, given a set of necessary restrictions so that the result achieved is viable and of better result possible.

Exposed the difficulty of choosing a viable offer that respects all the necessary restrictions for its execution, such as: market, employability, supply assertiveness, school education, characteristics of the locality, population, besides other factors, we opted for the use of the structured methodology in multi-criteria, so that through a well-qualified modeling of the criteria, we have as a result a proposal of courses offers with the greatest possibility of success, serving as an auxiliary tool in the decision making in the offer of these courses.

2 Decision Analysis

One of the definitions of Decision Making Process [1] is that it is a cognitive process that results in the selection of an option between several alternatives. It is widely used to include preference, inference, classification, and judgment, whether conscious or unconscious.

In management, decision making is the cognitive process by which a plan of action is chosen among several others (based on varied scenarios, environments, analyzes, and factors) for a problem situation. Every decision-making process produces a final choice. The output may be an action or an opinion of choice. That is, decision-making refers to the process of choosing the most appropriate path to the company, in a given circumstance.

Relevant points for structuring the decision-making process:

Level of importance within the organization:

- ✓ Highly important;
- ✓ Important;
- ✓ Fairly important;
- ✓ Less important;
- ✓ Not important.

Structuring:

✓ Structured;

✓ Not structured.

✓ Predictability:

✓ Routine or cyclical;

✓ Not routine or cyclical;

✓ Unpublished.

Any decision-making process in a company has a direct impact on its results, be they strategic, tactical or operational, so mechanisms should be created to support decision-makers, use of technology and good practices for mapping related variables to the problem has a direct influence on the quality of the decision.

According to [3], a potential action constitutes the object of the decision or that which is directed to support the decision, which is considered as potential when it is possible to apply it, or simply when it deserves some interest in the decision support process. Identified at the beginning of the decision process or over it and may become a solution to the problem being studied.

2.1 Multicriteria Support for the Decision

Multiple Criteria Decision Analysis (MCDA) consists of a set of methods and techniques to assist or support individuals and organizations in making decisions, given a multiplicity of criteria. Its best-known methods came in the 1970s.

The multicriteria allow structuring the control events that are being analyzed and to order them according to the degree of importance in the decision-making process of the search of the diagnosis, when of its application [2].

There are several approaches and characteristics that play a key role in the methodologies applied in the decision-making process. Here, we will mention some differences between American and European schools, where the former uses a more rational methodology and always seeks a more objective solution. The European defends humanity in the decision process, strengthened through subjectivity as a character of choice and the search for learning and a real understanding of the problem.

The multicriteria problems have the following classifications in Table 1:

Table 1. Classifications of multicriteria problems

Structured problems	Semi-structured problems	Non-structured problems
The solution is found in the logical well-defined processes	The problem is solved by logical process (structured part) and the final decision is delegated to the decision-maker (non-structured part)	These are qualitative problems, complex for the organization, formalization and numeric measurement. It is a subjective analyzes process

The Multiple Criteria Decision Analysis (MCDA) has as main objectives:

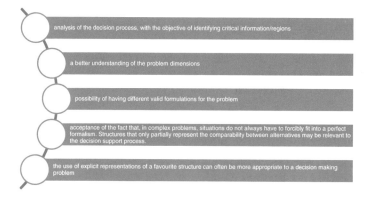

3   Cognitive Mapping

One of the definitions found is they are schematic structures or graphical representations of a set of ideas and concepts, arranged in a kind of network of propositions, in order to present more clearly the exposure of a knowledge.

The Conceptual/Cognitive Map was developed in the 1970s by the American researcher and navigator Joseph Novak, initially proposed as an administrative tool, to organize and represent knowledge in general, basically being an improvement of the well-known organizational chart, aimed to be used in teamwork and/or in a collegiate.

According to [4, 7, 8], it allows us to portray ideas, feelings, values and attitudes and their interrelationships, in a way that makes possible the study of a problem and its subsequent analysis, using a graphical representation.

According to Ackermann et al. [18], this tool is used to structure and analyze the problems, which can be mapped through verbal or documented interviews.

For the problem addressed in this work, we use cognitive mapping as a playful tool to visually represent the relevant criteria that the NCAS/Ceará decision makers team analyzes and studies for the choice of courses offered.

According to [5], the cognitive map makes it possible to know the problem variables from the point of view of several people directly involved in the activity. Knowing the constraints, criteria and objectives of the task studied.

During the creation of the cognitive map, it is possible to capture the decision maker information about goals and purposes and means objectives of the problem in order to obtain sufficient information about the decision criteria.

Cognitive mapping has several types of graphical representation [9, 14, 15]; I cite some more known for example Spider Web, Flowchart and Hierarchical. They are chosen according to the best way to display each problematic, we must reinforce that the visual form of the map has as the main purpose to indicate to the readers the whole context of the problem/objective in a way that is explicit and intuitive the relevance of the criteria and their relationships.

3.1   Construction of the Cognitive Map for Structuring a Multicriterial Model for Course Offering

In order to reach the objective proposed by the study, we developed a structured cognitive map in the multicriteria approach mapping the main criteria, variables, alternatives and relationships in the decision-making process of choosing a vocational course in NCAS/Ceará, aiming the creation of a viable model for assertiveness in this offer.

We followed a 4-step process:

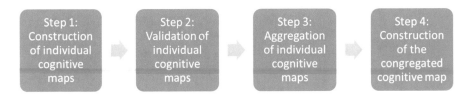

This process allowed us to capture the individuality of each technician responsible for the area of course development and then through the congregation of these ideas new criteria emerged because of the connections found between the individual mappings. In this sense, [6] emphasize the importance of interviews. At this step, he should avoid interfering with what they say (empathic approach), as this could misdirect the name inappropriately.

In step 1, we follow the following steps:

Step 1:   Defining a Label

The NCAS/Ceará has as strategic objective a reduction in costs and improve the assertiveness in the offer of its courses, since the erroneous offer of a course causes a significant damage to the company, as besides the costs of dissemination, elaboration, allocation of resources, among others, brings to the company a problem of credibility with its clients, because often due to the low demand is necessary to cancel the offer and the few clients who wish to carry out the courses are disappointed. Unit managers, technical consultants, and planning management are the decision makers. The label for the problem was defined "Assertiveness in the courses offering".

Step 2:   Definition of the Primary Assessment Elements (PAEs/EPAs)

The PAEs are the concepts that the decision maker considers relevant about the problem and transmits them to the facilitator at the time of the interview in the form of phrases.

Examples of PAEs for the problem of offering courses: high profitability, taking into account regional characteristics, diverse client profiles, professional qualification and inclusion in the labor market and courses for teams/companies and entrepreneurs.

Step 3:   Construction of Concepts

Moreover, each PAE should be modified by including some verbs for the infinitive suggesting that the PAE be action-oriented.

Example of modified EPAs: choose courses with high profitability, promote courses attending to regional characteristics, serve diverse client profiles, promote professional qualification and inclusion in the job market and perfect teams/companies and develop entrepreneurs;

Step 4: Construction of the Concept Hierarchy

In this stage, the classification was created by means of ordering on the concepts developed in the previous stages:

Furthermore, example of ordering: 1st: choosing courses with high profitability, 2nd: promoting professional qualification and inclusion in the labor market, 3rd: improving teams/companies, developing entrepreneurs, 4th: attend diverse profiles of client and 5th: promoting courses taking care of the regional characteristics;

Also, in step 2, the entire decision-making body was presented, the individual maps validated, and their concepts and variables validated for the objective, in this step were adjusted some points, so that we did not have very distant criteria from the intended purpose, we know that for each variable there can be "N" relationships, but we must choose wisely which ones are relevant to each scenario. We then review and validate the individual maps in the light of the main objective that is assertiveness in the course offering.

Moreover, in step 3, once the individual maps were validated, we entered the aggregation phase of these maps, aiming at the construction of a map in order to provide the vision of the process as a whole, allowing illustrating to the decision maker all the criteria that should be analyzed and prioritized for the decision-making process.

Also, in step 4, we developed an aggregate cognitive map with the whole group, taking into account all the procedures performed in the previous steps and aiming to reach the objective proposed by the work. Figure 1 shows the criteria that the group indicated as relevant for the decision making of the choice of courses:

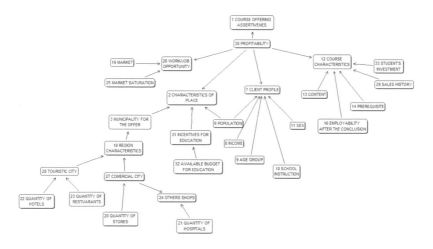

Fig. 1. Cognitive map for course offerings

The elaborated map indicated several essential characteristics for the decision making in the offer of courses. They reflect the complexity and the work required to define the course offering, such as characteristics of the course, characteristics of the population, characteristics of the locality, characteristics of the market, client profile, among others, are extremely important in this choice.

Once this survey has been made, it is necessary to establish which criteria are most relevant, so these can be used to help in choosing the best course offer. After analyzing the team, eleven items with the highest degree of importance for decision making are listed: Public Policies/Government Incentives, Improvements/Updates/Market Trends, Profitability (Courses must promote high profitability), Investment (Courses require high investment by the student), Employability (Courses promote a high rate of employability for the student), Business opportunity (Courses promote good business opportunity for the student), Regional (Courses must meet the regional characteristics of the places of offer), Customer Profile (Courses must be selected according to the client profile), Costs (High Cost Courses for SENAC), Infrastructure (Courses that need a specific infrastructure) and Competition (The courses are chosen taking into account the competition).

Questions like, what criteria have more weight or more priority, profitable courses? Courses targeted to the customer profile found? Is there any population investment power? What costs will be incurred for the offer of this course, is it advantageous? Is it necessary to meet the regional characteristics? By capturing these responses, it will be possible to qualify the criteria subjectively and prioritize the degree of importance of each topic in the decision-making process for the offer.

3.2 Prioritization of Criteria

After the elaboration of the cognitive map, the selection of the most relevant criteria and elaboration of the questions that must be answered for each criterion, we move on to the "ranking" phase of the criteria selected for the decision-making process.

We chose to apply a questionnaire that individually scored each criterion on a scale of 1 to 5, where 1 is no importance and five high importance. The intention was to individually measure each item, with the vision of each technician separately. The questionnaire template that was sent to the e-mail of those involved in the criteria prioritization process is published at the address:

https://docs.google.com/forms/d/1YDw76YYmkk9aoicd0tFlyGUPBk1LIFnuGZ-xUYMVtpVM/edit.

On the other hand, fourteen professionals in their various activities within the context of education answered the questionnaire, aiming at the miscegenation in several optics on the subject, among the actions we have: Technical Consulting of Educational Products, Coordination of Graduate Courses, Teachers, Management of Information Technology, Management of Professional Education Center, Education Unit Coordinator, Development Management and Educational Technology, Pedagogical Consulting, Coordination of the Bank of Opportunities, Educational Direction, Strategic Business Management and Administrative and Financial Directorate.

Moreover, as a result of the application of this questionnaire, we obtained the following classification of the descriptors in Table 2:

Table 2. Prioritization of the criteria - Result of the questionnaire applied

Descriptors	Average	Prioritization
Employability (Courses promote a liigli rate of employability for the student)	4,79	1st
Improvements/Updates/Market Trends	4,71	2nd
Business Opportunity (Courses promote good business opportunity for the student)	4,50	3rd
Client Profile (Courses must be selected using the client profile)	4,50	4th
Profitability (Courses should promote high profitability)	4,43	5th
Regional (Courses must meet the regional characteristics of places of supply)	4,29	6th
Infrastructure (Courses that need a specific infrastructure)	3,64	7th
Competition (Courses are chosen taking into account the competition)	3,57	8th
Public Policy/Government Incentive	3,50	9th
Costs (High-Cost Courses for SENAC)	3,07	10th
Investment (Courses require high investments by the student)	2,86	11th

Once categorized, descriptors will play a fundamental role in the decision-making process [10–13], it is up to the company to create mechanisms to capture information that subsidizes decision-makers when looking at each criterion, using ranking as a way of maximizing which approaches and which offer will be chosen for each need.

4 Results

As a result of this work, we have the application of the process of cognitive mapping where it was possible to raise all the criteria and relationships necessary for the decision making of vocational courses offering at NCAS/Ceará.

It was proposed a multicriteria structured model that aims, from the classification by means of priority ranking, which criteria should have greater weight in the decision of the course offering, aiming for a greater assertiveness in the sale, making use of a defined process that promotes the collection, analysis, tabulation and results, through various sources of data entry, whether they come from own systems, research sources, available research bases, BI, etc.

5 Conclusion

Cognitive mapping is an efficient way of capturing subjective perceptions. Its use allows several decision-makers to present their views, in a very personalized way and at the same time, through "negotiations" with the other decision-makers, searching a general consensus. It allows decision-makers learning and understanding other ways of approaching problematic situations, other than those they see themselves.

In decision-making contexts where conflicting interests of heterogeneous groups occur, it is a great way to address complex problems [16, 17]. For the scenario imposed by the work, it was a dynamic, playful, clear and objective approach, as it proved to be very efficient in the survey of several characteristics necessary for the cost-supply process. Ideas and thoughts that are often forgotten in the individuality of each one, have emerged and converged to a viable and validated model for all, a process that can now be executed with the support of computational resources, aiming for quality, assertiveness, and support in the processes of decision-making necessary for the institution's vitality.

We have a future work proposal, the application of this prioritization model in a real scenario, using real databases with the necessary characteristics for the model, courses mapping, data collection of a region served by NCAS/Ceará, client mappings, and all the possible factors pointed out as relevant in the cognitive mapping of this problem. The purpose of the work will be to indicate which courses should be offered to the chosen region, respecting the prioritization of the criteria established by the decision makers and their restrictions.

References

1. International Encyclopedia of the Social & Behavioral Sciences - Elsevier Science & Technology - Literati by Credo. corp.credoreference.com. Accessed 26 Nov 2016
2. Figueira, J., Greco, S., Ehrgott, M.: Multiple Criteria Decision Analysis: State of the Art Surveys. Springer, Boston, Dordrecht, London (2005)
3. Roy, B.: Paradigms, and challenges. In: Figueira, J., Greco, S., Ehrgott, M. (eds.) Multiple Criteria Decision Analysis: State of the Art Surveys Series: International Series in Operations Research and Management Science, vol. 78, XXXVI, pp. 3–24 (2005)
4. Gomes, L.F.A.M., Gomes, C.F.S., Almeida, A.T.: Tomada de Decisão Gerencial: Enfoque Multicritério, 2nd edn. Atlas, São Paulo (2006)
5. Ackermann, F., Eden, C., Cropper, S.: Getting started with cognitive mapping. Tutorial paper, 7th Young OR Conference. (Available from Banxia Software Ltd.) (1992)
6. Rieg, D.L., Araújo Filho, T.: Mapas cognitivos como ferramenta e resolução de problemas: O caso da Pró-Reitoria de extensão da UFSCar. Gestão e Produção, no. 2, pp. 145–162 (2003)
7. Cossette, P., Audet, M.: Mapping of an idiosyncratic schema. J. Manage. Stud. **29**(3), 321–347 (1992)
8. Pinheiro, P.R. de Souza, G.G.C.: A multicriteria model for production of a newspaper. In: The 17th International Conference on Multiple Criteria Decision Analysis, vol. 17, pp. 315–325. Simon Fraser University, British Columbia (2004)
9. Pinheiro, P.R., Souza, G.G.C., de Castro, A.K.A.: Estruturação do problema multicritério para produção de jornal. Pesquisa Oper. **28**, 203–216 (2008)
10. Bana e Costa, C.A.: Structuration, Construction et Exploitation d'un Modèle Multicritère d'Aide à la Décision. Tese de doutoramento, Instituto Superior Técnico, IST, Lisboa (1992)
11. Bana e Costa, C.A., Correa, E.C., Corte, J.M.D., Vansnick, J.C.: Facilitating bid evaluation in public call for tenders: a social-technical approach. OMEGA **30**, 227–242 (2002)
12. Bana e Costa, C.A., Beinat, E.: Model-structuring in public decision-aiding. Working Paper LSEOR 05.79. London School of Economics, London (2005)

13. Bana e Costa, C.A., Beinat, E., Vickerman, R.: Introduction, and Problem Definition. CEG-IST Working Paper no. 24 (2001)
14. Bastos, A.V.B.: Mapas cognitivos e a pesquisa organizacional: explorando aspectos metodológicos. Estudos de Psicologia, 7 (Número especial), 65–77 (2002)
15. Banxia Software, B.: Decision Explorer Software Package. University of Strathclyde, Glasgow (2002)
16. da Silva, O.B.: C3M – Gerenciamento de mudanças estruturado em uma metodologia de otimização multicritério. Dissertação de Mestrado. Universidade de Fortaleza, Fortaleza (2013)
17. da Silva, O.B., Holanda, R., Pinheiro, P.R.: A decision model for change management based on multicriteria methodology. In: 2013 Proceedings of the 8th International Workshop on Business-driven IT Management (BDIM), pp. 1241–1244 (2013)
18. Ackermann, F., Eden, C., Cropper., S.: Getting started with cognitive mapping (2004)

Algorithms for Congestion Control in LTE Mobile Networks

Teodor Iliev[1]([✉]) ⓘ, Tsviatko Bikov[1] ⓘ, Grigor Mihaylov[1] ⓘ, Elena Ivanova[1] ⓘ, Ivaylo Stoyanov[2] ⓘ, and Ventsislav Keseev[1] ⓘ

[1] Department of Telecommunication, University of Ruse, Ruse, Bulgaria
{tiliev,gmihaylov,epivanova,vkeseev}@uni-ruse.bg,
cbikov@gmail.com
[2] Department of Electrical Power Engineering, University of Ruse, Ruse, Bulgaria
stoyanov@uni-ruse.bg

Abstract. In the modern cellular data networks, with the tremendous amount of mobile devices, the constantly decreasing prices of the hardware and according the trends worldwide, provisioning of the data service and its quality will soon become an impossible task. Since current mobile cellular standards are unable to serve and meet the needs of this increase, the network will eventually become congested. As we are investigating data networks, they are based on the TCP protocol. TCP has various congestion control schemes used in both wired and wireless network, such as Cubic, Vegas, Reno, etc. Most modern cellular systems have already been migrated to 3G and 4G, so it could be very useful to identify which algorithm performs best in LTE (4G) wireless network. As most of the platforms for mobile devices, such as phones and tablets are unix/linux based and the current default algorithm in Linux is Cubic, the focus in this article will be on real-time modification and testing of the congestion control parameters of TCP CUBIC.

Keywords: Congestion control algorithm · Mobile networks · LTE

1 Introduction

During the last few decades, all technologies, including the mobile communication has evolved with enormous rates. The increasing demand of the consumers on telecommunication resources is pushing even further all technological developments in the mobile sphere. The multimedia applications like mobile TV services, live movies, VoD, video collaboration, VoIP, online gaming, etc. are only a few of all exciting applications currently running over the 4G. These applications are not subject of congestion control restriction enforcement so far. Because of this and the constantly increasing interest on these applications, the stability and the durability of the Internet could be eventually compromised if no measures are taken.

Five years ago, research and analysts firm named IDATE reported that the total number of mobile connections will eventually exceed 8 billion globally by 2019. In 2013, the researchers stated that the mobile subscribers will be increased with more than

© Springer International Publishing AG, part of Springer Nature 2019
R. Silhavy (Ed.): CSOC 2018, AISC 763, pp. 474–484, 2019.
https://doi.org/10.1007/978-3-319-91186-1_49

21% for the next 5 years. In present days, this number is already part of the reality. According to an online platforms which provides real-time data about the mobile connections, current mobile connections, including Machine to Machine (M2M), already exceeds 8.4 billion. Moreover, 5.1 billion of them are unique subscribers and the count is constantly growing. With this tremendous increasing in mobile connectivity and yet the continues rapidly growth demand for data, there is the outstanding question, whether the current equipment and network are able to deliver such bandwidth, which will comply these demands. Even the fixed internet subscribers are also supposed to be increased by 18%. If this is true, the Internet of Things (IoT) and M2M communications will also be expanded exponentially, as a side effect of the increasing in the mobile communications industry.

Anyway, to properly address this demands and the trends for the near future, the mobile operators and the network providers might put some effort and eventually take a step forward into migrating all the systems from 3G to 4G LTE networks, shutdown or restrict the 2G networks as much as possible and adjust the M2M Technology.

In telecommunications, the fourth generation of mobile wireless standard is named 4G. Its predecessors are the 3G and 2G families of standards. In 2009, ITU-R defined the International Mobile Telecommunications Advanced (IMT-Advanced) require-ments for the fourth generation standards, defining limitation requirements for 4G service at 100 Mbps cap speed for high mobility communication and 1 Gbps for low mobility communication [1, 2]. The 4th generation system is expected to deliver vaster and secure IP based mobile broadband solution to all kind of mobile and stationary network connected devices. Consuming it as ultra-broadband internet access, a lot of multimedia and entertainment applications and services could be delivered to the consumers. The developed IMT-Advanced compliant versions of LTE and WiMAX are "LTE-Advanced" and respectively "Wireless MAN-Advanced". On December, 2010, ITU recognized the LTE, WiMAX and some other evolved 3G technologies which not fulfill IMT-Advanced requirements, that could nevertheless be considered as '4G'. They had to provide at least same services and levels like their precursors to IMT-Advanced and in addition some significant levels of improvement in performance and features with respect to the prime third generation systems already deployed. As mentioned above, 4G mobile communication is based on entirely IP network. By design, core network is independent of specific wireless access network and could supply end to end IP service. In addition, the 4G systems are compatible with current core network and Public Switched Telephone Network (PSTN). Core network is open structure, containing three properties: Service, Control and Transport and they will be different from the previous generation's properties [3, 4].

2   Network Congestion Control in LTE

Despite all the positive aspects of the next generation of mobile networks, of course they also bring new problems that require attention. With the use of the digital technology and everything based on TCP/IP, their respective problems should be addressed as well with proper solutions. TCP Congestion control is one of these solutions. In simple words, congestion control is a set of behaviors defined by algorithms that each TCP performs to prevent network overload due to the vast amount of data passed through it [5].

Every signal or data could be transmitted through digital channels. There are two types of channels:

- Wired channel, where the signal is transmitted over a twisted-pair wire, coaxial cable and fiber optic cable.
- Wireless channel, where the information between two or more devices is transmitted without the need of any wired cables, but typically is done over radio frequencies (channels).

By design, TCP was planned for fixed end-platforms and wired cable networks. TCP is designed to assume congestion if detects dropped packets. As the TCP could not be fundamentally transformed due to the wide spreaded cable network installations, the mobility of the TCP must remain compatible [6].

In the wireless networks the TCP's performance is slightly different. The packet loss in wireless networks is typically due to bit errors, channel interruptions, hand-over, congestions, etc. Such errors could be received due to low a signal strength, interferences or high noise in the channel. These and other problems like delays, timeouts, throughput limitation, etc. could be considered as disadvantages of the wireless networks [7–9].

Building a wireless network will give a lot of space, flexibility and mobility of the users and will significantly help removing the copper wires required for creating a network, which relies entirely on wired channels. Of course there are a lot of other problems which has to be resolved, as the interference from other nearby wireless devices, delays and eventually the congestion of the network if the amount of devices grows. This does not mean that in wired network has not congestion. In wired networks, when host sends file to another host in remote area, the file is fragmented into hundreds of packets and it is send through numerous intermediate devices across all over the world. Each of this intermediates has its own buffer with pre-defined size. If the connections grows and the buffers becomes saturated, the speed will become reduced. Delays may occur and if the buffers get overflowed, the packets might be even dropped. These delays of the packets and the packet loss caused by enormous traffic in the networks are called congestion [10, 11]. When congestion on the network grows enormously, segments could be delayed or even dropped. This will cause retransmissions and even interruptions. Meanwhile, TCP will re-send the lost packets. Therefore, the network will be impacted and eventually exhausted if there are no implemented congestion control algorithms to handle it. Consequently, TCP congestion control schemes are developed to handle the congestions in TCP/IP.

The Congestion Control Schemes are decreasing the transmitting speed when congestion is about to happen in the network or when it is already happened. The most difficult part is to determine when the network is overloaded and even to avoid congestion before it happens. After that, the problem is how to slow down the transmission rate (on a congestion event or right before it appears) and when to increase it again. There are a lot of variations of TCP congestion control schemes and different environments where they can operate and where they could be manipulated. Such example is the operating system Linux, which is open source, so TCP congestion control schemes could be simply modified by commands and not only. As mentioned above, in this paper we will do exactly that. We will use Linux environment to manipulate with experimental intentions the parameters of the default TCP Congestion Control algorithm – CUBIC.

The International Mobile Telecommunications-2000 (IMT-2000) is another set of requirements developed by ITU, but targeting the family of standards for the 3rd Gen mobile communications.

To comply with these requirements, the systems are obliged to assure peak data rates of at least 200 Kbps. UMTS, released in 2001 and standardized by 3GPP. Typically the cell phones are UMTS and GSM hybrid combined. A few radio interfaces are provided, using one and same infrastructure:

- The original and most common radio interface is W-CDMA.
- The lately released UMTS interface – HSPA+ is capable to assure peak rates up to 56 Mbps in the downlink (in theory) and 22 Mbps in the uplink.
- The CDMA2000 system, first released in 2002, standardized by 3GPP2 and sharing same infrastructure with the IS-95 2nd Generation standard, as shown on Fig. 1.

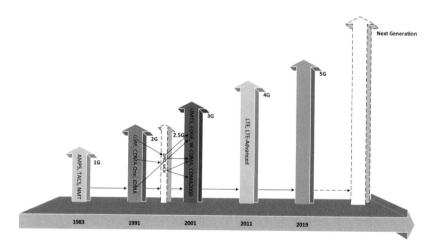

Fig. 1. Evolution of wireless technology.

Some of the 3G's limitations are:

- It is hard to achieve higher data rate in CDMA;
- It is hard to provide full range of multi-rate services with not fully integrated System;
- There are some propagation problems with the CDMA, related to its work in a multi path systems from private to public and indoor to wide areas [3, 12];

and respectively they are addressed in the future network communication standards as LTE and not only.

3 Mathematical Exposure

As today's internet connected devices requires fast transfer of large amount of data, but the traditional Additive Increase Multiplicative Decrease (AIMD) protocol algorithms perform badly because of to their slow response times in fast or long distance networks, TCP Cubic algorithm is developed to address this problem, while managing stability and compatibility with traditional TCP congestion control. Above all, the TCP Cubic is an algorithm created to improve an earlier TCP version known as TCP BIC and its weak spots.

The most noticeable difference between the TCP Cubic and the other AIMD-based algorithms is Cubic's fast windows growth capability. This algorithm allows very fast window amplification and acts like a high-speed TCP to address and resolve the issue of efficient TCP transport, in cases when the delays compared to the bandwidth are large.

As the name itself represents, the window growth function of TCP CUBIC is a cubic like and its shape is pretty similar to the function of BIC Algorithm. As mentioned above, Cubic is designed to simplify and improve the window control of BIC. The congestion window of TCP Cubic is determined below, as the target window size $-W$, is a function of elapsed time and the maximum window size reached prior to the last packet loss event $-W_{max}$:

$$W_{(t)} = C(t - K)^2 + W_{max}, \tag{1}$$

where C is the scaling parameter, t is the time passed from the last window attenuation, and:

$$K = \sqrt[3]{W_{max}\beta/C} \tag{2}$$

With this function, the flows with shorter trip times (RTT) will not grow as quickly as those with long RTTs, unlike the AIMD-based algorithms, or even Cubic's precursor - BIC. TCP Cubic's window growth could be depict as shown in Fig. 2.

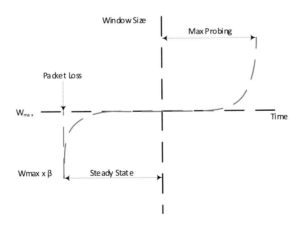

Fig. 2. TCP CUBIC growth behavior.

This algorithm contains three primary parameters, affecting the window growth:

- Beta (shown in Fig. 2 above as β). It controls the decreasing of the window. The default value of beta in TCP Cubic is 0.7, which means that the reduction of the window is by thirty percent on a packet loss event.
- Fast convergence, which modifies the value of maximum window size on a packet loss event occurrence. If the fast convergence is disabled, the maximum window size is set to the value of the congestion window (cwnd), at the time of the loss event. If the fast convergence is enabled and loss event occurs, the maximum windows size is scaled down to approximately eighty five percent of the cwnd.
- TCP friendliness, which permits Cubic to grow as a standard TCP in short RTT environments.
- To expose additionally the parameters above, alpha (α) is added. This helps scaling the value of W_{max} by setting:

$$W_{max} = W_{max} * \alpha \qquad (3)$$

This way allows TCPTuner to adjust and to manage how quickly the window grows by provoking a longer period of very fast growth after a reduction, or by leaving the window to remain reduced until Cubic reach the max probing state.

Based on these and some other functions described in more details in [13], experimental scenarios were defined and executed.

4   Testbed and Simulation Results

For the needs of the tests we used a specifically designed module for this purpose (freely distributed in github), named TCPTuner. It is designed to run on Linux kernel level and to be loaded as the main kernel module for congestion control. That it acts as the algorithm that manage the traffic on the machine. Because of this, we performed our tests on a virtual machine with installed Linux distribution based on Debian version 4.3.3-7. As the program was distributed as a source code, first all modules had to be built inside the OS and loaded into the machine's kernel. More detailed description of this procedure can be found in the README file of the TCPTuner project [13].

As the default algorithm/module for congestion control used in Linux is CUBIC and we want to create a real simulation, first thing that we did after compiling the TCP Tuner module is to load it as a primary with the `sysctl -w net.ipv4.tcp_congestion_control = tuner` command and then we check which is the currently used

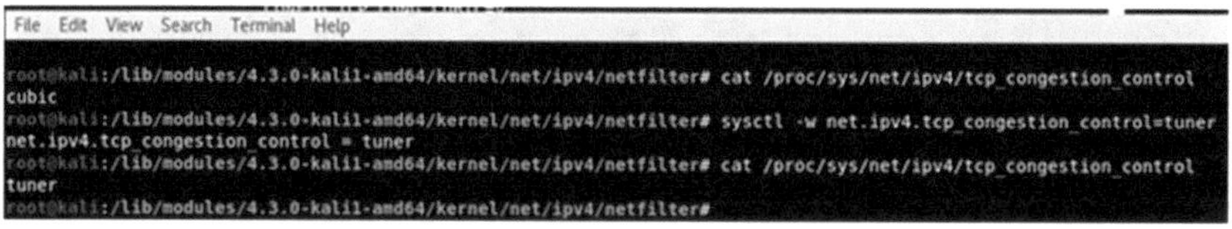

Fig. 3. Command window.

module with `cat/proc/sys/net/ipv4/tcp_congestion_control` command. The command window is shown on Fig. 3.

After that and a couple of other fine tunings of this module we start with real experimental scenarios. In the scope, the following scenarios were considered:

- Default set of parameters (Alpha = 512 and Beta = 717)
- Alpha = 654, Beta = 717;
- Alpha = 732, Beta = 717;
- Alpha = 512, Beta = 614;
- Alpha = 512, Beta = 124;
- Alpha = 512, Beta = 1024.

All described above scenarios were utilized with downloading the following four sample audio, text (shown in Table 1), video and picture files (described in Table 2) with approximate volume of 1 MB each:

- Text File - http://www.sample-videos.com/text/Sample-text-file-1000kb.txt;
- Audio File - http://www.sample-videos.com/audio/mp3/wave.mp3;
- Images File - http://www.sample-videos.com/img/Sample-jpg-image-1mb.jpg;
- Video File - http://www.sample-videos.com/video/mp4/720/big_buck_bunny_720p_1mb.mp4.

Table 1. Comparison results for downloading Text and Audio files with different alpha and beta congestion control parameters.

Alpha	Beta	Text file			Audio file		
		Volume (KB)	Speed (Kbps)	Time (s)	Volume (KB)	Speed (Kbps)	Time (s)
512	717	999.40	439	2.3	708.24	133	5.3
654	717	999.40	219	4.6	708.24	286	2.5
732	717	999.40	466	2.1	708.24	325	2.7
512	614	999.40	296	3.6	708.24	239	3.6
512	124	999.40	158	6.2	708.24	179	4.0
512	1024	999.40	298	3.4	708.24	314	2.3

We download the sample files using wget program and after each scenario we do removed all sample files to assure no delay or latency will occur because of the overriding the already downloaded files or other process. Additionally for better accuracy, each test were repeated twice and the average output data of the two tests was taken as a relevant.

As shown in both tables, despite the approximate volume of all downloaded files, the required time for download is quite different in each and every scenario in each and every different configuration for alpha and beta. So far and in this research no dependencies were examined or discovered, related to the required download time for the targeted files, described above.

Table 2. Comparison results for downloading Image and Audio files with different alpha and beta congestion control parameters.

Alpha	Beta	Image file			Video file		
		Volume (KB)	Speed (Kbps)	Time (s)	Volume (KB)	Speed (Kbps)	Time (s)
512	717	1009	487	2.1	1010	450	2.3
654	717	1009	219	5.0	1010	301	3.7
732	717	1009	545	1.9	1010	553	1.9
512	614	1009	251	5.0	1010	346	3.0
512	124	1009	203	5.1	1010	228	4.8
512	1024	1009	365	2.8	1010	402	2.6

More comparison details as graphical data from the TCP Tuner GUI are described on figures below (Fig. 4).

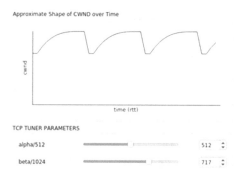

Fig. 4. TCP Tuner default settings and behavior.

With the default values of the congestion control algorithm, the results appears to be very balanced, but however in each and every independent simulation (downloading audio, video, etc. file), there was at least one of the other scenarios with better performance results (Fig. 5).

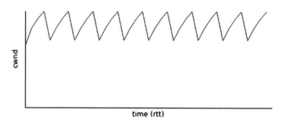

Fig. 5. CWND over Time with Alpha/654 and Beta/717.

Using this configuration, the performance is slightly degraded in almost every tested aspect, except in the audio download scenario (Fig. 6). More details could be found in Table 1.

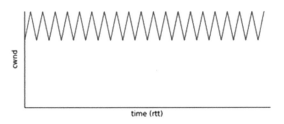

Fig. 6. CWND over Time with Alpha/732 and Beta/717.

In this scenario all tested simulations had better performance than the scenario with the default settings. This means that if the congestion control parameters for CUBIC algorithm are precisely tuned, even better performance could be achieved. In general this could be a time consuming exercise and it is not fully guarantee that the final results will be significantly better.

Compared to the default settings and behavior of the CUBIC congestion control algorithm, the scenario shown in Fig. 7 have notably worst output results in each and every simulation, except in the Audio download test, where every scenario have better performance results, compared to the default parameters.

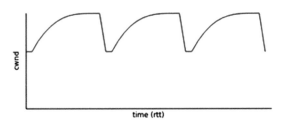

Fig. 7. CWND over Time with Alpha/512 and Beta/614.

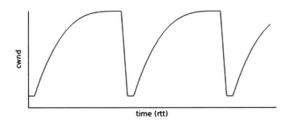

Fig. 8. CWND over Time with Alpha/512 and Beta/124.

The situation is the same also for Fig. 8, where even the download speed is simi-lar but also quite different for all tested file formats. In this figure is shown that with lower value for beta parameter, the cwnd depict in the graph is bigger, as well as the RTT.

5 Conclusion

Based on the results from the performed tests above, we can conclude that the used as a default in Linux OS congestion control algorithm – CUBIC is quite well balanced. During the tests we saw that changing only the alpha or beta parameters of CUBIC is instantly reflecting over the speed and respectively the time required to finish the download but indeed it was very variable and impermanent. Anyway, serious amount of time is required to be spend to find and choose the right optimal parameter configuration. Judging by the charts and the experimental data in this whitepaper (without doing more deeper investigation) it seems the CUBIC could even be more optimized with some small configuration fine-tunings. Our proposal is some time to be spend in that area and for example, alpha and beta parameters could be configured to be between 700 and 740 and more detailed and accurate tests to be performed.

References

1. Burst, K., Joiner, L., Grimes, G.: Delay based congestion detection and admission control for voice quality in enterprise or carrier controlled IP networks. IEEE Trans. Netw. Serv. Manag. **2**(1), 1–8 (2005)
2. Chang, C., Chen, B., Liu, T., Ren, F.: Fuzzy neural congestion control for integrated voice and data DS-CDMA/FRMA cellular networks. IEEE J. Sel. Areas Commun. **18**(2), 283–293 (2000)
3. Huang, S.: Evolution from 3G to 4G and beyond (5G). ITN-620 enterprise network design. In: International Conference on Information Technology, Las Vegas, Nevada (2005)
4. Wildstrom, S.: AT&T's Stephenson: the road to 4G (2009)
5. Fall, K., Stevens, W.: TCP/IP Illustrated, Volume 1: The Protocols, 2nd edn. Addison-Welsey, Boston (2011)
6. Prabhaker, M.: Mobile TCP CEG436: Mobile Computing
7. Rodriguez, E.: Wired vs. Wireless. http://www.skullbox.net/wiredvswireless.php. Accessed 02 Feb 2018
8. Kaur, N., Monga, S.: Comparisons of wired and wireless networks: a review. Int. J. Adv. Eng. Technol. **5**(2), 34–35 (2014)
9. Kozierok, C.: The TCP/IP Guide: A Comprehensive, Illustrated Internet Protocols Reference, 1st edn. No Starch Press, San Francisco (2005)
10. Abed, G., Ismail, M., Jumari, K.: Integrated approaches to enhance TCP performance over 4G wireless network. In: IEEE Symposium on Computers and Information, Penang, pp. 154–158 (2012)

11. Ghazaleh, H., Muhanna, M.: Enhancement of throughput time using MS-TCP transport layer protocol for 4G mobiles. In: Proceedings of the 5th International Multi-Conference on Systems, Signals and Devices, pp. 1–5. IEEE, Amman (2008)
12. Agilent. http://www.agilent.com/cm/wireless/pdf/3GSeminar2001_02.pdf
13. GitHub. https://github.com/Gasparila/TCPTuner/blob/master/README.md. Accessed 01 Feb 2018

Correction to: Computing Importance Value of Medical Data Parameters in Classification Tasks and Its Evaluation Using Machine Learning Methods

Andrea Peterkova, Martin Nemeth, German Michalconok, and Allan Bohm

Correction to:
Chapter "Computing Importance Value of Medical Data Parameters in Classification Tasks and Its Evaluation Using Machine Learning Methods" in: R. Silhavy (Ed.): *Software Engineering and Algorithms in Intelligent Systems,* **AISC 763, https://doi.org/10.1007/978-3-319-91186-1_41**

In the original version of the book, acknowledgement should be included in chapter "Computing Importance Value of Medical Data Parameters in Classification Tasks and Its Evaluation Using Machine Learning Methods". The correction chapter and the book have been updated with the change.

The updated version of this chapter can be found at
https://doi.org/10.1007/978-3-319-91186-1_41

Author Index

Printed in the United States
By Bookmasters